Internal Combustion Engines

Internal Combustion Engines

V. GANESAN
Professor of Mechanical Engineering
Indian Institute of Technology
Madras

McGraw-Hill, Inc.

New York   San Francisco   Washington, D.C.   Auckland   Bogotá   Caracas
Lisbon   London   Madrid   Mexico   Milan   Montreal   New Delhi
San Juan   Singapore   Sydney   Tokyo   Toronto

Library of Congress Cataloging-in-Publication Data

Ganesan, V.
    Internal combustion engines / V. Ganesan.

        p.   cm.
    ISBN 0-07-462122-X
    1. Internal combustion engines.   I. Title.
    TJ755.G36   1995
    621.43—dc20   95-23419
                                          -        CIP

First published © 1994,
Tata McGraw-Hill Publishing Company Limited

Copyright © 1996 by McGraw-Hill, Inc. All rights reserved. Printed in
the United States of America. Except as permitted under the United States
Copyright Act of 1976, no part of this publication may be reproduced or
distributed in any form or by any means, or stored in a data base or
retrieval system, without the prior written permission of the publisher.

1 2 3 4 5 6 7 8 9 0   DOC/DOC   9 0 9 8 7 6 5

ISBN 0-07-462122-X

Printed and bound by R. R. Donnelley and Sons Company.

I dedicate this book to my mother, Mrs. L. Seetha Ammal, who devoted herself, heart and soul to bringing up her two children and shielding them from the repercussions of the premature demise of their father.

FOREWORD

Focusing on the need of a first level text book for the under-graduates, postgraduates and a professional reference book for practising engineers, the author of this work Dr. V. Ganesan has brought forth this volume using his extensive teaching and research experience in the field of internal combustion engineering. It is a great pleasure to write a foreword to such a book which satisfies a long-felt requirement.

For selfish reasons alone, I wish that this book would have come out much earlier for the benefit of several teachers like me who have finished their innings a long time ago. For me, this would have been just the required text book for my young engineering students and engineers in the transportation and power fields. The style of the book reflects the teaching culture of premier engineering institutions like IITs, since a vast topic has to be covered in a comprehensive way in a limited time. Each chapter is presented with elegant simplicity requiring no special prerequisite knowledge of supporting subjects. Self-explanatory sketches, graphs, line schematics of processes and tables have been generously used to curtail long and wordy explanations. Numerous illustrated examples, exercises and problems at the end of each chapter serve as a good source material to practise the application of the basic principles presented in the text. SI system of units has been used throughout the book which is not so readily available in the currently-used books.

It is not a simple task to bring out a comprehensive book on an all-encompassing subject like internal combustion engines. Over a century has elapsed since the discovery of the diesel and gasoline engines. Excluding a few developments of rotary combustion engines, the IC engine has still retained its basic anatomy. As a descendent of the steam engine, it is still crystallized into a standard piston-in-cylinder mechanism, reciprocating first in order to rotate finally. The attendent kinematics requiring numerous moving parts are still posing dynamic problems of vibration, friction losses and mechanical noise. Empiricism has been the secret of its evolution in its yester years.

As our knowledge of engine processes has increased, these engines have continued to develop on a scientific basis. The present day engines have to satisfy the strict environmental constraints and fuel economy standards in addition to meeting the competitiveness of the world market. Today, the IC engine has synthesized the basic knowledge of many disciplines— thermodynamics, fluid flow, combustion, chemical kinetics and heat transfer as applied to a system with both spatial and temporal variations in a state of non-equilibrium. With the availability of sophisticated computers, multi-dimensional mathematical modelling the electronic instrumentation have added new refinements to the engine design. From my personal knowledge, Dr. Ganesan has himself made many original contributions in these intricate areas. It is a wonder for me how he has modestly kept out these details from the text as it is beyond the scope of this book. However, the reader is not denied the benefits of these investigations. Skilfully the overall findings and updated information have been summarized as is reflected in topics on combustion and flame propagation, engine heat transfer, scavenging processes and engine emissions—to name a few examples. Indeed, it must have been a difficult task to summarize the best of the wide ranging results of combustion engine research and compress them in an elegant and simple way in this book. The author has also interacted with the curriculum development cell so that the contents of the book will cater to the needs of any standard accredited university.

I congratulate the author, Dr. V. Ganesan on bringing out this excellent book for the benefit of students in IC engines. While many a student will find it rewarding to follow this book for his class work, I also hope that it will motivate a few of them to specialize in some key areas and take up combustion engine research as a career. With great enthusiasm, I recommend this book to students and practising engineers.

B.S. MURTHY
Former Professor, IIT Madras

PREFACE

Keeping in view the increasing importance of IC engines, various Universities are introducing courses on the subject as an intrinsic part of the thermal engineering curriculum. However, the lack of a suitable textbook has created difficulties in fully appreciating the basic principles and applications of IC engines. This book has been written to fulfill this need.

I have endeavoured to explain the various topics right from the fundamentals so that even a beginner can understand the exposition. Keeping this in view, chapters 1 to 15 are framed so that the book will be useful to both undergraduate and postgraduate students as well as to practising engineers.

In writing this book, I have kept in mind the tremendous amount of ground which the student and the practising engineer of today is expected to cover. On this account, the work has been organized to form, it is hoped, a continuous logical narrative.

SI units have been consistently used throughout the book. The book includes a large number of typical worked out examples and several illustrative figures for an easier understanding of the subject. Exercises have also been provided in various chapters so that the inquisitive student may solve these problems and compare with the answers given.

Care has been taken to minimise the errors and typing mistakes. I would be obliged to the readers for finding out any such error and mistake, and would be grateful for any constructive criticism for the improvement of various topics in the book.

It would be impossible to refer in detail to the many persons who have been consulted in the compilation of this work. I am thankful to all my Punctilious and highly Devoted scholars, past and present, who have helped me in various ways in bringing out this book. I may be excused for not naming them individually.

I am particularly grateful for the help I have received from my colleagues at I.C. Engines Laboratory, Indian Institute of Technology, Madras and especially to Prof. P. Srinivasa Rao, who spared a great amount of his valuable time in going through the manuscript and discussing various points in each chapter. I wish to thank the Centre for Continuing Education of the Indian Institute of Technology, Madras for their help in the preparation of this book.

I would like to express my gratitude and offer my sincere thanks to Vijayashree, G. Venkatasubramanian and V. Satish Kumar, who were instrumental in composing and formatting the entire book. I will be failing in my duty if I do not acknowledge P. Chandrasekaran who has done all the drawings excellently.

Last but not the least my sincere thanks are due to my dear wife P. Rajalakshmi, my son G. Venkatasubramanian, and daughter G. Aparna who have borne with me for the last three and a half years and provided me with a quiet home where so much of this work was written.

V. GANESAN

CONTENTS

Chapter 2   Review of Basic Principles

Chapter 5 Actual Cycles and Their Analysis

Chapter 6 Fuels

Chapter 7 Carburetion

Chapter 8   Injection

Chapter 9   Ignition

Chapter 10   Combustion and Combustion Chambers

Chapter 12 Heat Rejection and Cooling

Chapter 13   Measurements and Testing

Chapter 15   Two-Stroke Engines

NOMENCLATURE

A

a_1	constant
$amep$	mean effective pressure required to drive the auxiliary components
A	piston area [Chp.1]
A	area of heat transfer [Chp.12]
A	average projected area of each particles [Chp.13]
A	TEL in ml/gal of fuel [Chp.6]
A_1	cross-sectional area at inlet of the carburettor
A_2	cross-sectional area at venturi of the carburettor
A_{act}	actual amount of air in kg for combustion per kg of fuel
A_f	area of cross-section of the fuel nozzle [Chp.7]
A_f	area of fin [Chp.12]
A_e	effective area
A_{th}	theoretical amount of air in kg per kg of fuel
A/F	air-fuel ratio

B

b_1	constant
bp	brake power
bhp	brake horsepower
$bmep$	brake mean effective pressure
$bsfc$	brake specific fuel consumption
BDC	Bottom Dead Centre

C

$cmep$	mean effective pressure required to drive the compressor or scavenging pump
C	velocity [Chp.6]
C_d	coefficient of discharge for the orifice [Chp.8]

C_{da}	coefficient of discharge for the venturi
C_{df}	coefficient of discharge for fuel nozzle
C_f	fuel velocity at the nozzle exit
C_p	specific heat of gas at constant pressure
C_{rel}	relative charge
C_v	specific heat at constant volume
CV	calorific value of the fuel

D

d	cylinder bore diameter [Chp.1]
d	diameter of orifice [Chp.8]
D	brake drum diameter

E

e	expansion ratio
E	stored energy [Chp.2]
E	enrichment [Chp.7]
EVC	Exhaust Valve Closing
EVO	Exhaust Valve Opening

F

f	coefficient of friction
$fmep$	frictional mean effective pressure
fp	frictional power
F	force
F/A	fuel-air ratio
F_R	relative fuel-air ratio

G

g	acceleration due to gravity
g_c	gravitational constant
gp	gross power

H

h	specific enthalpy
h	pressure difference [Chp.8]
h	convective heat transfer coefficient [Chp.12]
H	enthalpy

I

ip	indicated power
$imep$	indicated mean effective pressure
$isfc$	indicated specific fuel consumption
I	intensity
IDC	Inner Dead Centre
IVC	Inlet Valve Closing
IVO	Inlet Valve Opening

K

k	thermal conductivity of gases
k_1	constant [Chp.4]
K	number of cylinders
K_{ac}	optical absorption coefficient

L

l	characteristic length
l	distance [Chp.13]
L	stroke
L_ℓ	length of the light path

M

m	mass
m	exponent [Chp.12]
mep	mean effective pressure
$mmep$	mechanical mean effective pressure
$\dot{m}_a$	mass flow rate of air
$\dot{m}_{act}$	actual mass flow rate of air
$\dot{m}_{th}$	theoretical mass flow rate of air
M_{del}	mass of fresh air delivered
M_f	molecular weight of the fuel
M	molecular weight
M_{ref}	reference mass

N

n	number of power strokes
n	polytropic index [Chp.2]
n	number of soot particles per unit volume [Chp.13]
N	speed in revolutions per minute
N_i	number of injections per minute [Chp.8]

O

ODC	Outer Dead Centre
ON	Octane Numbers

P

p	pressure
$pmep$	charging mean effective pressure
pp	pumping power
p_{ar}	pure air ratio
p_{bm}	brake mean effective pressure
p_e	exhaust pressure
p_i	inlet pressure
p_{im}	indicated mean effective pressure
p_m	mean effective pressure
P_{cyl}	pressure of charge inside the cylinder
P_{inj}	fuel pressure at the inlet to injector
P_l	pressure loss coefficient
P_s	specific power output
PN	performance number

Q

q	heat transfer
$\dot{q}$	rate of heat transfer
Q_R	heat rejected
Q_S	heat supplied

R

r	compression ratio
rpn	relative performance number
r_c	cut-off ratio
r_p	pressure ratio
R	length of the moment arm
R	delivery ratio [Chp.15]
$\overline{R}$	universal gas constant
R_{del}	delivery ratio

S

$\overline{s}_p$	mean piston speed
sfc	specific fuel consumption
S	spring scale reading

T

T	absolute temperature
T	torque [Chp.14]
TDC	Top Dead Centre
T_b	black body temperature
T_f	friction torque
T_g	mean gas temperature
T_l	load torque

U

u	specific internal energy
U	internal energy
U_c	chemical energy
U_s	stored energy

V

V	volume
V_{ch}	volume of cylinder charge
V_{cp}	volume of combustion products
V_{del}	volume of air delivered
V_f	fuel jet velocity
V_{pure}	volume of pure air
V_{ref}	reference volume
V_{res}	volume of residual gas
V_{ret}	volume of retained air or mixture
V_s	displacement volume
V_{short}	short circuiting air
V_C	clearance volume
V_T	volume at bottom dead centre

W

w	specific weight
w	work transfer [Chp.7]
W	net work
W	weight [Chp.13]
W	number of quartz windows [Chp.13]
W	load [Chp.11]
WOT	Wide Open Throttle
W_x	external work

Z

z	height of the nozzle exit [Chp.7]
z_2	datum height [Chp.2]
Z	constant

GREEK

α	air coefficient
γ	ratio of specific heats
ΔT	temperature difference between the gas and the wall
ϵ	heat exchanger efficiency
η	efficiency
η_{sc}	scavenging efficiency
$\eta_{air\ std}$	air standard efficiency
η_{bth}	brake thermal efficiency
η_{ith}	indicated thermal efficiency
η_m	mechanical efficiency
η_{rel}	relative efficiency
η_{sc}	scavenging efficiency
η_{th}	thermal efficiency
η_{trap}	trapping efficiency
η_v	volumetric efficiency
θ	crank angle [Chp.10]
θ	specific absorbance per particle [Chp.13]
λ	wave length [Chp.13]
λ	excess air factor [Chp.15]
μ	kinematic viscosity of gases
ν	dynamic viscosity
ρ	density
ρ_f	density of fuel
ϕ	equivalence ratio
ψ	magnetic field strength
ω	angular velocity

1

INTRODUCTION

1.1 ENERGY CONVERSION

The distinctive feature of our civilization today, one that makes it different from all others, is the wide use of mechanical power. At one time, the primary source of power for the work of peace or war was chiefly man's muscles. Later, animals were trained to help and afterwards the wind and the running stream were harnessed. But, the great step was taken in this direction when man learned the art of energy conversion from one form to another. The machine which does this job of energy conversion is called an engine.

1.1.1 Definition of 'Engine'

An engine is a device which transforms one form of energy into another form. However, while transforming energy from one form to another, the efficiency of conversion plays an important role. Normally, most of the engines convert thermal energy into mechanical work and therefore they are called 'heat engines'.

1.1.2 Definition of 'Heat Engine'

Heat engine is a device which transforms the chemical energy of a fuel into thermal energy and utilizes this thermal energy to perform useful work. Thus, thermal energy is converted to mechanical energy in a heat engine.

Heat engines can be broadly classified into two categories :

(i) Internal Combustion Engines (IC Engines)
(ii) External Combustion Engines (EC Engines)

1.1.3 Classification and Some Basic Details of Heat Engines

Engines whether Internal Combustion or External Combustion are of two types, viz.,

(i) Rotary engines

(ii) Reciprocating engines

A detailed classification of heat engines is given in Fig.1.1. Of the various types of heat engines, the most widely used ones are the reciprocating internal combustion engine, the gas turbine and the steam turbine. The steam engine is rarely used nowadays. The reciprocating internal combustion engine enjoys some advantages over the steam turbine due to the absence of heat exchangers in the passage of the working fluid (boilers and condensers in steam turbine plant). This results in a considerable mechanical simplicity and improved efficiency of the internal combustion engine.

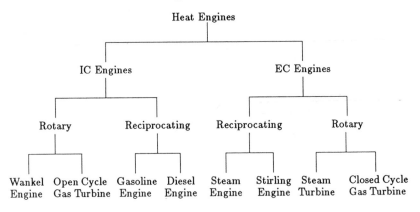

Fig.1.1 Classification of Heat Engines

Another advantage of the reciprocating internal combustion engine over the other two types is that all its components work at an average temperature which is much below the maximum temperature of the working fluid in the cycle. This is because the high temperature of the working fluid in the cycle persists only for a very small fraction of the cycle time. Therefore, very high working fluid temperatures can be employed resulting in higher thermal efficiency.

Further, in internal combustion engines, higher thermal efficiency can be obtained with moderate maximum working pressure of the fluid in the cycle, and therefore, the weight to power ratio is less than that of the steam turbine plant. Also, it has been possible to develop reciprocating internal combustion engines of very small power output (power output of even a fraction of a kilowatt) with reasonable thermal efficiency and cost.

The main disadvantage of this type of engine is the problem of vibration caused by the reciprocating components. Also, it is not possible to use a variety of fuels in these engines. Only liquid or gaseous fuels of given specification can be efficiently used. These fuels are relatively more expensive.

Considering all the above factors the reciprocating internal combustion engines have been found suitable for use in automobiles, motor-cycles and scooters, power boats, ships, slow speed aircraft, locomotives and power units of relatively small output.

1.1.4 External Combustion and Internal Combustion Engines

External combustion engines are those in which combustion takes place outside the engine whereas in internal combustion engines combustion takes place within the engine. For example, in a steam engine or a steam turbine, the heat generated due to the combustion of fuel is employed to generate high pressure steam which is used as the working fluid in a reciprocating engine or a turbine.

In case of gasoline or diesel engines, the products of combustion generated by the combustion of fuel and air within the cylinder form the working fluid.

1.2 BASIC ENGINE COMPONENTS AND NOMENCLATURE

Even though reciprocating internal combustion engines look quite simple, they are highly complex machines. There are hundreds of components which have to perform their functions satisfactorily to produce output power. Before going through the working principle of this complex machine a brief description of the important engine components and the nomenclature associated with an engine are more appropriate and are given in the following sections.

1.2.1 Engine Components

A cross section of a single cylinder spark-ignition engine with side valves is shown in Fig.1.2. The major components of the engine and their functions are briefly described below.

Cylinder : As the name implies it is a cylindrical vessel or space in which the piston makes a reciprocating motion. The varying volume created in the cylinder during the operation of the engine is filled with the working fluid and subjected to different thermodynamic processes. The cylinder is supported in the cylinder block.

Piston : It is a cylindrical component fitted into the cylinder forming the moving boundary of the combustion system. It fits perfectly (snugly) into the cylinder providing a gas-tight space with the piston rings and the lubricant. It forms the first link in transmitting the gas forces to the output shaft.

Combustion Chamber : The space enclosed in the upper part of the cylinder, by the cylinder head and the piston top during the

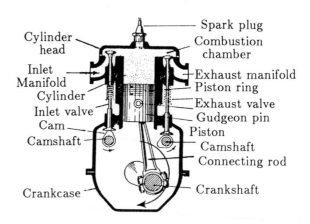

Fig.1.2 Cross-section of a Spark Ignition Engine

combustion process, is called the combustion chamber. The combustion of fuel and the consequent release of thermal energy results in the building up of pressure in this part of the cylinder.

Inlet Manifold : The pipe which connects the intake system to the inlet valve of the engine and through which air or air-fuel mixture is drawn into the cylinder is called the inlet manifold.

Exhaust Manifold : The pipe which connects the exhaust system to the exhaust valve of the engine and through which the products of combustion escape into the atmosphere is called the exhaust manifold.

Inlet and Exhaust Valves : Valves are commonly mushroom shaped poppet type. They are provided either on the cylinder head or on the side of the cylinder for regulating the charge coming into the cylinder (inlet valve) and for discharging the products of combustion (exhaust valve) from the cylinder.

Spark Plug : It is a component to initiate the combustion process in Spark Ignition (SI) engines and is usually located on the cylinder head.

Connecting Rod : It interconnects the piston and the crankshaft and transmits the gas forces from the piston to the crankshaft.

Crankshaft : It converts the reciprocating motion of the piston into useful rotary motion of the output shaft. In the crankshaft of a single cylinder engine there are a pair of crank arms and balance weights. The balance weights are provided for static and dynamic balancing of the rotating system. The crankshaft is enclosed in a crankcase.

Piston Rings : Piston rings, fitted into the slots around the piston, provide a tight seal between the piston and the cylinder wall thus preventing leakage of combustion gases.

Gudgeon Pin : It forms the link between the small end of the connecting rod and the piston.

Camshaft : The camshaft and its associated parts control the opening and closing of the two valves. The associated parts are push rods, rocker arms, valve springs and tapets. This shaft also provides the drive to the ignition system. The camshaft is driven by the crankshaft through timing gears.

Cams : These are made as integral parts of the camshaft and are designed in such a way to open the valves at the correct timing and to keep them open for the necessary duration.

Fly Wheel : The net torque imparted to the crankshaft during one complete cycle of operation of the engine fluctuates causing a change in the angular velocity of the shaft. In order to achieve a uniform torque an inertia mass in the form of a wheel is attached to the output shaft and this wheel is called the flywheel. The variation of net torque decreases with increase in the number of cylinders in the engine and thereby the size of the flywheel also becomes smaller. This means that a single cylinder engine will have a larger flywheel whereas a multi-cylinder engine will have a smaller flywheel.

1.2.2 Nomenclature

Cylinder Bore (d) : The nominal inner diameter of the working cylinder is called the cylinder bore and is designated by the letter d and is usually expressed in millimeter (mm).

Piston Area (A) : The area of a circle of diameter equal to the cylinder bore is called the piston area and is designated by the letter A and is usually expressed in square centimeter (cm^2).

Stroke (L) : The nominal distance through which a working piston moves between two successive reversals of its direction of motion is called the stroke and is designated by the letter L and expressed is usually in millimeter (mm).

Dead Centre : The position of the working piston and the moving parts which are mechanically connected to it, at the moment when the direction of the piston motion is reversed at either end of the stroke is called the dead centre. There are two dead centres in the engine as indicated in Fig.1.3. They are :

(i) Top Dead Centre   (ii) Bottom Dead Centre

(i) *Top Dead Centre* (TDC) : It is the dead centre when the piston is farthest from the crankshaft. It is designated as TDC. It is also called the Inner Dead Centre (IDC).

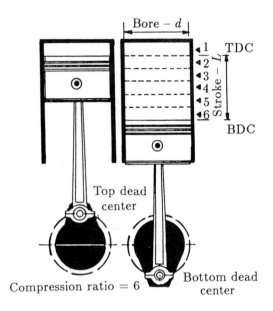

Fig.1.3 Top and Bottom Dead Centres

(ii) *Bottom Dead Centre* (*BDC*) : It is the dead centre when the piston is nearest to the crankshaft. It is designated as *BDC*. It is also called the Outer Dead Centre (*ODC*).

Displacement or Swept Volume (*V_s*) : The nominal volume swept by the working piston when traveling from one dead centre to the other is called the displacement volume. It is expressed in terms of cubic centimeter (cc) and given by

$$V_s \;\; = \;\; A \times L \;\; = \;\; \frac{\pi}{4} d^2 L \tag{1.1}$$

Clearance Volume (*V_C*) : The nominal volume of the combustion chamber above the piston when it is at the top dead centre is the clearance volume. It is designated as V_C and expressed in cubic centimeter (cc).

Compression Ratio (*r*) : It is the ratio of the total cylinder volume when the piston is at the bottom dead centre, V_T, to the clearance volume, V_C. It is designated by the letter *r*.

$$r \;\; = \;\; \frac{V_T}{V_C} \;\; = \;\; \frac{V_C + V_s}{V_C} \;\; = \;\; 1 + \frac{V_s}{V_C} \tag{1.2}$$

1.3 THE WORKING PRINCIPLE OF ENGINES

If an engine is to work successfully then it has to follow a cycle of operations in a sequential manner. The sequence is quite rigid and cannot be changed. In the following sections the working principle of both SI and CI engines is described. Even though both engines have much in common there are certain fundamental differences.

The credit of inventing the spark-ignition engine goes to Nicolaus A. Otto (1876) whereas compression-ignition engine was invented by Rudolf Diesel (1892). Therefore, they are often referred to as Otto engine and Diesel engine.

1.3.1 Four-Stroke Spark Ignition Engine

In a four-stroke engine, the cycle of operations is completed in four strokes of the piston or two revolutions of the crankshaft. During the four strokes, there are five events to be completed, viz., suction, compression, combustion, expansion and exhaust. Each stroke consists of 180° of crankshaft rotation and hence a four-stroke cycle is completed through 720° of crank rotation. The cycle of operation for an ideal four-stroke SI engine consists of the following four strokes :

(i)   Suction or intake stroke

(ii) Compression stroke

(iii) Expansion or power stroke

(iv) Exhaust stroke

The details of various processes of a four-stroke Spark Ignition engine with side valves are shown in Fig.1.4 (a-d). The ideal indicator diagram, showing the *p-V* plot for the four-stroke SI engine is shown in Fig.1.5.

(i) *Suction or Intake Stroke :* Suction stroke 0→1 (Fig.1.5) starts when the piston is at the top dead centre and about to move downwards. The inlet valve is open at this time and the exhaust valve is closed, Fig 1.4(a). Due to the suction created by the motion of the piston towards the bottom dead centre, the charge consisting of fuel-air mixture is drawn into the cylinder. When the piston reaches the bottom dead centre the suction stroke ends and the inlet valve closes.

(ii) *Compression Stroke :* The charge taken into the cylinder during the suction stroke is compressed by the return stroke of the piston 1→2, (Fig.1.5). During this stroke both inlet and exhaust valves are in the closed position, Fig.1.4(b). The mixture which fills the entire cylinder volume is now compressed into the clearance volume. At the end of the compression stroke the mixture is ignited with the help of an electric spark between the electrodes

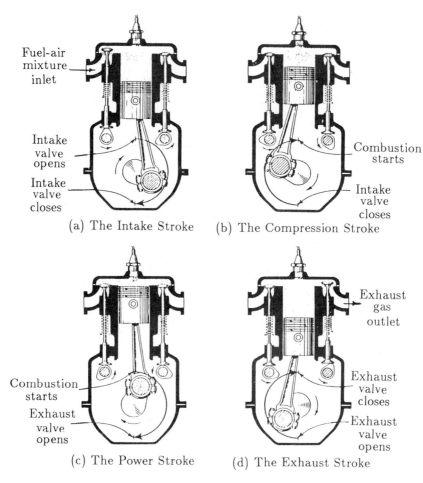

Fuel-air
mixture
inlet

Intake
valve
opens

Intake
valve
closes

(a) The Intake Stroke

Combustion
starts

Intake
valve
closes

(b) The Compression Stroke

Combustion
starts

Exhaust
valve
opens

(c) The Power Stroke

Exhaust
gas
outlet

Exhaust
valve
closes

Exhaust
valve
opens

(d) The Exhaust Stroke

Fig.1.4 Working Principle of a Four-Stroke SI Engine

of a spark plug located on the cylinder head. Burning takes place almost instantaneously when the piston is at the top dead centre and hence the burning process can be approximated as heat addition at constant volume. During the burning process the chemical energy of the fuel is converted into heat energy producing a temperature rise of about 2000 °C (Process 2→3) (Fig.1.5). The pressure at the end of the combustion process is considerably increased due to the heat release.

(iii) *Expansion or Power Stroke :*   The high pressure of the burnt gases forces the piston towards the *BDC*, stroke 3→4 (Fig.1.5),

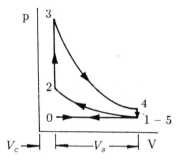

Fig.1.5 Ideal Indicator Diagram of a Four-Stroke SI Engine

with both the inlet and exhaust valves remaining closed, Fig.1.4(c). Thus, power is obtained during this stroke. Both pressure and temperature decrease during expansion.

(iv) *Exhaust Stroke :* At the end of the expansion stroke the exhaust valve opens and the inlet valve remains closed, Fig.1.4(d). The pressure falls to atmospheric level as a part of the burnt gases escape. The piston moves from the bottom dead centre to top dead centre (stroke 5→0), Fig.1.5 and sweeps the burnt gases out from the cylinder almost at atmospheric pressure. The exhaust valve closes at the end of the exhaust stroke and some residual gases trapped in the clearance volume remain in the cylinder.

These residual gases mix with the fresh charge coming in during the following cycle, forming its working fluid. Each cylinder of a four-stroke engine completes the above four operations in two engine revolutions, one revolution of the crankshaft occurs during the suction and compression strokes and the second revolution during the power and exhaust strokes. Thus for one complete cycle there is only one power stroke while the crankshaft turns by two revolutions.

1.3.2 Four-Stroke Compression Ignition Engine

The four-stroke CI engine is similar to the four-stroke SI engine but it operates at a much higher compression ratio. The compression ratio of an SI engine varies from 6 to 10 while for a CI engine it is from 16 to 20. In the CI engine during suction stroke, air, instead of a fuel-air mixture, is inducted. Due to the high compression ratio employed, the temperature at the end of the compression stroke is sufficiently high to self ignite the fuel which is injected into the combustion chamber. In CI engines, a high pressure fuel pump and an injector are provided to inject the fuel into the combustion chamber. The carburettor and

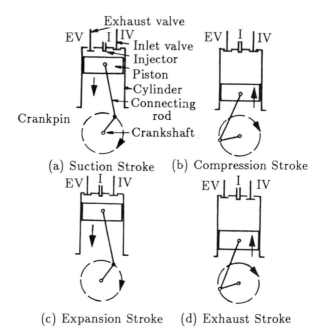

(a) Suction Stroke   (b) Compression Stroke

(c) Expansion Stroke   (d) Exhaust Stroke

Fig.1.6 Cycle of Operation of a CI Engine

ignition system necessary in the SI engine are not required in the CI engine.

The ideal sequence of operations for the four-stroke CI engine is as follows :

(i) *Suction Stroke :* Air alone is inducted during the suction stroke. During this stroke intake valve is open and exhaust valve is closed, Fig.1.6(a).

(ii) *Compression Stroke :* Air inducted during the suction stroke is compressed into the clearance volume. Both valves remain closed during this stroke, Fig.1.6(b).

(iii) *Expansion Stroke :* Fuel injection starts nearly at the end of the compression stroke. The rate of injection is such that the combustion maintains the pressure constant inspite of the piston movement on its expansion stroke increasing the volume. Heat is assumed to have been added at constant pressure. After the injection of fuel is completed (i.e. after cut-off) the products of combustion expand. Both the valves remain closed during the expansion stroke, Fig.1.6(c).

(iv) *Exhaust Stroke :* The piston traveling from BDC to TDC pushes out the products of combustion. The exhaust valve is open and the intake valve is closed during this stroke, Fig.1.6(d).

Due to higher pressures in the cycle of operations the CI engine has to be more sturdy than a spark-ignition engine for the same output. This results in a CI engine being heavier than the SI engine. However, it has a higher thermal efficiency on account of the high compression ratio (of about 18 as against about 8 in SI engines) used.

1.3.3 Comparison of SI and CI Engines

In four-stroke engines, there is one power stroke for every two revolutions of the crankshaft. There are two non-productive strokes of exhaust and suction which are necessary for flushing the products of combustion from the cylinder and filling it with the fresh charge. If this purpose could be served by an alternative arrangement, without the movement of the piston, it is possible to obtain a power stroke for every revolution of the crankshaft increasing the output of the engine. However, in both SI and CI engines operating on four-stroke cycle, power can be obtained only in every two revolution of the crankshaft.

Since both SI and CI engines have much in common, it is worthwhile to compare them based on important parameters like basic cycle of operation, fuel induction, compression ratio etc. The detailed comparison is given in Table 1.1.

1.3.4 Two-Stroke Engine

As already mentioned, if the two unproductive strokes, viz., the suction and exhaust could be served by an alternative arrangement, especially without the movement of the piston then there will be a power stroke for each revolution of the crankshaft. In such an arrangement, theoretically the power output of the engine can be doubled for the same speed compared to a four-stroke engine. Based on this concept, Dugald Clark (1878) invented the two-stroke engine.

In two-stroke engines the cycle is completed in one revolution of the crankshaft. The main difference between two-stroke and four-stroke engines is in the method of filling the fresh charge and removing the burnt gases from the cylinder. In the four-stroke engine these operations are performed by the engine piston during the suction and exhaust strokes respectively. In a two-stroke engine, the filling process is accomplished by the charge compressed in crankcase or by a blower. The induction of the compressed charge moves out the product of combustion through exhaust ports. Therefore, no piston strokes are required for these two operations. Two strokes are sufficient to complete the cycle, one for compressing the fresh charge and the other for expansion or power stroke.

Table 1.1 *Comparison of SI and CI Engines*

Description	SI Engine	CI Engine
Basic cycle	Otto cycle or constant volume heat addition cycle	Diesel cycle or constant pressure heat addition cycle.
Fuel	Gasoline, a highly volatile fuel. Self-ignition temperature is high.	Diesel oil, a non-volatile fuel. Self-ignition temperature is comparatively low.
Introduction of fuel	A gaseous mixture of fuel and air is introduced during the suction stroke. A carburettor is necessary to provide the mixture.	Fuel is injected directly into the combustion chamber at high pressure at the end of the compression stroke. A fuel pump and injector are necessary.
Load control	Throttle controls the quantity of mixture introduced.	The quantity of fuel is regulated in the pump. Air quantity is not controlled.
Ignition	Requires an ignition system with spark plug in the combustion chamber. Primary voltage is provided by a battery or a magneto.	Self-ignition occurs due to high temperature of air because of the high compression. Ignition system and spark plug are not necessary.
Compression ratio	6 to 10. Upper limit is fixed by antiknock quality of the fuel.	16 to 20. Upper limit is limited by weight increase of the engine.
Speed	Due to light weight and also due to homogeneous combustion, they are high speed engines	Due to heavy weight and also due to heterogeneous combustion, they are low speed engines
Thermal efficiency	Because of the lower CR, the maximum value of thermal efficiency that can be obtained is lower.	Because of higher CR, the maximum value of thermal efficiency that can be obtained is higher.
Weight	Lighter due to lower peak pressures.	Heavier due to higher peak pressures.

Figure 1.7 shows one of the simplest two-stroke engines, viz., the crankcase scavenged engine. Figure 1.8 shows the ideal indicator diagram of the crankcase scavenged two-stroke engine. The air or charge is inducted into the crankcase through the spring loaded inlet valve when the pressure in the crankcase is reduced due to upward motion of the piston during compression stroke. After the compression and ignition, expansion takes place in the usual way.

During the expansion stroke the charge in the crankcase is compressed. Near the end of the expansion stroke, the piston uncovers the exhaust ports and the cylinder pressure drops to atmospheric pressure as the combustion products leave the cylinder. Further movement of the piston uncovers the transfer ports, permitting the slightly compressed charge in the crankcase to enter the engine cylinder. The top of the piston has usually a projection to deflect the fresh charge towards the top of the cylinder before flowing to the exhaust ports. This serves the double purpose of scavenging the upper part of the cylinder of the combustion products and preventing the fresh charge from flowing directly to the exhaust ports.

The same objective can be achieved without piston deflector by proper shaping of the transfer port. During the upward motion of the piston from BDC the transfer ports close first and then the exhaust ports close when compression of the charge begins and the cycle is repeated.

1.3.5 Comparison of Four-Stroke and Two-Stroke Engines

The two-stroke engine was developed to obtain a greater output from the same size of the engine. The engine mechanism also eliminates the valve arrangement making it mechanically simpler. Almost all two-stroke engines have no conventional valves but only ports (some have an exhaust valve). This simplicity of the two-stroke engine makes it cheaper to produce and easy to maintain. Theoretically a two-stroke engine develops twice the power of a comparable four-stroke engine because of one power stroke every revolution (compared to one power stroke every two revolutions of a four-stroke engine). This makes the two-stroke engine more compact than a comparable four-stroke engine. In actual practice power output is not exactly doubled but increased by only about 30% because of

(i) reduced effective expansion stroke and

(ii) increased heating caused by increased number of power strokes which limits the maximum speed.

The other advantages of the two-stroke engine are more uniform torque on crankshaft and comparatively less exhaust gas dilution. However, when applied to the spark-ignition engine the two-stroke

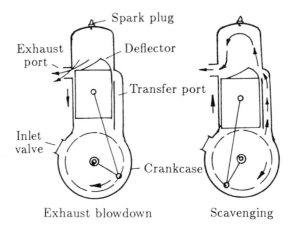

Exhaust blowdown   Scavenging

Fig.1.7 Crank Case Scavenged Two-Stroke Engine

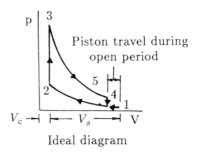

Ideal diagram

Fig.1.8 Ideal Indicator Diagram of a
Two-Stroke SI Engine

cycle has certain disadvantages which have restricted its application
to only small engines suitable for motor cycles, scooters, lawn mowers,
outboard engines etc. In the SI engine, the incoming charge consists
of fuel and air. During scavenging, as both inlet and exhaust ports are
open simultaneously for some time, there is a possibility that some
of the fresh charge containing fuel escapes with the exhaust. This
results in high fuel consumption and lower thermal efficiency. The
other drawback of two-stroke engine is the lack of flexibility, viz., the
capacity to operate with the same efficiency at all speeds. At part
throttle operating condition, the amount of fresh mixture entering
the cylinder is not enough to clear all the exhaust gases and a part of
it remains in the cylinder to contaminate the charge. This results in
irregular operation of the engine.

The two-stroke diesel engine does not suffer from these defects. There is no loss of fuel with exhaust gases as the intake charge in diesel engine is only air. The two-stroke diesel engine is used quite widely. Many of the high output diesel engines work on this cycle. A disadvantage common to all two-stroke engines, gasoline as well as diesel, is the greater cooling and lubricating requirements due to one power stroke in each revolution of the crankshaft. Consumption of lubricating oil is also high in two-stroke engines due to higher temperature. Table 1.2 gives the comparison of two-stroke and four-stroke engines.

1.4 ACTUAL ENGINES

Actual engines differ from the ideal engines because of various constraints in their operation. The indicator diagram also differs considerably from the ideal indicator diagrams. The details of the actual indicator diagrams for the four-stroke and the two-stroke SI engines are given in Figs.1.9(a) and 1.9(b) respectively.

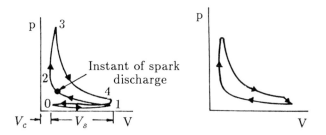

(a) Four-stroke Engine   (b) Two-stroke Engine

Fig.1.9 Actual Indicator Diagrams of a Four-Stroke
and Two-Stroke SI Engine

1.5 CLASSIFICATION OF IC ENGINES

Internal combustion engines are usually classified on the basis of the thermodynamic cycle of operation, type of fuel used, method of charging the cylinder, type of ignition, type of cooling and the cylinder arrangement. Details are given in Fig.1.10.

1.5.1 Number of Strokes

SI and CI engines are classified into four-stroke and two-stroke engines depending on the number of piston strokes for completing a cycle of operation. Accordingly they are classified as

(i) Four-stroke engines   (ii) Two-stroke engines

Table 1.2 *Comparison of Four and Two-Stroke Cycle Engines*

Four-Stroke Engine	Two-Stroke Engine
The thermodynamic cycle is completed in four strokes of the piston or in two revolutions of the crankshaft. Thus, one power stroke is obtained in every two revolutions of the crankshaft.	The thermodynamic cycle is completed in two strokes of the piston or in one revolution of the crankshaft. Thus one power stroke is obtained in each revolution of the crankshaft.
Because of the above, turning moment is not so uniform and hence a heavier flywheel is needed.	Because of the above, turning moment is more uniform and hence a lighter flywheel can be used.
Again, because of one power stroke for two revolutions, power produced for same size of engine is less, or for the same power the engine is heavier and bulkier.	Because of one power stroke for every revolution, power produced for same size of engine is more (theoretically twice; actually about 1.3 times), or for the same power the engine is lighter and more compact.
Because of one power stroke in two revolutions lesser cooling and lubrication requirements. Lower rate of wear and tear.	Because of one power stroke in one revolution greater cooling and lubrication requirements. Higher rate of wear and tear.
The four-stroke engine contains valves and valve actuating mechanisms to open and close the valves.	Two-stroke engines have no valves but only ports (some two-stroke engines are fitted with conventional exhaust valve or reed valve).
Because of the heavy weight and complicated valve mechanism, the initial cost of the engine is more.	Because of light weight and simplicity due to the absence of valve mechanism, initial cost of the engine is less.
Volumetric efficiency is more due to more time for induction.	Volumetric efficiency is low due to lesser time for induction.
Thermal efficiency is higher; part load efficiency is better than two-stroke cycle engine.	Thermal efficiency is lower; part load efficiency is poor compared to a four-stroke cycle engine.
Used where efficiency is important, viz., in cars, buses, trucks, tractors, industrial engines, aeroplanes, power generation etc.	Used where low cost, compactness and light weight are important, viz., in mopeds, scooters, motorcycles, hand sprayers etc.

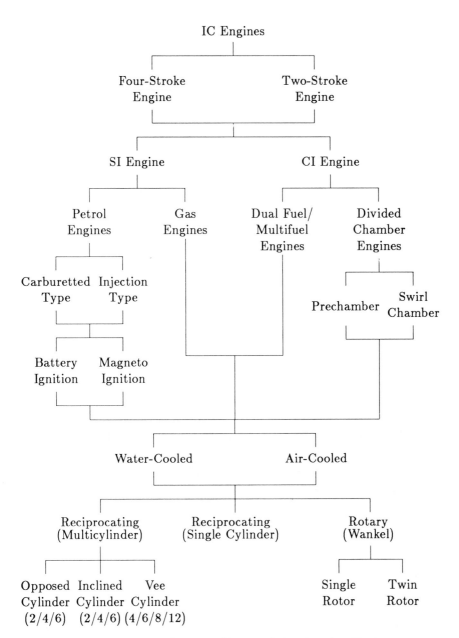

Fig.1.10 Classification of Internal Combustion Engines

1.5.2 Cycle of Operation

According to the cycle of operation, IC engines are basically classified into two categories

(i) Constant volume heat addition cycle engine or Otto cycle engine. It is also called a Spark Ignition engine, SI engine or Gasoline engine.

(ii) Constant-pressure heat addition cycle engine or Diesel cycle engine. It is also called a Compression Ignition engine, CI engine or Diesel engine.

1.5.3 Type of Fuel Used

Based on the type of fuel used engines are classified as

(i) Engines using volatile liquid fuels like gasoline, alcohol, kerosene, benzene etc.

The fuel is generally mixed with air to form a homogeneous charge in a carburettor outside the cylinder and drawn into the cylinder in its suction stroke. The charge is ignited near the end of the compression stroke by an externally applied spark and therefore these engines are called Spark Ignition engines.

(ii) Engines using gaseous fuels like natural gas, petroleum gas, blast furnace gas, town gas and biogas etc.

The gas is mixed with air and the mixture is introduced into the cylinder during the suction process. Working of this type of engine is similar to that of the engines using volatile liquid fuels (SI gas engine).

(iii) Engine using solid fuels like charcoal, powdered coal etc.

Solid fuels are generally converted into gaseous fuels outside the engine in a separate gas producer and the engine works as a gas engine.

(iv) Engines using viscous (low volatility at normal atmospheric temperatures) liquid fuels like heavy and light diesel oils.

The fuel is generally introduced into the cylinder in the form of minute droplets by a fuel injection system near the end of the compression process. Combustion of the fuel takes place due to its coming into contact with the high temperature compressed air in the cylinder. Therefore, these engines are called Compression Ignition engines.

(v) Engines using two fuels (Dual-fuel engines)

A gaseous fuel or a highly volatile liquid fuel is supplied along with air during the suction stroke or during the initial part of compression through a gas valve in the cylinder head and the other fuel (a viscous liquid fuel) is injected into the combustion space near the end of the compression stroke (dual-fuel engines).

1.5.4 Method of Charging

According to the method of charging, the engines are classified as

(i) Naturally aspirated engines : Admission of air or fuel-air mixture at near atmospheric pressure.

(ii) Supercharged Engines : Admission of air or fuel-air mixture under pressure, i.e., above atmospheric pressure.

1.5.5 Type of Ignition

Spark-ignition engines require an external source of energy for the initiation of spark and thereby the combustion process. A high voltage spark is made to jump across the spark plug electrodes. In order to produce the required high voltage there are two types of ignition systems which are normally used. They are :

(i) battery ignition system

(ii) magneto ignition system

They derive their name based on whether a battery or a magneto is used as the primary source of energy for producing the spark.

In the case of CI engines there is no need for an external means to produce the ignition. Because of high compression ratio employed, the resulting temperature at the end of the compression process is high enough to self-ignite the fuel when injected. However, the fuel should be atomized into very fine particles. For this purpose a fuel injection system is normally used.

1.5.6 Type of Cooling

Cooling is very essential for the satisfactory running of an engine. There are two types of cooling systems in use and accordingly, the engines is classified as

(i) air-cooled engine

(ii) water-cooled engine

1.5.7 Cylinder Arrangements

Another common method of classifying reciprocating engines is by the cylinder arrangement. The cylinder arrangement is only applicable to multi cylinder engines. Two terms used in connection with cylinder arrangements must be defined first.

(i) *Cylinder Row :* An arrangement of cylinders in which the centre-line of the crankshaft journals is perpendicular to the plane containing the centrelines of the engine cylinders.

(ii) *Cylinder Bank* : An arrangement of cylinders in which the centre-line of the crankshaft journals is parallel to the plane containing the centrelines of the engine cylinders.

A number of cylinder arrangements popular with designers are described below. The details of various cylinder arrangements are shown in Fig.1.11.

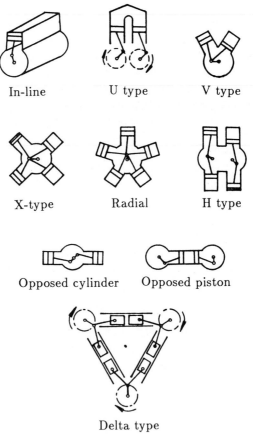

Fig.1.11 Engine Classification by Cylinder Arrangements

In-line Engine :   The in-line engine is an engine with one cylinder bank, i.e. all cylinders are arranged linearly, and transmit power to a single crankshaft. This type is quite common with automobile engines. Four and six cylinder in-line engines are popular in automotive applications.

'V' *Engine* :   In this engine there are two banks of cylinders (i.e., two in line engines) inclined at an angle to each other and with one crankshaft. Most of the high powered automobiles use the 8 cylinder 'V' engine, four in line on each side of the 'V'. Engines with more than six cylinders generally employ this configuration.

Opposed Cylinder Engine :   This engine has two cylinder banks located in the same plane on opposite sides of the crankshaft. It can be visualized as two 'in-line' arrangements 180 degrees apart. It is inherently a well balanced engine and has the advantages of a single crankshaft. This design is used in small aircrafts.

Opposed Piston Engine :   When a single cylinder houses two pistons, each of which driving a separate crankshaft, it is called an opposed piston engine. The movement of the pistons is synchronized by coupling the two crankshafts. Opposed piston arrangement, like opposed cylinder arrangement, is inherently well balanced. Further, it has the advantage of requiring no cylinder head. By its inherent features, this engine usually functions on the principle of two-stroke engines.

Radial Engine :   Radial engine is one where more than two cylinders in each row are equally spaced around the crankshaft. The radial arrangement of cylinders is most commonly used in conventional air-cooled aircraft engines where 3, 5, 7 or 9 cylinders may be used in one bank and two to four banks of cylinders may be used. The odd number of cylinders is employed from the point of view of balancing. Pistons of all the cylinders are coupled to the same crankshaft.

'X' *Type Engine* :   This design is a variation of 'V' type. It has four banks of cylinders attached to a single crankshaft.

'H' *Type Engine* :   The 'H' type is essentially two 'Opposed cylinder' type utilizing two separate but interconnected crankshafts.

'U' *Type Engine* :   The 'U' type is a variation of opposed piston arrangement.

Delta Type Engine :   The delta type is essentially a combination of three opposed piston engine with three crankshafts interlinked to one another.

In general, automobile engines and general purpose engines utilize the 'in-line' and 'V' type configuration or arrangement. The 'radial' engine was used widely in medium and large aircrafts till it was replaced by the gas turbine. Small aircrafts continue to use either the 'opposed cylinder' type or 'in-line' or 'V' type engines. The 'opposed piston' type engine is widely used in large diesel installations. The 'H' and 'X' types do not presently find wide application, except in some diesel installations. A variation of the 'X' type is referred to as the 'pancake' engine.

1.6 APPLICATION OF IC ENGINES

The most important application of IC engines is in transport on land, sea and air. Other applications include industrial power plants and as prime movers for electric generators. Table 1.3 gives, in a nutshell, the applications of both IC and EC engines.

Table 1.3 *Application of Engines*

IC Engine		EC Engine	
Type	Application	Type	Application
Gasoline Engines	Automotive Marine Aircraft	Steam Engines	Locomotives Marine
Gas Engines	Industrial Power	Stirling Engines	Experimental Space Vehicles
Diesel Engines	Automotive Locomotive Power Marine	Steam Turbines	Power Large Marine
Gas Turbines	Power Aircraft Industrial Marine	Closed Cycle Gas Turbine	Power Marine

1.6.1 Two-Stroke Gasoline Engines

Small two-stroke gasoline engines are used where simplicity and low cost of the prime mover are the main considerations. In such applications a little higher fuel consumption is acceptable. The smallest engines are used in mopeds (50 cc engine) and lawn mowers. Scooters and motor cycles, the commonly used two wheeler transport, have generally 100-150 cc, two-stroke gasoline engines developing a maximum brake power of about 5 kW at 5500 rpm. High powered motor cycles have generally 250 cc two-stroke gasoline engines developing a maximum brake power of about 10 kW at 5000 rpm. Two-stroke gasoline engines may also be used in very small electric generating sets, pumping sets, and outboard motor boats.

However, their specific fuel consumption is higher due to the loss of fuel-air charge in the process of scavenging and because of high speed of operation for which such small engines are designed.

1.6.2 Two-Stroke Diesel Engines

Very high power diesel engines used for ship propulsion are commonly two-stroke diesel engines. In fact all engines between 400 to 900 mm bore are two-stroke engines, uniflow with exhaust valves or loop scavenged. The brake power on a single crankshaft can be upto 37000 kW. Nordberg, 12 cylinder 800 mm bore and 1550 mm stroke, two-stroke diesel engine develops 20000 kW at 120 rpm. This speed allows the engine to be directly coupled to the propeller of a ship without the necessity of gear reducers.

1.6.3 Four-Stroke Gasoline Engines

The most important application of small four-stroke gasoline engines is in automobiles. A typical automobile is powered by a four-stroke four cylinder engine developing an output in the range of 30-60 kW at a speed of about 4500 rpm. American automobile engines are much bigger and have 6 or 8 cylinder engines with a power output upto 185 kW. However, the oil crisis and air pollution from automobile engines have reversed this trend towards smaller capacity cars.

Four-stroke gasoline engines were also used for buses and trucks. They were generally 4000 cc, 6 cylinder engines with maximum brake power of about 90 kW. However, in this application gasoline engines have been practically replaced by diesel engines. The four-stroke gasoline engines have also been used in big motor cycles with side cars. Another application of four-stroke gasoline engine is in small pumping sets and mobile electric generating sets.

Small aircraft generally use radial four-stroke gasoline engines. Engines having maximum power output from 400 kW to 4000 kW have been used in aircraft. An example is the Bristol Contours 57, 18 cylinder two row, sleeve valve, air-cooled radial engine developing, a maximum brake power of about 2100 kW.

Modern large aircraft use gas turbine such as turboprop engine or turbo jet engine and these have replaced the radial gasoline engine except in small sizes (200 kW range).

1.6.4 Four-Stroke Diesel Engines

The four-stroke diesel engine is one of the most efficient and versatile prime movers. It is manufactured in sizes from 50 mm to more than 1000 mm of cylinder diameter and with engine speeds ranging from 100 to 4500 rpm while delivering outputs from 1 to 35000 kW.

Small diesel engines are used in pump sets, construction machinery, air compressors, drilling rigs and many miscellaneous applications. Tractors for agricultural application use about 30 kW diesel engines whereas jeeps, buses and trucks use 40 to 100 kW diesel engines. Generally, the diesel engines with higher outputs than about 100 kW are supercharged. Earth moving machines use supercharged diesel engines in the output range of 200 to 400 kW. Locomotive applications require outputs of 600 to 4000 kW. Marine applications, from fishing vessels to ocean going ships use diesel engines from 100 to 35000 kW. Diesel engines are used both for mobile and stationary electric generating plants of varying capacities.

Development is going on in the use of diesel engines in personal automobiles. This is mainly because, compared to gasoline engines, diesel engines are more efficient. However, the vibrations from the engine and the unpleasant odour in the exhaust are the main drawbacks.

1.7 THE FIRST LAW ANALYSIS OF ENGINE CYCLE

Before a detailed thermodynamic analysis of the engine cycle is done, it is desirable to have a general picture of the energy flow or energy balance of the system so that one becomes familiar with the various performance parameters. Figure 1.12 shows the energy flow through the reciprocating engine and Fig.1.13 shows its block diagram as an open system.

According to the first law of thermodynamics, energy can neither be created nor destroyed. It can only be converted from one form to another. Therefore, there must be an energy balance of input and output to a system.

In the reciprocating internal combustion engine the fuel is fed into the combustion chamber where it burns in air converting chemical energy of the fuel into heat. The liberated heat energy cannot be totally utilized for driving the piston as there are losses through the engine exhaust, to the coolant and due to radiation. The heat energy which is converted to power at this stage is called the indicated power, *ip* and it is utilized to drive the piston. The energy represented by the gas forces on the piston passes through the connecting rod to the crankshaft. In this transmission there are energy losses due to bearing friction, pumping losses etc. In addition, a part of the energy available is utilized in driving the auxiliary devices like feed pump, valve mechanisms, ignition systems etc. The sum of all these losses, expressed in power units is termed as frictional power, *fp*. The remaining energy is the useful mechanical energy and is termed as the brake power, *bp*. In energy balance, generally, frictional power is not shown separately because ultimately this energy is accounted in exhaust, cooling water, radiation, etc.

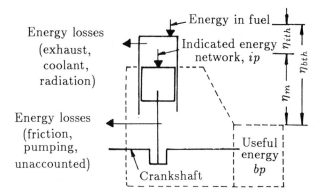

Fig.1.12 Energy Flow through the Reciprocating Engine

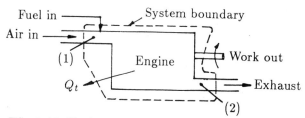

Fig.1.13 Reciprocating Engine as an Open System

1.8 ENGINE PERFORMANCE PARAMETERS

The engine performance is indicated by the term *efficiency*, η. Five important engine efficiencies and other related engine performance parameters are given below :

(i)	Indicated thermal efficiency	(η_{ith})
(ii)	Brake thermal efficiency	(η_{bth})
(iii)	Mechanical efficiency	(η_m)
(iv)	Volumetric efficiency	(η_v)
(v)	Relative efficiency or Efficiency ratio	(η_{rel})
(vi)	Mean effective pressure	(p_m)
(vii)	Mean piston speed	$(\bar{s}_p)$
(viii)	Specific power output	(P_s)
(ix)	Specific fuel consumption	(sfc)
(x)	Fuel-air or air-fuel ratio	$(F/A \text{ or } A/F)$
(xi)	Calorific value of the fuel	(CV)

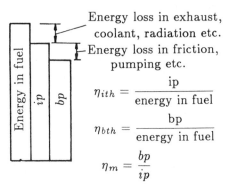

$$\eta_{ith} = \frac{ip}{\text{energy in fuel}}$$

$$\eta_{bth} = \frac{bp}{\text{energy in fuel}}$$

$$\eta_m = \frac{bp}{ip}$$

Fig.1.14 Energy Distribution

Figure 1.14 shows the diagrammatic representation of energy distribution in an IC engine.

1.8.1 Indicated Thermal Efficiency (η_{ith})

Indicated thermal efficiency is the ratio of energy in the indicated power, *ip*, to the input fuel energy in appropriate units.

$$\eta_{ith} = \frac{ip \ [\text{kJ/s}]}{\text{energy in fuel per second} \ [\text{kJ/s}]} \qquad (1.3)$$

$$= \frac{ip}{\text{mass of fuel/s} \times \text{calorific value of fuel}} \qquad (1.4)$$

1.8.2 Brake Thermal Efficiency (η_{bth})

Brake thermal efficiency is the ratio of energy in the brake power, *bp*, to the input fuel energy in appropriate units.

$$\eta_{bth} = \frac{bp}{\text{mass of fuel/s} \times \text{calorific value of fuel}} \qquad (1.5)$$

1.8.3 Mechanical Efficiency (η_m)

Mechanical efficiency is defined as the ratio of brake power (delivered power) to the indicated power (power provided to the piston).

$$\eta_m \quad = \quad \frac{bp}{ip} \quad = \quad \frac{bp}{bp + fp} \qquad (1.6)$$

$$fp \quad = \quad ip - bp \qquad (1.7)$$

It can also be defined as the ratio of the brake thermal efficiency to the indicated thermal efficiency.

1.8.4 Volumetric Efficiency (η_v)

The engine output is limited by the maximum amount of air that can be taken in during the suction stroke because only a certain amount of fuel can be burned effectively with this quantity of air. Volumetric efficiency is an indication of the breathing ability of the engine and is defined as the ratio of the volume of air actually inducted at ambient conditions to the swept volume of the engine. The volumetric efficiency can be calculated considering mass or volume of air. However, it is preferable to use mass basis.

$$\eta_v = \frac{\text{mass of charge actually inducted}}{\substack{\text{mass of charge represented by swept volume} \\ \text{at ambient temperature and pressure}}} \qquad (1.8)$$

$$= \frac{\substack{\text{volume of charge aspirated per stroke} \\ \text{at ambient conditions}}}{\text{swept volume}} \qquad (1.9)$$

1.8.5 Relative Efficiency or Efficiency Ratio (η_{rel})

Relative efficiency or efficiency ratio is the ratio of thermal efficiency of an actual cycle to that of the ideal cycle. The efficiency ratio is a very useful criterion which indicates the degree of development of the engine.

$$\eta_{rel} = \frac{\text{Actual thermal efficiency}}{\text{Air-standard efficiency}} \qquad (1.10)$$

1.8.6 Mean Effective Pressure (p_m)

Mean effective pressure is the average pressure inside the cylinders of an internal combustion engine based on the calculated or measured power output. It increases as manifold pressure increases. For any particular engine, operating at a given speed and power output, there will be a specific indicated mean effective pressure, *imep*, and a corresponding brake mean effective pressure, *bmep*. They are derived from the indicated and brake power respectively. For derivation see Chapter 14. Indicated power can be shown to be

$$ip \;=\; \frac{p_{im}LAnK}{60 \times 1000} \qquad (1.11)$$

then, the indicated mean effective pressure can be written as

$$p_{im} \;=\; \frac{60000 \times ip}{LAnK} \qquad (1.12)$$

Similarly, the brake mean effective pressure is given by

$$p_{bm} \;=\; \frac{60000 \times bp}{LAnK} \qquad (1.13)$$

where ip =   indicated power (kW)

p_{im} =   indicated mean effective pressure (N/m^2)

L =   length of the stroke (m)

A =   area of the piston (m^2)

N =   speed in revolutions per minute (rpm)

n =   Number of power strokes
$N/2$ for 4-stroke and N for 2-stroke engines

K =   number of cylinders

Another way of specifying the indicated mean effective pressure p_{im} is from the knowledge of engine indicator diagram (p-V diagram). In this case, p_{im}, may be defined as

$$p_{im} \quad = \quad \frac{\text{Area of the indicator diagram}}{\text{Length of the indicator diagram}}$$

where the length of the indicator diagram is given by the difference between the total volume and the clearance volume.

1.8.7 Mean Piston Speed $(\bar{s}_p)$

An important parameter in engine applications is the mean piston speed, $\bar{s}_p$. It is defined as

$$\bar{s}_p \quad = \quad 2LN$$

where L is the stroke and N is the rotational speed of the crankshaft in rpm. It may be noted that $\bar{s}_p$ is often a more appropriate parameter than crank rotational speed for correlating engine behaviour as a function of speed.

Resistance to gas flow into the engine or stresses due to the inertia of the moving parts limit the maximum value of $\bar{s}_p$ to within 8 to 15 m/s. Automobile engines operate at the higher end and large marine diesel engines at the lower end of this range of piston speeds.

1.8.8 Specific Power Output (P_s)

Specific power output of an engine is defined as the power output per unit piston area and is a measure of the engine designer's success in using the available piston area regardless of cylinder size. The specific power can be shown to be proportional to the product of the mean effective pressure and mean piston speed.

$$\text{Specific power output, } P_s \quad = \quad bp/A \qquad (1.14)$$

$$= \quad \text{constant} \times p_{bm} \times \bar{s}_p \quad (1.15)$$

Thus the specific power output consists of two elements, viz., the force available to work and the speed with which it is working. Thus, for the same piston displacement and *bmep*, an engine running at a higher speed will give a higher output.

It is clear that the output of an engine can be increased by increasing either the speed or the *bmep*. Increasing the speed involves increase in the mechanical stresses of various engine components. For increasing the *bmep* better heat release from the fuel is required and this will involve more thermal load on engine cylinder.

1.8.9 Specific Fuel Consumption (sfc)

The fuel consumption characteristics of an engine are generally expressed in terms of specific fuel consumption in kilograms of fuel per kilowatt-hour. It is an important parameter that reflects how good the engine performance is. It is inversely proportional to the thermal efficiency of the engine.

$$\text{sfc} \quad = \quad \frac{\text{Fuel consumption per unit time}}{\text{Power}} \qquad (1.16)$$

Brake specific fuel consumption and indicated specific fuel consumption, abbreviated as *bsfc* and *isfc*, are the specific fuel consumptions on the basis of *bp* and *ip* respectively.

1.8.10 Fuel-Air (F/A) or Air-Fuel Ratio (A/F)

The relative proportions of the fuel and air in the engine are very important from the standpoint of combustion and the efficiency of the engine. This is expressed either as a ratio of the mass of the fuel to that of the air or vice versa.

In the SI engine the fuel-air ratio practically remains a constant over a wide range of operation. In CI engines at a given speed the air flow does not vary with load; it is the fuel flow that varies directly with load. Therefore, the term fuel-air ratio is generally used instead of air-fuel ratio.

A mixture that contains just enough air for complete combustion of all the fuel in the mixture is called a chemically correct or stoichiometric fuel-air ratio. A mixture having more fuel than that in a chemically correct mixture is termed as rich mixture and a mixture that contains less fuel (or excess air) is called a lean mixture. The ratio of actual fuel-air ratio to chemically correct fuel-air ratio is called equivalence ratio and is denoted by ϕ.

$$\phi = \frac{\text{Actual fuel-air ratio}}{\text{Stoichiometric fuel-air ratio}} \qquad (1.17)$$

Accordingly, $\phi = 1$ means chemically correct mixture, $\phi < 1$ means lean mixture and $\phi > 1$ means rich mixture.

1.8.11 Calorific Value (CV)

Calorific value of a fuel is the thermal energy released per unit quantity of the fuel when the fuel is burned completely and the products of combustion are cooled back to the initial temperature of the combustible mixture. Other terms used for the calorific value are heating value and heat of combustion.

When the products of combustion are cooled to 25 °C practically all the water vapour resulting from the combustion process is condensed. The heating value so obtained is called the higher calorific value or gross calorific value of the fuel. The lower or net calorific value is the heat released when water vapour in the products of combustion is not condensed and remains in the vapour form.

1.9 DESIGN AND PERFORMANCE DATA

Engine ratings usually indicate the highest power at which manufacturers expect their products to give satisfactory economy, reliability, and durability under service conditions. Maximum torque, and the speed at which it is achieved, is usually given also. Since both of these quantities depend on displaced volume, for comparative analysis between engines of different displacements in a given engine category normalized performance parameters are more useful.

Typical design and performance data for SI and CI engines used in different applications are summarized in Table 1.4. The four-stroke cycle dominates except in the smallest and largest engines. The larger engines are turbocharged or supercharged. The maximum rated engine speed decreases as engine size increases, maintaining the maximum mean piston speed in the range of about 8 to 15 m/s. The maximum brake mean effective pressure for turbocharged and supercharged engines is higher than for naturally aspirated engines. Because the maximum fuel-air ratio for SI engines is higher than for CI engines, their naturally aspirated maximum *bmep* levels are higher. As the engine size increases, brake specific fuel consumption decreases and fuel conversion efficiency increases due to the reduced heat losses and friction. For the large CI engines, brake thermal efficiencies of about 40% and indicated thermal efficiencies of about 50% can be obtained in modern engines.

Table 1.4 *Typical Design and Performance Data for Modern Internal Combustion Engines*

	Operating cycle (Stroke)	Compression ratio	Bore (m)	Stroke/ bore ratio	Rated Maximum		Weight/ Power ratio (kg/kW)	Approx. best bsfc (g/kW h)
					Speed (rev/min)	bmep (atm)		
Spark-ignition engines								
Small (e.g. motorcycles)	2/4	6–10	0.05–0.085	1.2–0.9	4500–7500	4–10	5.5–2.5	350
Passenger cars	4	8–10	0.07–0.1	1.1–0.9	4500–6500	7–10	4–2	270
Trucks	4	7–9	0.09–0.13	1.2–0.7	3600–5000	6.5–7	6.5–2.5	300
Large gas engines	2/4	8–12	0.22–0.45	1.1–1.4	300–900	6.8–12	23–35	200
Wankel engines	4	≈ 9	0.57 dm³ per chamber		6000–8000	9.5–10.5	1.6–0.9	300
Compression-ignition engines								
Passenger cars	4	16–20	0.075–0.1	1.2–0.9	4000–5000	5–7.5	5–2.5	250
Trucks	4	16–20	0.1–0.15	1.3–0.8	2100–4000	6–9	7–4	210
Locomotive	4/2	16–18	0.15–0.4	1.1–1.3	425–1800	7–23	6–18	190
Large engines	2	10–12	0.4–1	1.2–3.0	110–400	9–17	12–50	180

Worked out Examples

1.1. A four-cylinder, four-stroke, spark-ignition engine has a bore of 80 mm and stroke of 80 mm. The compression ratio is 8. Calculate the cubic capacity of the engine and the clearance volume of each cylinder. What type of engine is this?

SOLUTION:

$$\text{Swept volume, } V_s \;=\; \frac{\pi}{4}d^2 L \;=\; \frac{\pi}{4} \times 8^2 \times 8$$

$$=\; 402.1 \text{ cc}$$

$$\text{Cubic capacity of the engine} \;=\; \text{Number of cylinders} \times V_s$$

$$=\; 4 \times 402.1$$

$$=\; \mathbf{1608.4 \text{ cc}} \qquad \underset{\displaystyle =}{\text{Ans}}$$

$$\text{Compression ratio, } r \;=\; \frac{V_s + V_C}{V_C} \;=\; 1 + \frac{V_s}{V_C}$$

$$\text{Clearance volume, } V_C \;=\; \left(\frac{1}{r-1}\right) V_s$$

$$=\; \frac{1}{8-1} \times 402.1$$

$$=\; \mathbf{57.4 \text{ cc}} \qquad \underset{\displaystyle =}{\text{Ans}}$$

Since the bore and stroke are equal the engine is called a square engine.

1.2. A four-stroke, compression-ignition engine with four cylinders develops an indicated power of 125 kW and delivers a brake power of 100 kW. Calculate (i) frictional power (ii) mechanical efficiency of the engine.

SOLUTION:

$$\text{Frictional power, } fp \;=\; ip - bp \;=\; 125 - 100$$

$$=\; \mathbf{25 \text{ kW}} \qquad \underset{\displaystyle =}{\text{Ans}}$$

$$\text{Mechanical efficiency, } \eta_m \;=\; \frac{bp}{ip} \times 100 \;=\; \frac{100}{125} \times 100$$

$$=\; \mathbf{80\%} \qquad \underset{\displaystyle =}{\text{Ans}}$$

1.3. An engine with 80 per cent mechanical efficiency develops a brake power of 30 kW. Find its indicated power and frictional power. If frictional power is assumed to be constant, what will be the mechanical efficiency at half load.

SOLUTION:

$$\text{Indicated power, } ip \;=\; \frac{bp}{\eta_m} \;=\; \frac{30}{0.8}$$

$$=\; \textbf{37.5 kW} \qquad\qquad \underset{\Longleftarrow}{\text{Ans}}$$

$$\text{Frictional power, } fp \;=\; ip - bp \;=\; 37.5 - 30$$

$$=\; \textbf{7.5 kW} \qquad\qquad \underset{\Longleftarrow}{\text{Ans}}$$

$$\text{Brake power at half load} \;=\; 0.5 \times 30$$

$$=\; \textbf{15 kW} \qquad\qquad \underset{\Longleftarrow}{\text{Ans}}$$

$$\text{Mechanical efficiency, } \eta_m \;=\; \frac{bp}{bp + fp} \times 100$$

$$=\; \frac{15}{15 + 7.5} \times 100$$

$$=\; \textbf{66.7\%} \qquad\qquad \underset{\Longleftarrow}{\text{Ans}}$$

1.4. A single-cylinder, compression-ignition engine with a brake thermal efficiency of 30% uses high speed diesel oil having a calorific value of 42000 kJ/kg. If its mechanical efficiency is 80 per cent, calculate (i) *bsfc* in kg/kW h (ii) *isfc* in kg/kW h

SOLUTION:

$$\text{Brake thermal efficiency, } \eta_{bth} \;=\; \frac{bp}{\dot{m}_f \times CV}$$

where $\dot{m}_f$ is fuel consumption in kg/s

$$\text{Brake sp. fuel consumption, } bsfc \;=\; \frac{3600}{\eta_{bth} \times CV}$$

$$=\; \frac{3600}{0.3 \times 42000}$$

$$=\; \textbf{0.286 kg/kW h} \qquad\qquad \underset{\Longleftarrow}{\text{Ans}}$$

Ind. sp. fuel consumption, $isfc$ $=$ $bsfc \times \eta_m$ $=$ 0.286×0.8

$$= \quad 0.229 \text{ kg/kW h} \qquad \underline{\text{Ans}}$$

1.5. A petrol engine uses a fuel of calorific value of 42000 kJ/kg and has a specific gravity of 0.75. The brake thermal efficiency is 24 per cent and mechanical efficiency is 80 per cent. If the engine develops a brake power of 29.44 kW, calculate (i) volume of the fuel consumed per second (ii) indicated thermal efficiency

SOLUTION:

Brake thermal efficiency, η_{bth} $= \dfrac{bp}{\dot{m}_f \times CV}$

Fuel flow rate, $\dot{m}_f$ $= \dfrac{bp}{\eta_{bth} \times CV}$ $= \dfrac{29.44}{0.24 \times 42000}$

$$= \quad 2.92 \times 10^{-3} \text{ kg/s} \qquad \underline{\text{Ans}}$$

Density of fuel $= \quad 0.75 \times 1000 \quad = \quad 750 \text{ kg/m}^3$

Volume rate of fuel, $\dot{V}_f$ $= \dfrac{2.92 \times 10^{-3}}{750}$

$$= 3.89 \times 10^{-6} \text{ m}^3/\text{s}$$

η_{ith} $= \dfrac{\eta_{bth}}{\eta_m} \times 100$ $= \dfrac{0.24}{0.80} \times 100$

$$= \quad 30\% \qquad \underline{\text{Ans}}$$

1.6. A single-cylinder, four-stroke diesel engine having a displacement volume of 790 cc is tested at 300 rpm. When a braking torque of 49 Nm is applied, analysis of the indicator diagram gives a mean effective pressure of 980 kPa. Calculate the brake power and mechanical efficiency of the engine.

SOLUTION:

$$bp \quad = \quad \frac{2\pi NT}{60000} \quad = \quad \frac{2\pi \times 300 \times 49}{60000}$$

$$= \quad 1.54 \text{ kW} \qquad \underline{\text{Ans}}$$

$$\text{Indicated power, } ip \quad = \quad \frac{P_m\ LA\ n}{60000} \quad = \quad \frac{P_{im}V_s n}{60}$$

$$= \quad 980000 \times 790 \times 10^{-6} \times \frac{300}{2} \times \frac{1}{60}$$

$$= \quad 1.94 \text{ kW}$$

$$\eta_m \quad = \quad \frac{bp}{ip} \times 100 \quad = \quad \frac{1.54}{1.94} \times 100$$

$$= \quad \mathbf{79.4} \qquad\qquad \underset{\Longleftarrow}{\text{Ans}}$$

1.7. A four-stroke SI engine delivers a brake power of 441.6 kW with a mechanical efficiency of 85 per cent. The measured fuel consumption is 160 kg of fuel in one hour and air consumption is 410 kg during one sixth of an hour. The heating value of the fuel is 42000 kJ/kg. Calculate (i) indicated power (ii) frictional power (iii) air-fuel ratio (iv) indicated thermal efficiency (v) brake thermal efficiency

SOLUTION:

$$\text{Indicated power, } ip \quad = \quad \frac{bp}{\eta_m} \quad = \quad \frac{441.6}{0.85}$$

$$= \quad \mathbf{519.5 \text{ kW}} \qquad\qquad \underset{\Longleftarrow}{\text{Ans}}$$

$$\text{Frictional power, } fp \quad = \quad ip - bp \quad = \quad 519.5 - 441.6$$

$$= \quad \mathbf{77.9 \text{ kW}} \qquad\qquad \underset{\Longleftarrow}{\text{Ans}}$$

$$\text{Air-fuel ratio, } A/F \quad = \quad \frac{0.683}{0.044}$$

$$= \quad \mathbf{15.5} \qquad\qquad \underset{\Longleftarrow}{\text{Ans}}$$

$$\text{Ind. th. efficiency,} \eta_{ith} \quad = \quad \frac{ip}{\dot{m}_f \times CV}$$

$$= \quad \frac{519.5}{0.044 \times 42000}$$

$$= \quad 0.281$$

$$= \quad \mathbf{28.1\%} \qquad\qquad \underset{\Longleftarrow}{\text{Ans}}$$

$$\text{Brake thermal efficiency, } \eta_{bth} \quad = \quad \eta_{ith} \times \eta_m$$

$$= \quad 0.281 \times 0.85 \quad = \quad 0.239$$

$$= \quad \textbf{23.9\%} \qquad \underset{\Longleftarrow}{\textbf{Ans}}$$

1.8. A two-stroke cycle CI engine delivers a brake power of 368 kW while 73.6 kW is used to overcome the friction losses. It consumes 180 kg/h of fuel at an air-fuel ratio of 20:1. The heating value of the fuel is 42000 kJ/kg. Calculate (i) indicated power (ii) mechanical efficiency; (iii) Air consumption (iv) indicated thermal efficiency (v) brake thermal efficiency

SOLUTION:

$$\text{Indicated power, } ip \quad = \quad bp + fp \quad = \quad 368 + 73.6$$

$$= \quad \textbf{441.6 kW} \qquad \underset{\Longleftarrow}{\textbf{Ans}}$$

$$\text{Mechanical efficiency, } \eta_m \quad = \quad \frac{bp}{ip} \times 100$$

$$= \quad \frac{368}{441.6} \times 100$$

$$= \quad \textbf{83.3\%} \qquad \underset{\Longleftarrow}{\textbf{Ans}}$$

$$\text{Air consumption} \quad = \quad \frac{A}{F} \times \dot{m}_f \quad = \quad 20 \times 0.05$$

$$= \quad \textbf{1 kg/s} \qquad \underset{\Longleftarrow}{\textbf{Ans}}$$

$$\text{Ind. th. efficiency, } \eta_{ith} \quad = \quad \frac{ip}{\dot{m}_f \times CV} \times 100$$

$$= \quad \frac{441.6}{0.05 \times 42000} \times 100$$

$$= \quad \textbf{21\%} \qquad \underset{\Longleftarrow}{\textbf{Ans}}$$

$$\text{Brake thermal efficiency, } \eta_{bth} \quad = \quad \eta_{ith} \times \eta_m$$

$$= \quad 0.21 \times 0.833 \quad = \quad 0.175$$

$$= \quad \textbf{17.5\%} \qquad \underset{\Longleftarrow}{\textbf{Ans}}$$

1.9. A four-stroke petrol engine delivers a brake power of 36.8 kW with a mechanical efficiency of 80%. The air-fuel ratio is 15:1 and the fuel consumption is 0.4068 kg/kW h. The heating value of the fuel is 42000 kJ/kg. Calculate (i) indicated power (ii) frictional power(iii) brake thermal efficiency (iv) indicated thermal efficiency (v) total fuel consumption (vi) air consumption/second

SOLUTION:

$$\text{Indicated power, } ip \quad = \quad \frac{bp}{\eta_m} = \frac{36.8}{0.8}$$

$$= \quad \textbf{46 kW} \qquad \underline{\underline{\textbf{Ans}}}$$

$$\text{Frictional power, } fp \quad = \quad ip - bp$$

$$= \quad 46 - 36.8$$

$$= \quad \textbf{9.2 kW}$$

$$\text{Brake thermal efficiency, } \eta_{bth} \quad = \quad \frac{bp}{\dot{m}_f \times CV}$$

$$bsfc \quad = \quad \frac{0.4068}{3600}$$

$$= \quad 1.13 \times 10^{-4} \text{ kg/kW s}$$

$$\eta_{bth} \quad = \quad \frac{1}{1.13 \times 10^{-4} \times 42000}$$

$$= \quad \textbf{21\%} \qquad \underline{\underline{\textbf{Ans}}}$$

$$\text{Ind. th. efficiency, } \eta_{ith} \quad = \quad \frac{\eta_{bth}}{\eta_m} \times 100$$

$$= \quad \frac{0.21}{0.8} \times 100$$

$$= \quad \textbf{26.25\%} \qquad \underline{\underline{\textbf{Ans}}}$$

$$\text{Fuel flow rate, } \dot{m}_f \quad = \quad bsfc \times bp$$

$$= \quad 1.13 \times 10^{-4} \times 36.8$$

$$= \quad \textbf{0.0042 kg/s} \qquad \underline{\underline{\textbf{Ans}}}$$

$$\text{Air consumption, } \dot{m}_a \;=\; \dot{m}_f \times \text{Air-fuel ratio}$$

$$=\; 0.0042 \times 15$$

$$=\; \mathbf{0.063 \; kg/s} \qquad \underset{\displaystyle =}{\text{Ans}}$$

1.10. A spark-ignition engine has a fuel-air ratio of 0.067. How many kgs of air per hour are required for a brake power output of 73.6 kW at an overall brake thermal efficiency of 20%? How many m^3 of air are required per hour if the density of air is 1.15 kg/m^3. If the fuel vapour has a density four times that of air, how many m^3 per hour of the mixture is required? The calorific value of the fuel is given as 42000 kJ/kg.

SOLUTION:

$$\text{Fuel flow rate, } \dot{m}_f \;=\; \frac{bp}{\eta_{bth} \times CV}$$

$$=\; \frac{73.6}{0.2 \times 42000} \times 3600$$

$$=\; \mathbf{31.54 \; kg/h} \qquad \underset{\displaystyle =}{\text{Ans}}$$

$$\text{Air consumption/h} \;=\; \frac{\dot{m}_f}{(F/A)} \;=\; \frac{31.54}{0.067}$$

$$=\; \mathbf{470.75 \; kg/h} \qquad \underset{\displaystyle =}{\text{Ans}}$$

$$\text{Volume flow rate of air, } \dot{V}_a \;=\; \frac{470.75}{\rho_{air}} \;=\; \frac{470.75}{1.15}$$

$$=\; \mathbf{409.35 \; m^3/h} \qquad \underset{\displaystyle =}{\text{Ans}}$$

$$\text{Volume flow rate of fuel,} \dot{V}_f \;=\; \frac{\dot{m}_f}{\rho_{fuel}} \;=\; \frac{31.54}{1.15 \times 4}$$

$$=\; \mathbf{6.86 \; m^3/h} \qquad \underset{\displaystyle =}{\text{Ans}}$$

$$\text{Mixture flow rate, } \dot{V}_m \;=\; \dot{V}_a + \dot{V}_f \;=\; 409.35 + 6.86$$

$$=\; \mathbf{416.21 \; m^3/h} \qquad \underset{\displaystyle =}{\text{Ans}}$$

Questions :-

1.1 Define the following :
    (i) engine
    (ii) heat engine

1.2 How are heat engines classified?

1.3 Explain the basic difference in their work principle?

1.4 Give examples of EC and IC engines.

1.5 Compare EC and IC engines.

1.6 What are the important basic components of an IC engine? Explain them briefly.

1.7 Draw the cross-section of a single cylinder spark-ignition engine and mark the important parts.

1.8 Define the following :
    (i) bore
    (ii) stroke
    (iii) displacement volume
    (iv) clearance Volume
    (v) compression ratio
Mention the units in which they are normally measured.

1.9 What is meant by TDC and BDC? In a suitable sketch mark the two dead centres.

1.10 Classify the internal combustion engine with respect to
    (i) cycle of operation
    (ii) types of fuels used
    (iii) method of charging the cylinder
    (iv) types of ignition
    (v) cylinder arrangements

1.11 What is meant by cylinder row and cylinder bank?

1.12 With neat sketches explain the working principle of four-stroke spark-ignition engine.

1.13 In what respects four-stroke cycle CI engine differ from that of an SI engine?

1.14 What is the main reason for the development of two-stroke engines and what are the two main types of two-stroke engines?

1.15 Describe with a neat sketch the working principle of a crankcase scavenged two-stroke engine.

1.16 Draw the ideal and actual indicator diagrams of a two-stroke SI engine. How are they different from that of a four-stroke cycle engine?

1.17 Compare four-stroke and two-stroke cycle engines. Bring out clearly their relative merits and demerits.

1.18 Compare SI and CI engines with respect to
   (i) basic cycle
   (ii) fuel used
   (iii) introduction of fuel
   (iv) ignition
   (v) compression ratio
   (vi) speed
   (vii) efficiency
   (viii) weight

1.19 Discuss in detail the application of various types of internal combustion engines.

1.20 Give an account of the first law analysis of an internal combustion engine.

1.21 Show by means of a diagram the energy flow in a reciprocating internal combustion engine.

1.22 Define the following efficiencies :
   (i) indicated thermal efficiency
   (ii) brake thermal efficiency
   (iii) mechanical efficiency
   (iv) volumetric efficiency
   (v) relative efficiency

1.23 Explain briefly
   (i) mean effective pressure
   (ii) specific output
   (iii) specific fuel consumption
   (iv) fuel-air ratio
   (v) heating value of the fuel

1.24 What is meant by mean piston speed? Explain its importance.

1.25 Discuss briefly the design performance data of SI and CI engines.

Exercise:-

1.1. A diesel engine has a brake thermal efficiency of 30 per cent. If the calorific value of the fuel is 42000 kJ/kg. Find its brake specific fuel consumption. *Ans*: 0.2857 kg/kW h

1.2. A gas engine having a cylinder 250 mm bore and 450 mm stroke has a volumetric efficiency of 80%. Air-gas ratio equals 9:1 calorific value of fuel 21000 kJ per m^3 at NTP. Calculate the heat supplied to the engine per working cycle. If the compression ratio is 5:1, what is the heat value of the mixture per working stroke per m^3 of total cylinder volume? *Ans*: 1348.62 kJ

1.3. A certain engine at full load delivers a brake power of 36.8 kW. It requires a frictional power of 7.36 kW to rotate the engine without fuel at the same speed. Calculate its mechanical efficiency. Assuming that the mechanical losses remain constant what will be the mechanical efficiency at (i) half load and (ii) quarter load.
Ans: (i) 83.3% (ii) 71.4% (iii) 55.5%

1.4. An engine used for pumping water develops a brake power of 3.68 kW. Its indicated thermal efficiency is 30%, mechanical efficiency is 80% calorific value of the fuel is 42,000 kJ/kg and its specific gravity = 0.875. Calculate (i) the fuel consumption of the engine in (a) kg/h (b) litres/h (ii) indicated specific fuel consumption and (iii) brake specific fuel consumption.
Ans: (i) (a) 1.31 kg/h (b) 1.5 litres/h
(ii) 0.285 kg/kW h (iii) 0.356 kg/kW h

1.5. A two-stroke CI engine develops a brake power of 368 kW while its frictional power is 73.6 kW. Its fuel consumption is 180 kg/h and works with an air-fuel ratio of 20:1. The heating value of the fuel is 42000 kJ/kg. Calculate (i) indicated power (ii) mechanical efficiency (iii) air consumption per hour (iv) indicated thermal efficiency and (v) brake thermal efficiency.
Ans: (i) 441.6 kW (ii) 83.3%
(iii) 3600 kg/h (iv) 21.1% (v) 17.6%

1.6. Compute the brake mean effective pressure of a four-cylinder, four-stroke diesel engine having 150 mm bore and 200 mm stroke which develops a brake power of 73.6 kW at 1200 rpm.
Ans: 5.206 bar

1.7. Compute the brake mean effective pressure of a four-cylinder, two-stroke engine, 100 mm bore 125 mm stroke when it develops a torque of 490 Nm.   *Ans*: 7.84 bar

1.8. Find the brake thermal efficiency of an engine which consumes 7 kg of fuel in 20 minutes and develops a brake power of 65 kW. The fuel has a heating value of 42000 kJ/kg.   *Ans*: 26.5%

1.9. Find the mean piston speed of a diesel engine running at 1500 rpm. The engine has a 100 mm bore and L/d ratio is 1.5.

Ans: 7.5 m/s

1.10. An engine is using 5.2 kg of air per minute while operating at 1200 rpm. The engine requires 0.2256 kg of fuel per hour to produce an indicated power of 1 kW. The air-fuel ratio is 15:1. Indicated thermal efficiency is 38% and mechanical efficiency is 80%. Calculate (i) brake power and (ii) heating value of the fuel.

Ans: (i) 73.7 kW (ii) 42000 kJ/kg

2

REVIEW OF BASIC PRINCIPLES

2.1 INTRODUCTION

The main objectives in studying the theory of IC engines can be summarized as

(i) To have a better understanding of the various processes taking place and the conditions prevailing in the engine cylinder.

(ii) To predict the changes in power, fuel consumption and reliability resulting from the changes in the operating conditions or changes in the design features of a given engine.

(iii) To predict the operating characteristics of a new design from test results on a similar engine of a different size.

Basic knowledge and the familiarity of the principles of thermodynamics, physics and chemistry is a must for achieving the above objectives. However, mere familiarity alone will not serve the purpose. The familiarity must be supplemented by the ability to apply these principles correctly. This requires an understanding of fundamentals used to derive the important formulae. The main aim of this chapter is to place before the reader the basic principles of thermodynamics, physics, chemistry and other relevant information for the understanding of the various details discussed in the subsequent chapters. Although a fairly exact treatment has been attempted, the reader is advised to go through other relevant textbooks to get more detailed account of the various basic principles.

2.2 BASIC AND DERIVED QUANTITIES

In this section we will briefly recall the definition of basic quantities like length, time, mass, weight, acceleration due to gravity, force and pressure. These quantities will be used to derive other quantities.

2.2.1 Length

The unit of length is metre. Originally, metal bar standards were used to define length. They have been abandoned now in favour of a standard length which can be reproduced in any laboratory in the world. Length is now defined in terms of the atomic standards. The orange light emitted by the gas krypton under special experimental conditions is used to define length. One metre is defined to be 1,650,763.73 wavelengths of the orange light Krypton-86. However, for convenience in engine applications, millimeter (mm) and centimeter (cm) may also be used to represent length.

2.2.2 Time

The unit of time is second. Time is also defined in terms of atomic standards. The time standard is based on the duration of the periods of the radiation of the atom cesium-133.

$$1 \text{ second} \quad = \quad 9,192,631,770 \text{ periods}$$

The second, minute, hour, day and year are used throughout the world. Their interrelation requires no description here.

2.2.3 Mass and Weight

The mass of a body, m is the amount of matter in the body. The mass of a given body therefore is the same anywhere in the universe. The unit of mass is kilogram (kg). However, for engine applications, gram (gm) may also be used as the unit of mass.

The weight of a body, w is the force with which the earth attracts the body. The weight of a given body will change with the value of acceleration due to gravity, g.

The weight, w is always equal to the number of kg mass, m times the local value of g or

$$w \quad = \quad mg \tag{2.1}$$

2.2.4 Gravitational Constant

The force of attraction by the earth experienced by any body is called the gravitational force. Since, the mass of most of the bodies under consideration are negligible compared to that of earth this force effects only the other bodies. This results in bodies not able to float and a body without support experiences a motion in the direction of the gravitational force. The acceleration of such freely falling body is called acceleration due to gravity. This is found to vary from place to place on the surface of the earth and hence an average value is used for engineering calculations, viz., 9.81 m/s^2 or 32 ft/s^2.

In the derivation of unit of force it is required to prescribe mass and acceleration. In prescribing the acceleration, two conventions are used, one to prescribe unit acceleration and the other to prescribe the acceleration due to gravity. In the four methods of systems of units so far evolved two use the former and two later.

F.P.S.	:	1 pound	=	1 slug	×	32	ft/s^2
C.G.S.	:	1 dyne	=	1 gm	×	1	m/s^2
M.K.S.	:	1 kgf	=	1 kg	×	9.81	m/s^2
SI	:	1 Newton	=	1 kg	×	1	m/s^2

In using M.K.S. system care should be taken to differentiate between kgf and kg since 1 kgf is actually equal to 9.81 Newtons. This is done by defining a gravitational constant, g_c which is numerically equal to the acceleration due to gravity but without dimensions. Since it is proposed to use SI units in this book use of g_c is not elaborated further.

2.2.5 Force

In this book the unit of force will be Newton and is abbreviated as N. A Newton is the force required to give an acceleration of $1 \ m/s^2$ to a mass of one kg of matter. This definition is adopted because it is convenient to metrologists; its adoption is rendered possible by the universal validity of Newton's Second Law of Motion.

2.2.6 Pressure

Fluids, gas or liquid exert forces on its boundaries. These forces are not concentrated at particular points, but are distributed. It is therefore useful to define the quantity pressure as the force exerted on a surface which is normal to the force, divided by the area of the surface. The units of pressure is N/m^2. However, it may also be expressed in bar, Pascal and standard atmosphere, (atm).

The inter relation between bar, Pascal and standard atmosphere with N/m^2 is given below.

1 Pascal	(Pa)	=	$1 \ N/m^2$
1 bar	(bar)	=	10^5 Pascal
1 standard atmosphere	(atm)	≈	$1.01325 \times 10^5 \ N/m^2$

2.2.7 Temperature

The unit of temperature is Kelvin. The Kelvin is $1/273.16$ of the thermodynamic temperature of the triple point of water, viz., the point at which liquid water, water vapour and ice are in equilibrium.

2.3 THE GAS LAWS

During the various strokes of the engine different pressures and temperatures exist in the engine cylinder and all gases encountered in
engine operation are assumed to obey at all these pressures and temperatures the simple gas law,

$$pV = N\overline{R}T \qquad (2.2)$$

where

p	=	Pressure of the gas	(N/m^2)
V	=	Volume occupied by the gas	(m^3)
N	=	Number of moles of gas	$(kmol)$
$\overline{R}$	=	Universal gas constant	$(8.314 \text{ kJ/kmol K})$
T	=	Absolute temperature	(K)

The above equation is an approximation sufficiently close for engineering purposes. This law is very simple and 'always' applies to
'any' gas, regardless of the process to which the gas is subjected, as
long as the gas is not near its liquefaction.

2.3.1 The Mole

A kilomol of any gas is a quantity of the gas whose weight is equal to
the molecular weight in kg. Thus

$$N = \frac{m}{M} \qquad (2.3)$$

where m = Mass of gas (kg)

M = Molecular weight of the gas, (g/mol)

Combining Eqns.2.2 and 2.3, the gas law may be written as

$$pV = \frac{m}{M}\overline{R}T = m\left(\frac{\overline{R}}{M}\right)T = mRT \qquad (2.4)$$

The quantity $\overline{R}/M$ is called the characteristic gas constant and
is usually denoted by R. Since M is different for different gases, the
characteristic gas constant will also be different for different gases.

2.3.2 Specific Volume and Density

The volume occupied by unit mass of a substance is known as its
specific volume (m^3/kg) and is denoted by v.

The inverse of the specific volume is the density (kg/m^3) and is
denoted by ρ.

The specific volume and density of the gas depend upon the pressure and temperature.

2.3.3 Simplification of the Gas Laws

When the number of moles m/M of gas under consideration is fixed, both N and R are constant. Then from Eqns.2.2 and 2.4,

$$\frac{p_1 v_1}{T_1} = \frac{p_2 v_2}{T_2} = \cdots = \frac{p_n v_n}{T_n} = \text{Constant} \quad (2.5)$$

This simple relation is always true when applied to a constant number of moles of a gas and makes it possible to calculate the effect of changing any two of the variables (pressure, volume or temperature) upon the remaining one.

2.3.4 Mixture of Gases

During the expansion and exhaust strokes (also during suction and compression) the cylinder is occupied by two or more gases. Each of these gases diffuses and fills up the entire space, obeying the gas law just as though the other gases were not present. The principle is called Dalton's Law. For example, let three gases 1, 2 and 3 occupy the cylinder at a particular instant. Let the volume of the cylinder at that instant be V. If the temperature of the mixture is T, then the pressure exerted on the walls of the cylinder by gas 1 is, by the gas law,

$$p_1 = \frac{N_1 \overline{R} T}{V}$$

and the pressure exerted by gas 2 is

$$p_2 = \frac{N_2 \overline{R} T}{V}$$

The total pressure within the container is

$$p_t = p_1 + p_2 + p_3 + \ldots = \frac{(N_1 + N_2 + N_3 + \ldots)\overline{R}T}{V} = \frac{N_t \overline{R} T}{V} \quad (2.6)$$

From the above expression for p_1 and p_2, it is seen that

$$\frac{p_1}{p_2} = \frac{N_1}{N_2}$$

and from Eqn.2.6

$$\frac{p_1}{p_t} = \frac{N_1}{N_t} \quad (2.7)$$

The pressures p_1, p_2 etc. which go to make up the total pressure are called partial pressure.

2.4 FORMS OF ENERGY

An engine takes in chemical energy in the fuel and changes most of this into sensible energy during the combustion process, and then transforms part of the sensible energy into mechanical work on the piston head.

During the suction stroke the atmosphere does work on the mixture as it forces it into the cylinder. During the exhaust stroke the products of combustion do work on the atmosphere as they leave the cylinder. In order to understand these processes, it is necessary to study the various forms of energy in some detail.

Energy is defined as the capacity to do work. If a machine or body, because of its position, temperature or velocity is capable of doing work on another body, it is said to have energy. Work is done whenever motion takes place against a resistance. The amount of work done is the distance moved, times the magnitude of the resisting force in the direction of motion.

$$\text{Work} \quad = \quad \text{Force} \times \text{distance}$$

Work and energy are expressed in the same units, viz., Nm or Joule.

When energy is being transferred from one body to another by virtue of a temperature difference, it is called heat and is usually measured in Joules.

Work and heat are related by the mechanical equivalent of heat J. The relation is

$$W \quad = \quad JH \tag{2.8}$$

where W is the work, H is the heat and $J = 427$ kgf m/kcal. Thus work and heat can be expressed in any units of convenience with the help of J.

2.4.1 Transfer of Energy

Heat and work are both energy in transit. When two bodies at different temperatures are brought in contact, heat flows from a body at higher temperature to a body at lower temperature. Similarly, when work is done, the system doing work loses energy and the system upon which the work is being done gains exactly this amount of energy.

2.4.2 Stored Energy

In the two cases above, when the process is completed the energy gained by the second body is not called heat or work, since, heat is no longer flowing and work is no longer being done. The additional energy which is now stored in the second body may be called stored energy, E, which includes many types of energies like kinetic energy, potential energy, surface tension, electrical energy etc. depending upon the manner in which they are stored.

2.4.3 Potential Energy

It is the energy contained in the system by virtue of its elevation with reference to an arbitrarily chosen datum, usually the sea level. Or alternatively it is equivalent to energy required to raise the system from an arbitrary datum.

2.4.4 Kinetic Energy

Any body of mass, m kg, moving with a velocity of C metres per second will do $\frac{1}{2}mC^2$ Joules of work before coming to rest. In this state of motion, the body is said to have kinetic energy equal to $\frac{1}{2}mC^2$ Joules.

2.4.5 Internal Energy

A body may possess energy due to the motion or position or the attraction between the particles of which it is made. In a permanent gas (i.e. a gas which is far from its liquefaction point), the internal energy, E, will usually be in the form of

(i) translational motion of the molecules of the gas and

(ii) motion of the atoms within the molecules

These two kinds of internal energy are known as sensible internal energy, although only that part which is due to molecular translation can actually be felt as warmth. For a given quantity of a particular permanent gas, the amount of sensible internal energy present is fixed by the temperature of the gas alone. The sensible internal energy is denoted by U_s. The gas under consideration may also contain chemical energy which is usually denoted by U_c.

The sum of the sensible internal energy and the chemical energy of the fuel present and is called the total internal energy, U, i.e.,

$$U \;=\; U_s + U_c \tag{2.9}$$

If during any combustion process no heat is allowed to escape from the gases and no work is permitted to be done by the gases, then at any time during the process the increase in sensible energy due to combustion is exactly equal to the loss in chemical energy as the fuel is used up. The total internal energy therefore remains unchanged.

2.5 FIRST LAW OF THERMODYNAMICS

Whenever a system undergoes a cyclic change, the algebraic sum of work transfer is proportional to the algebraic sum of heat transfer as work and heat are mutually convertible from one form into the other.

For a closed system, a change in the energy content is the algebraic difference between the heat supply, Q, and the work done, W, during any change in the state. Mathematically,

$$dE \;\; = \;\; \delta Q - \delta W \tag{2.10}$$

The energy E may include many types of energies like kinetic energy, potential energy, electrical energy, surface tension etc., but from the thermodynamic point of view these energies are ignored and the energy due to rise in temperature alone is considered. It is denoted by U and the first law is written as:

$$dU \;\; = \;\; \delta Q - \delta W \tag{2.11}$$

or during a process 1-2, dU can be denoted as

$$U_2 - U_1 \;\; = \;\; \int_1^2 (\delta Q - \delta W) \tag{2.12}$$

2.6 PROCESS

A change in the condition or state of a substance is called a process. The process may consist of heating, flow from one place to another, expansion etc. In general, a process may be divided into non-flow or flow processes.

2.6.1 Non-flow Processes

If there is no flow of material into or out of a system during a process, it is called a non-flow process. This is the simplest kind of process, and much can be learned about it by applying the principle of conservation of energy.

2.7 ANALYSIS OF NON-FLOW PROCESSES

The purpose of the analysis is to apply the First Law of Thermodynamics to process in which a non-flow system changes from one state to the other and also to develop some useful relations.

2.7.1 Constant Volume or Isochoric Process

This process is usually encountered in the analysis of air-standard Otto, Diesel and Dual cycles. Figure 2.1 shows the constant volume process on a *p-V* diagram.

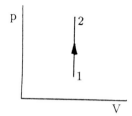

Fig.2.1 Constant Volume Process

As there is no change in volume, the work $\int (pdV)$ is zero. Hence, according to the first law for the constant volume process the change in internal energy is equal to the heat transfer, i.e.,

$$dU \quad = \quad \delta Q \quad = \quad mC_v dT \quad = \quad mC_v(T_2 - T_1) \qquad (2.13)$$

For unit mass,

$$du \quad = \quad \delta q \quad = \quad C_v dT$$

$$C_v \quad = \quad \left(\frac{du}{dT}\right)_V \qquad\qquad (2.14)$$

i.e., specific heat at constant volume is the rate of change of internal energy with respect to absolute temperature.

2.7.2 Constant Pressure or Isobaric Process

Figure 2.2 shows a system that changes from state 1 to state 2 at constant pressure. Application of first law yields,

$$\delta Q \quad = \quad dU + pdV \quad = \quad d(U + pV) \quad = \quad dH \qquad (2.15)$$

where H is known as the enthalpy.

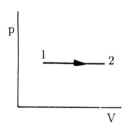

Fig.2.2 Constant Pressure Process

Thus, during constant pressure process, heat transfer is equal to change in enthalpy or

$$dH \quad = \quad \delta Q \quad = \quad mC_p dT \qquad (2.16)$$

For unit mass,

$$\delta q \quad = \quad dh \quad = \quad C_p dT \qquad (2.17)$$

$$C_p \quad = \quad \left(\frac{dh}{dT}\right)_p \qquad (2.18)$$

i.e., specific heat at constant pressure is the rate of change of specific enthalpy with respect to absolute temperature.

2.7.3 Constant Temperature or Isothermal Process

The isothermal process on a *p-V* diagram is illustrated in Fig.2.3. As there is no temperature change during this process, there will not be any change in internal energy i.e., $dU = 0$, then according to the first law

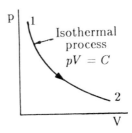

Fig.2.3 Constant Temperature Process

$$\delta Q \quad = \quad \delta W \qquad (2.19)$$

or

$$Q_{1-2} \quad = \quad \int_1^2 pdV \quad = \quad p_1 V_1 \log_e \left(\frac{V_2}{V_1}\right) \qquad (2.20)$$

2.7.4 Reversible Adiabatic or Isentropic Process

If a process occurs in such a way that there is no heat transfer between the surroundings and the system, but the boundary of the system moves giving displacement work, the process is said to be adiabatic. Such a process is possible if the system is thermally insulated from the surroundings. Hence, $\delta Q = 0$, therefore,

$$\delta W = -\delta U = -mC_v dT \qquad (2.21)$$

Reversible adiabatic process is also known as isentropic process. Let $pV^\gamma = C$ be the law of the isentropic process. For unit mass flow,

$$q_{1-2} = 0 = w_{1-2} + u_2 - u_1 \qquad (2.22)$$

or

$$w_{1-2} = -(u_2 - u_1)$$

In other words, work is done at the expense of internal energy

$$W_{1-2} = \int_1^2 pdV = \int_1^2 \frac{C}{V^\gamma} dV \qquad (2.23)$$

$$= \frac{\left[CV^{1-\gamma}\right]_{V1}^{V2}}{1-\gamma}$$

$$= \frac{CV_2^{1-\gamma} - CV_1^{1-\gamma}}{1-\gamma}$$

when $C = p_1 V_1^\gamma = p_2 V_2^\gamma$,

$$W_{1-2} = \frac{p_2 V_2^\gamma V_2^{1-\gamma} - p_1 V_2^\gamma V_2^{1-\gamma}}{1-\gamma}$$

$$= \frac{p_1 V_1 - p_2 V_2}{\gamma - 1}$$

Using $pV = RT$ for unit mass flow, we have,

$$w_{1-2} = \frac{R(T_1 - T_2)}{\gamma - 1}$$

$$= C_v(T_1 - T_2) = -(u_2 - u_1)$$

$$\frac{p_1 V_1}{T_1} = \frac{p_2 V_2}{T_2}$$

therefore,

$$\frac{T_2}{T_1} = \frac{p_2 V_2}{p_1 V_1} = \left(\frac{V_1}{V_2}\right)^\gamma \left(\frac{V_2}{V_1}\right)$$

$$= \left(\frac{V_1}{V_2}\right)^{(\gamma-1)} \tag{2.24}$$

$$\frac{T_2}{T_1} = \frac{p_2 V_2}{p_1 V_1} = \left(\frac{p_1}{p_2}\right)^{\frac{1}{\gamma}}\left(\frac{p_2}{p_1}\right)$$

$$= \left(\frac{p_2}{p_1}\right)^{\left(1-\frac{1}{\gamma}\right)} = \left(\frac{p_2}{p_1}\right)^{\left(\frac{\gamma-1}{\gamma}\right)} \tag{2.25}$$

$$\frac{T_2}{T_1} = \left(\frac{V_1}{V_2}\right)^{(\gamma-1)} = \left(\frac{p_2}{p_1}\right)^{\left(\frac{\gamma-1}{\gamma}\right)} \tag{2.26}$$

2.7.5 Reversible Polytropic Process

In polytropic process, both heat and work transfers take place. It is denoted by the general equation $pV^n = C$, where n is the polytropic index. The following equations can be written by analogy to the equations for the reversible adiabatic process which is only a special case of polytropic process with $n = \gamma$. Hence, for a polytropic process

$$p_1 V_1^n = p_2 V_2^n \tag{2.27}$$

$$\frac{T_1}{T_2} = \left(\frac{V_2}{V_1}\right)^{n-1}$$

$$\frac{T_1}{T_2} = \left(\frac{p_1}{p_2}\right)^{\left(\frac{n-1}{n}\right)} \tag{2.28}$$

and

$$w_{1-2} = \frac{p_1 V_1 - p_2 V_2}{n - 1} \tag{2.29}$$

2.7.6 Heat Transfer During Polytropic Process

Heat transfer per unit mass,

$$q_{1-2} = (u_2 - u_1) + \int_1^2 p\,dV \tag{2.30}$$

$$= C_v(T_2 - T_1) + \frac{R(T_2 - T_1)}{1 - n}$$

$$= \left(C_v + \frac{R}{1 - n}\right)(T_2 - T_1)$$

$$= \left(C_v + \frac{C_p - C_v}{1 - n}\right)(T_2 - T_1)$$

$$= \left(\frac{C_p - nC_v}{1 - n}\right)(T_2 - T_1)$$

$$= \left(\frac{C_v}{1 - n}\right)\left(\frac{C_p}{C_v} - n\right)(T_2 - T_1)$$

$$= \left(\frac{C_v}{1 - n}\right)(\gamma - n)(T_2 - T_1)$$

$$= \left(\frac{\gamma - n}{1 - n}\right)C_v(T_2 - T_1) \tag{2.31}$$

Hence,

$$q_{1-2} = C_n(T_2 - T_1)$$

where,

$$C_n = \left(\frac{\gamma - n}{1 - n}\right)C_v \tag{2.32}$$

Table 2.1 gives the formulae for various process relations for easy reference.

2.8 ANALYSIS OF FLOW PROCESS

Consider a device shown in the Fig.2.4 through which a fluid flows at uniform rate and which absorbs heat and does work, also at a uniform rate. In engines, the gas flow, heat flow and work output vary throughout each cycle, but if a sufficiently long time interval (such as a minute) be chosen, then, even an engine can be considered to be operating under steady-flow conditions.

Applying the principle of conservation of energy to such a system during the chosen time interval, the total energy in the mass of fluid which enters the machine across boundary 1-1 plus the heat added to the fluid through the walls of the machine, minus the work done by the machine, must equal the total energy left in the equal mass of fluid crossing boundary 2-2 and leaving the machine. Expressed in the mathematical form,

$$Q - W_x = m\left[\left(h_2 + \frac{C_2^2}{2} + gz_2\right) - \left(h_1 + \frac{C_1^2}{2} + gz_1\right)\right]$$

Table 2.1 *Summary of Process Relations for a Perfect Gas*

Process	Index n	p, V, T relation	Heat Transfer	Work $\int p\,dV$	Work $\int V\,dp$	ΔU
Constant Volume	∞	$T_1/T_2 = p_1/p_2$	$mC_v(T_2 - T_1)$	0	$V(p_2 - p_1)$	$mC_v(T_2 - T_1)$
Constant Pressure	0	$T_1/T_2 = V_1/V_2$	$mC_p(T_2 - T_1)$	$p(V_2 - V_1)$	0	$mC_v(T_2 - T_1)$
Isothermal	1	$p_1V_1 = p_2V_2$	$p_1V_1 \log_e \left(\dfrac{V_2}{V_1}\right)$	$p_1V_1 \log_e \left(\dfrac{V_2}{V_1}\right)$	$p_1V_1 \log_e \left(\dfrac{V_1}{V_2}\right)$	0
Isentropic	γ	$p_1V_1^{\gamma} = p_2V_2^{\gamma}$ $\dfrac{T_1}{T_2} = \left(\dfrac{V_2}{V_1}\right)^{\gamma-1}$ $= \left(\dfrac{p_1}{p_2}\right)^{\frac{\gamma-1}{\gamma}}$	0	$\dfrac{p_1V_1 - p_2V_2}{\gamma - 1}$	$\dfrac{\gamma}{\gamma-1}(p_2V_2 - p_1V_1)$	$mC_v(T_2 - T_1)$
Polytropic	n	$p_1V_1^{n} = p_2V_2^{n}$ $\dfrac{T_1}{T_2} = \left(\dfrac{V_2}{V_1}\right)^{n-1}$ $= \left(\dfrac{p_1}{p_2}\right)^{\frac{n-1}{n}}$	$\left(\dfrac{\gamma-n}{1-n}\right) mC_v(T_2 - T_1)$	$\dfrac{p_1V_1 - p_2V_2}{n - 1}$	$\dfrac{n}{n-1}(p_2V_2 - p_1V_1)$	$mC_v(T_2 - T_1)$

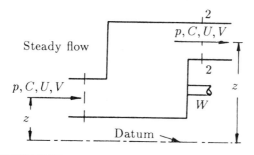

p	$\rightarrow$	Pressure	V	$\rightarrow$	Volume	C	$\rightarrow$	Velocity
U	$\rightarrow$	Internal Energy	H	$\rightarrow$	Enthalpy	z	$\rightarrow$	Datum Height

Fig.2.4 Illustration of Steady Flow Process

or

$$\frac{Q - W_x}{m} = \Delta\left(h + \frac{C^2}{2} + gz\right) \qquad (2.33)$$

In the above equation, m denotes mass. It is known as the general energy equation and is the one which illustrates the first law of thermodynamics.

Here, W_x is the external work and given by

$$W = W_x \quad mp_1V_1 + mp_2V_2 \qquad (2.34)$$

where W is the net work and V stands for volume.

2.8.1 Flow Work

The term pV in the above equation is called flow work and must be considered whenever flow takes place. It is the work necessary either to push certain quantity of mass into or out of the system.

2.9 WORK, POWER AND EFFICIENCY

In this section let us define work, power and efficiency and also the relation between them.

2.9.1 Work

Work might be defined as the energy expended when motion takes place in opposition to a force. Work thus represents energy being transferred from one place to another.

2.9.2 Power

Power is the rate of doing work. It is generally useless to do a certain piece of work unless it can be done within a reasonable length of time. When work is done at the rate of 736 Nm/s or 75 kgf m/s, the quantity of power being developed is known as one horse power. In SI units power is expressed in kW = 1 kN m/s.

2.9.3 Efficiency

It is common practice in engineering to establish a figure of merit for a device by comparing the actual performance of the device with the performance it would have had under some arbitrary set of ideal conditions. The ratio of actual performance to the ideal performance is called the efficiency of the device. Since the ideal performance is usually unattainable, the efficiency is usually less than 1. In internal combustion engines, one of the most important efficiencies is the thermal efficiency, which is defined as

$$\text{Thermal efficiency} = \eta_{th} = \frac{\text{Work delivered by the engine}}{\text{Chemical energy in the fuel}}$$

$$= \frac{\text{Work output}}{\text{Fuel energy input}}$$

Obviously, the output and input may be measured over any convenient interval such as one cycle or one minute. The output and input energies must be expressed in same kind of units, since η_{th} is a number without units.

Let W be the work delivered by the engine (Nm) and m_f be the mass of fuel (kg) required and CV be the calorific value of the fuel (kJ/kg), then the efficiency η_{th} is given by

$$\eta_{th} = \frac{W}{m_f \, CV} \tag{2.35}$$

If W is the indicated work then η_{th} is called the indicated thermal efficiency, η_{ith}, and if W is the work delivered at the output shaft then it is called the brake thermal efficiency, η_{bth}. The ratio of the brake thermal efficiency to the indicated thermal efficiency is called the mechanical efficiency, η_m. These efficiencies have already been defined in Chapter 1.

Worked out Examples

2.1. A closed system undergoes a thermodynamic cycle consisting of four separate and distinct processes. The work and heat transfer for each of these processes is given below

Process	Heat Transfer (J/s)	Work Transfer (Nm/s)
1–2	7000	5300
2–3	−3500	9100
3–4	17500	8700
4–1	0	−2100

Show that the above data is in accordance with the first law of thermodynamics and determine : (i) Net rate of work output (ii) Efficiency of the cycle (iii) Change in internal energy in each process

SOLUTION:
For the above closed cycle 1–2–3–4–1,

$$\oint \delta Q = 7000 - 3500 + 17500 + 0 = 21000 \text{ J/s}$$

$$\oint \delta W = 5300 + 9100 + 8700 - 2100$$

$$= 21000 \text{ Nm/s} = 21000 \text{ J/s}$$

As $\oint \delta Q = \oint \delta W$, the above cycle satisfies the first law of thermodynamics

$$\text{Power} = \text{Net rate of work} = 21000 \text{ J/s}$$

$$= \textbf{21 kW} \qquad \underset{\Leftarrow}{\text{Ans}}$$

$$\text{Thermal efficiency} = \frac{\text{Net work}}{\text{Heat input}} = \frac{21000}{7000 + 17500} \times 100$$

$$= \textbf{85.71\%} \qquad \underset{\Leftarrow}{\text{Ans}}$$

Change in internal energy of each process can be estimated by using the relation, $\delta Q = du + \delta W$

Process	δQ J/s	δW Nm/s	$du = \delta Q - \delta W$ J/s
1-2	7000	5300	**1700**
2-3	−3500	9100	**−12600**
3-4	17500	8700	**8800**
4-1	0	−2100	**2100**

2.2. Air at the rate of 10 kg/min flows steadily through an air compressor in which the inlet and discharge pipe lines are at the same level. Following data regarding the fluid are available:

	At inlet	At outlet
Fluid velocity	5 m/s	10 m/s
Fluid pressure	1 bar	8 bar
Specific volume	0.5 m^3/kg	0.20 m^3/kg

The internal energy of the air leaving the compressor was 250 kJ/kg greater than that entering and during the process the system lost 140 kJ/s of energy dissipated as heat to the cooling water and to the environment, Find (i) Rate of shaft work input to the air (ii) Ratio of inlet pipe diameter to outlet pipe diameter

SOLUTION:

Consider 1 kg/min of mass flow :

$$\text{Heat lost to cooling water and environment} \quad = \quad \frac{140 \times 60}{10}$$

$$= \quad 840 \text{ kJ/kg}$$

Applying steady flow energy equation, (the potential energy changes are neglected as the inlet and outlet pipes are at the same level).

$$U_1 + p_1V_1 + \frac{C_1^2}{2} + q \quad = \quad U_2 + p_2V_2 + \frac{C_2^2}{2} + w$$

therefore,

$$
\begin{aligned}
w \quad &= \quad (U_1 - U_2) + (p_1V_1 - p_2V_2) + \frac{C_1^2 - C_2^2}{2} + q \\[2mm]
&= \quad -250 \times 10^3 + (1 \times 10^5 \times 0.5 - 8 \times 10^5 \times 0.20) \\[2mm]
&\quad + \frac{5^2 - 10^2}{2} - 140 \times 10^3 \\[2mm]
&= \quad -500037.5 \text{ J/kg} \quad \approx \quad -500 \text{ kJ/kg} \\[2mm]
W \quad &= \quad mw \quad = \quad 10 \times (-500) \\[2mm]
&= \quad -5000 \text{ kJ/min} \quad = \quad -83.33 \text{ kJ/s}
\end{aligned}
$$

Power input $\quad = \quad$ **83.33 kW** $\qquad$ **Ans**

From continuity equation, i.e., $A_1 C_1 \rho_1 \quad = \quad A_2 C_2 \rho_2$

Since, the specific volume, $v_1 = \frac{1}{\rho_1}$. We have,

$$\frac{A_1 C_1}{v_1} = \frac{A_2 C_2}{v_2}$$

$$\frac{A_1}{A_2} = \frac{v_1 C_2}{v_2 C_1} = \frac{0.5 \times 10}{0.2 \times 5} = 5$$

$$\frac{D1}{D2} = \sqrt{5}$$

$$= \mathbf{2.236} \qquad \underset{=}{\text{Ans}}$$

2.3. A tank of volume 0.1 m^3 contains 4 kg nitrogen, 1.5 kg oxygen and 0.75 kg carbon dioxide. If the temperature of the mixture is 20 °C, determine : (i) the total pressure of the mixture, (ii) the gas constant of the mixture. (Given that $R_{N_2} = 296.8$ J/kg K, $R_{O_2} = 259.83$ J/kg K and $R_{CO_2} = 188.9$ J/kg K).

SOLUTION:

N_2 : $m_1 = 4.00$ kg; $t_1 = t = 20$ °C; $V_1 = V = 0.1$ m^3
O_2 : $m_2 = 1.50$ kg; $t_2 = t = 20$ °C; $V_2 = V = 0.1$ m^3
CO_2 : $m_3 = 0.75$ kg; $t_3 = t = 20$ °C; $V_3 = V = 0.1$ m^3

Assumptions :

(i) the component gases and the mixture behave like an ideal gas
(ii) the mixture obeys the Gibbs-Dalton law

For N_2 :

$$p_1 = \frac{m_1 R_{N_2} T}{V}$$

$$= \frac{4 \times 296.8 \times (273 + 20)}{0.1} = 34.78 \times 10^5 \text{ N/m}^2$$

For O_2 :

$$p_2 = \frac{m_2 R_{O_2} T}{V}$$

$$= \frac{1.5 \times 259.83 \times (273 + 20)}{0.1} = 11.42 \times 10^5 \text{ N/m}^2$$

For CO_2 :

$$p_3 = \frac{m_3 R_{CO_2} T}{V}$$

$$= \frac{0.75 \times 188.9 \times (273 + 20)}{0.1} = 4.15 \times 10^5 \text{ N/m}^2$$

Then, total pressure, P

$$p = p_1 + p_2 + p_3 = (34.78 + 11.42 + 4.15) \times 10^5$$

$$= 50.35 \times 10^5 \text{ N/m}^2 \qquad \text{Ans}$$

$$R = \frac{1}{m}(m_{N_2} R_{N_2} + m_{O_2} R_{O_2} + m_{CO_2} R_{CO_2})$$

$$= \frac{1}{4 + 1.5 + 0.75}(4 \times 296.8 + 1.5 \times 259.83 + 0.75 \times 188.92)$$

$$= 274.98 \text{ J/kg K} \qquad \text{Ans}$$

.4. Determine the work done by the air which enters an evacuated bottle from the atmosphere when the cock is opened. The atmospheric pressure is 1.013×10^5 N/m^2 and 0.3 m^3 of air (measured at atmospheric conditions) enter.

SOLUTION:

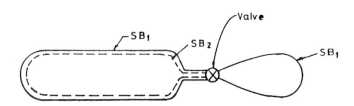

No work is done by the part of the boundary in contact with the bottle. Only the moving external part need to be considered. Over this part pressure is uniform at 1.013×10^5 N/m^2, therefore,

$$W_d = \int_{\substack{free\text{-}air \\ boundary}} pdV + \int_{Bottle} pdV$$

$$=. P_{atm} \int dV + 0 = 1.013 \times 10^5(-0.3)$$

$$= -30390 \text{ Nm} \qquad \text{Ans}$$

The work is negative as the boundary is contracting.

2.5. A balloon of flexible material is to be filled with air from a storage bottle until it has a volume of 1 m^3. The atmospheric pressure is 1.01×10^5 N/m^2. Determine the work done by the system comprising the air initially in the bottle, given that the balloon is light and requires no stretching.

SOLUTION:

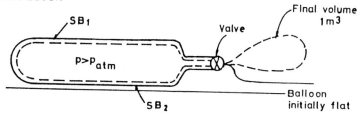

Initially the system boundary coincides with the inner surface of the storage bottle. At the end of the process the boundary also encloses the 1 m³ content of the balloon.

The displacement work which is the only work in this process, is obtained by taking the summation of the values of $\int pdV$ for each part of the boundary. As there is no change in the volume of the bottle, dV is zero for the part of the boundary which is in contact with the bottle surface. Hence the pressure inside the cylinder is not necessary in the calculation.

Therefore,

$$W_{displacement} = \int_{Balloon} pdV + \int_{Bottle} pdV$$

$$= p_{atm} \int dV + 0 = 1.01 \times 10^5 \times 1.0$$

$$= 1.01 \times 10^5 \text{ Nm} \qquad \underset{\Longleftarrow}{\text{Ans}}$$

Questions :-

2.1 What are the main objectives in studying the theory of I.C. engines?

2.2 What are the important fundamental quantities used in the application and analysis of I.C. engines? Explain them briefly.

2.3 What is meant by mole?

2.4 For a mixture of gases show that $p_1/p_t = N_1/N_t$.

2.5 What are the various forms of energy normally used in engine applications. Briefly explain them.

2.6 What are isochoric and isobaric processes? Explain them.

2.7 For an isentropic process, show that (i) $T_2/T_1 = (V_1/V_2)^{\gamma-1}$ and (ii) $T_2/T_1 = (p_1/p_2)^{(\gamma-1)/\gamma}$.

2.8 Derive an expression for heat transfer during a polytropic process.

2.9 Explain how an internal combustion engine can be analysed from the point of view of steady flow process.

2.10 Define: (i) work (ii) power and (iii) efficiency.

Exercise:-

2.1. The piston of an oil engine, of area 50 cm², moves downwards 10 cm and draws in 300 cc of fresh air from the atmosphere. The pressure in the cylinder is uniform during the process at 0.8 bar, while the atmospheric pressure is 1.013 bar. The difference in the pressure is accounted for by flow resistance in the induction pipe and inlet valve. Determine the displacement work done by the air finally in the cylinder.   *Ans:* −9.61 J

2.2. Air at 600 K and 1.2 bar undergoes an adiabatic expansion. The new value of pressure is 0.6 bar. The mass of the air, which may be treated as an ideal gas, is 0.12 kg. Determine (i) the work done by the air and (ii) the change in the internal energy of the air.   *Ans:* (i) 9.285 kJ (ii) −9.285 kJ

2.3. An open tank is filled to the brim with a liquid of density 1100 kg/m³. A spherical balloon 0.5 m in diameter is immersed in the liquid with its centre 2.5 m below the free liquid surface. Gas from a storage vessel is used to inflate the balloon thereby causing the tank to overflow. The atmospheric pressure is 10^5 N/m². Evaluate the work done by the gas in the balloon on the storage vessel as the balloon diameter increases to 1 m. Assume $g = 9.81$ m/s².   *Ans:* 58.2 $\times 10^3$ Nm

2.4. A gas flows steadily through a rotary compressor. The gas enters the compressor at a temperature of 16°, a pressure of 10^5 N/m² and an enthalpy of 391.2 KJ/kg. The gas leaves the compressor at a temperature of 245 °C, a pressure of 6×10^5 N/m² and an enthalpy of 534.5 KJ/kg. There is no net heat transfer to or from the gas as it flows through the compressor.

(i) Evaluate the external work done per unit mass of gas assuming the gas velocities at entry and exit to be negligible.

(ii) Evaluate the external work done per unit mass of gas when the gas velocity at entry is 80 m/s and that at exit is 160 m/s.
                    Ans: (i) 143.3 KJ/kg (ii) 152.9 KJ/kg

2.5. The relation between the properties of gaseous oxygen gas may be expressed over a restricted range by :

$$Pv = 260t + 71 \times 10^3 \quad and \quad t = 1.52u - 273$$

where P is in N/m², v in m³/kg, t in °C and u in kJ/kg.

(i) Evaluate the specific heat at constant volume and specific heat at constant pressure in kJ/kg K.

(ii) Show that for any process executed by unit mass of oxygen, $\Delta u = C_v \Delta t$ and $\Delta h = C_p \Delta t$

Ans: (i) 0.658 kJ/kg K (ii) 0.918 kJ/kg K

2.6. Two kg of a gas at 10 bar expands adiabatically and reversibly till the pressure falls to 5 bar. During the process 170 kJ of non-flow work is done by the system, and the temperature drops from 377 °C to 257 °C. Calculate (i) the value of the index of expansion and (ii) characteristic gas constant.

Ans: (i) 1.416 (ii) 294 J/kg K

2.7. Show that for a Vander Waals gas which has the equation of state

$$\left(P + \frac{a}{V_2} \ = \ RT \right)$$

The work done at constant temperature per unit mass of gas is

$$RT \ln \frac{V_2 - b}{V_1 - b} - a \left(\frac{1}{V_1} - \frac{1}{V_2} \right)$$

where V_1 and V_2 denote the initial and final specific volumes respectively.

2.8. In an air-standard cycle, heat is supplied at constant volume resulting in an increase in temperature of air from T_1 to T_2. The air is then expanded isentropically till its temperature falls to T_1. Finally, it is returned to its original state by a reversible isothermal compression process.

Show that the efficiency of the cycle is given by

$$\eta \ = \ 1 - \frac{T_1}{T_2 - T_1} \ln \frac{T_2}{T_1}$$

2.9. A volume of 5 m³ of air at a pressure of 1 bar and 27 °C is compressed adiabatically to 5 bar. The compressed air is then expanded isothermally to original volume. Find :

(i) The final pressure of the air after expansion

(ii) The quantity of heat added from the beginning of compression to the end of expansion

(iii) The quantity of heat that must be added or subtracted to reduce the air after expansion to the original state of pressure, volume and temperature.

Ans: (i) 1.58 bar (ii) 910.09 kJ (iii) −729.8 kJ

2.10. A perfect gas of molecular weight 30 has the ratio of specific heats 1.3. One kg of the gas at 1 bar and 27 °C is compressed adiabatically inside a cylinder to 7 times its initial pressure. Heat is now extracted from the system at constant pressure until it reaches a state (temperature 27 °C) such that the gas when expanded isothermally from this state will reach the state before adiabatic compression. Calculate the heat transfer during constant pressure cooling and the work done during isothermal expansion. What is the amount of work transfer during the cycle and in which direction does it take place. $R = 0.287$ kJ/kg K.

Ans: (i) -204 kJ/kg (ii) 161.8 kJ/kg (iii) −42.2 kJ/kg

2.11. A perfect gas flows steadily through a horizontal cooler. The mass flow rate through the cooler is 1 kg/s. The pressure and temperature are 2 bar and 127 °C at entry and 1.5 bar and 7 °C at exit. The cross sectional areas at entry and exit are each 80 cm^2. Calculate the velocities of the gas at entry and exit of the cooler and the heat transfer rate in the cooler. Take for the gas $C_v = 0.70$ kJ/kg K and $R = 160$ J/kg K.

Ans: (i) 40 m/s (ii) 37.5 m/s (iii) −103.3 kJ/s

2.12. Air inside a cylinder is compressed from the same initial state, without friction such that the compression ratio, $\frac{V_2}{V_1}$ is 15. In one case it is compressed isothermally and in the other case it is compressed polytropically with polytropic index, $n = 1.3$. Calculate the ratios of work done and heat transfer in two cases. What will be the ratio of final pressure in the two cases.

Ans: (i) 0.649 (ii) 2.59 (iii) 0.444

<h1 align="center">3</h1>

AIR-STANDARD CYCLES
AND
THEIR ANALYSIS

3.1 INTRODUCTION

The operating cycle of an internal combustion engine can be broken down into a sequence of separate processes viz., intake, compression, combustion, expansion and exhaust. The internal combustion engine does not operate on a thermodynamic cycle as it involves an open system i.e., the working fluid enters the system at one set of conditions and leaves at another. However, it is often possible to analyze the open cycle as though it were a closed one by imagining one or more processes that would bring the working fluid at the exit conditions back to the condition of the starting point.

The accurate analysis of internal combustion engine processes is very complicated. In order to understand them it is advantageous to analyze the performance of an idealised closed cycle that closely approximates the real cycle. One such approach is the air-standard cycle, which is based on the following assumptions :

(i) The working medium is assumed to be a perfect gas and follows the relation $pV = mRT$ or $p = \rho RT$.

(ii) There is no change in the mass of the working medium.

(iii) All the processes that constitute the cycle are reversible.

(iv) Heat is assumed to be supplied from a constant high temperature source and not from chemical reactions during the cycle.

(v) Some heat is assumed to be rejected to a constant low temperature sink during the cycle.

(vi) It is assumed that there are no heat losses from the system to the surroundings.

(vii) The working medium has constant specific heats throughout the cycle.

(viii) The physical constants viz., C_p, C_v, γ and M of working medium are the same as those of air at standard atmospheric conditions. For example in S.I. units,

$$C_p = 1.005 \text{ kJ/kg K} \quad M = 29 \text{ kg/kmol}$$
$$C_v = 0.717 \text{ kJ/kg K} \quad \gamma = 1.4$$

Due to these assumptions, the analysis becomes over-simplified and the results do not agree with those of the actual engine. Work output, peak pressure, peak temperature and thermal efficiency based on air-standard cycles will be the maximum that can be attained and will differ considerably from those of the actual engine. It is often used, mainly because of the simplicity in getting approximate answers to the complicated processes in internal combustion engines.

In this chapter, we will review the various cycles and also derive the equations for work output, mean effective pressure, efficiency etc. Also, comparison will be made between Otto, Dual and Diesel cycles to see which cycle is more efficient under a set of given operating conditions.

3.2 THE CARNOT CYCLE

Sadi Carnot, a French engineer, proposed a reversible cycle in 1894, in which the working medium receives heat at a higher temperature and rejects heat at a lower temperature. The cycle will consist of two isothermal and two reversible adiabatic processes as shown in Fig.3.1. Carnot cycle is represented as a standard of perfection and engines can be compared with it to judge the degree of perfection. It gives the concept of maximizing work output between two temperature limits.

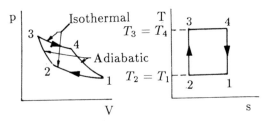

Fig.3.1 Carnot Engine

The working of an engine based on the Carnot cycle can be explained referring to Fig.3.2 which shows a cylinder and piston arrangement working without friction. The walls of cylinder are assumed to

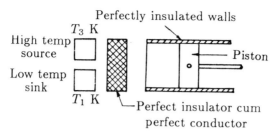

Fig.3.2 Working Principle of a Carnot Engine

be perfect insulators. The cylinder head is so arranged that it can be a perfect heat conductor as well as a perfect heat insulator.

First the heat is transferred from a high temperature source, (T_3), to the working medium in the cylinder and as a result the working medium expands. This is represented by the isothermal process $3{\rightarrow}4$ in Fig.3.1. Now the cylinder head is sealed and it acts as a perfect insulator. The working medium in the cylinder is now allowed to expand further from state 4 to state 1 and is represented by reversible adiabatic process $4{\rightarrow}1$ in p-V and T-s diagrams in Fig.3.1. Now the system is brought into contact with a constant low temperature sink, (T_1), as the cylinder head is now made to act as a perfect heat conductor. Some heat is rejected to the sink without altering the temperature of sink and as a result the working medium is compressed from state 1 to 2 which is represented by isothermal line $1{\rightarrow}2$. Finally the cylinder head is made again to act as a perfect insulator and the working medium is compressed adiabatically from state 2 to 3 which is represented by process $2{\rightarrow}3$. Thus the cycle is completed.

Analysing the cycle thermodynamically the efficiency of the cycle can be written as

$$\eta_{Carnot} = \frac{\text{Work done by the system during the cycle } (W)}{\text{Heat supplied to the system during the cycle } (Q_S)}$$

According to the first law of thermodynamics,

$$\text{Work done} = \text{Heat supplied} - \text{Heat rejected}$$

$$W = Q_S - Q_R \qquad (3.1)$$

Considering the isothermal processes $1{\rightarrow}2$ and $3{\rightarrow}4$, we get

$$Q_R = mRT_1 \log_e \frac{V_1}{V_2} \qquad (3.2)$$

$$Q_S = mRT_3 \log_e \frac{V_4}{V_3} \tag{3.3}$$

Considering adiabatic processes $2{\rightarrow}3$ and $4{\rightarrow}1$

$$\frac{V_3}{V_2} = \left(\frac{T_2}{T_3}\right)^{\left(\frac{1}{\gamma-1}\right)} \tag{3.4}$$

and

$$\frac{V_4}{V_1} = \left(\frac{T_1}{T_4}\right)^{\left(\frac{1}{\gamma-1}\right)} \tag{3.5}$$

Since $T_1 = T_2$ and $T_4 = T_3$ we have,

$$\frac{V_4}{V_1} = \frac{V_3}{V_2}$$

or

$$\frac{V_4}{V_3} = \frac{V_1}{V_2} = r \quad (say) \tag{3.6}$$

then,

$$\eta_{Carnot} = \frac{mRT_3 \log_e r - mRT_1 \log_e r}{mRT_3 \log_e r} \tag{3.7}$$

$$= \frac{T_3 - T_1}{T_3} = 1 - \frac{T_1}{T_3} \tag{3.8}$$

The lower temperature i.e., sink temperature, T_1, is normally the atmospheric temperature or the cooling water temperature and hence fixed. So the increase in thermal efficiency can be achieved only by increasing the source temperature. In other words, the upper temperature is required to be maintained as high as possible, to achieve maximum thermal efficiency. Between two fixed temperatures Carnot cycle (and other reversible cycles) has the maximum possible efficiency compared to other air-standard cycles. Inspite of this advantage, Carnot cycle does not provide a suitable basis for the operation of an engine using a gaseous working fluid because the work output from this cycle will be quite low.

Mean effective pressure, p_m, is defined as that hypothetical constant pressure acting on the piston during its expansion stroke producing the same work output as that from the actual cycle. Mathematically,

$$p_m = \frac{\text{Work Output}}{\text{Swept Volume}} \qquad (3.9)$$

It can be shown as

$$p_m = \frac{\text{Area of indicator diagram}}{\text{Length of diagram}} \times \text{constant} \qquad (3.10)$$

The constant depends on the mechanism used to get the indicator diagram and has the units, bar/m. These formulae are quite often used to calculate the performance of an internal combustion engine. If the work output is the indicated output then it is called indicated mean effective pressure, p_{im}, and if the work output is the brake output then it is called brake mean effective pressure, p_{bm}.

3.3 THE STIRLING CYCLE

The Carnot cycle has a low mean effective pressure because of its very low work output. Hence, one of the modified forms of the cycle to produce higher mean effective pressure whilst theoretically achieving full Carnot cycle efficiency is the Stirling cycle. It consists of two isothermal and two constant volume processes. The heat rejection and addition take place at constant temperature. The p-V and T-s diagrams for the Stirling cycle are shown in Figs.3.3(a) and 3.3(b) respectively.

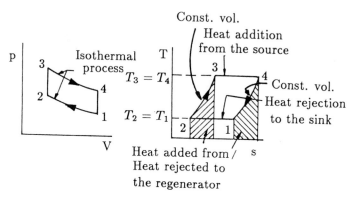

Fig.3.3 Stirling Cycle

It is clear from Fig.3.3(b) that the amount of heat addition and rejection during constant volume processes is same. Hence, the efficiency of the cycle is given as

$$\eta_{Stirling} = \frac{RT_3 \log_e \left(\frac{V_4}{V_3}\right) - RT_1 \log_e \left(\frac{V_1}{V_2}\right)}{RT_3 \log_e \left(\frac{V_4}{V_3}\right)} \tag{3.11}$$

But $V_3 = V_2$ and $V_4 = V_1$

$$\eta_{Stirling} = \frac{T_3 - T_1}{T_3} \tag{3.12}$$

same as Carnot efficiency

The Stirling cycle was used earlier for hot air engines and became obsolete as Otto and Diesel cycles came into use. The design of Stirling engine involves a major difficulty in the design and construction of heat exchanger to operate continuously at very high temperatures. However, with the development in metallurgy and intensive research in this type of engine, the Stirling engine has staged a come back in practical appearance. In practice, the heat exchanger efficiency cannot be 100%. Hence the Stirling cycle efficiency will be less than Carnot efficiency and can be written as

$$\eta = \frac{R(T_3 - T_1) \log_e r}{RT_3 \log_e r + (1 - \epsilon)C_v(T_3 - T_1)} \tag{3.13}$$

where ϵ is the heat exchanger effectiveness.

3.4 THE ERICSSON CYCLE

The other modified form of Carnot cycle for higher mean effective pressure is the Ericsson cycle. The Ericsson cycle consists of two isothermal processes and two constant pressure processes. The heat addition and rejection take place at constant pressure.

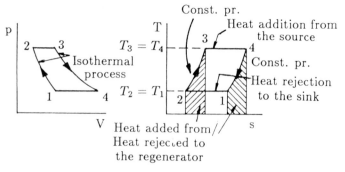

Fig.3.4 Ericsson Cycle

The cycle is shown on p-V and T-s diagrams in Fig.3.4(a) and 3.4(b) respectively. Comparing the Carnot, Stirling and Ericsson cycles all operating between the same source and sink temperatures, it is seen that the efficiency is the largest for the Carnot cycle. The advantage of the Ericsson cycle over the Carnot and Stirling cycles is its smaller pressure ratio for a given ratio of maximum to minimum specific volume.

The Ericsson cycle does not find practical application in piston engines but is approached by a gas turbine employing a large number of stages with heat exchangers, insulators and reheaters.

3.5 THE OTTO CYCLE

The main drawback of the Carnot cycle is its impracticability due to high pressure and high volume ratios employed with comparatively low mean effective pressure. Nicolaus Otto (1876), proposed a constant-volume heat addition cycle which forms the basis for the working of today's spark-ignition engines. The cycle is shown on p-V and T-s diagrams in Fig.3.5(a) and 3.5(b) respectively.

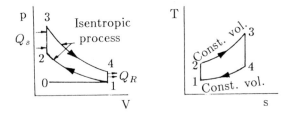

Fig.3.5 Otto Cycle

When the engine is working on full throttle, the processes $0 \rightarrow 1$ and $1 \rightarrow 0$ on the p-V diagram represents suction and exhaust processes and their effect is nullified. The process $1 \rightarrow 2$ represents isentropic compression of the air when the piston moves from bottom dead centre to top dead centre. During the process $2 \rightarrow 3$ heat is supplied reversibly at constant volume. This process corresponds to spark-ignition and combustion in the actual engine. The processes $3 \rightarrow 4$ and $4 \rightarrow 1$ represent isentropic expansion and constant volume heat rejection respectively.

3.5.1 Thermal Efficiency

The thermal efficiency of Otto cycle can be written as

$$\eta_{Otto} = \frac{Q_S - Q_R}{Q_S} \tag{3.14}$$

Considering constant volume processes 2→3 and 4→1, the heat supplied and rejected of air can be written as

$$Q_S = mC_v(T_3 - T_2) \tag{3.15}$$

$$Q_R = mC_v(T_4 - T_1) \tag{3.16}$$

$$\eta_{Otto} = \frac{m(T_3 - T_2) - m(T_4 - T_1)}{m(T_3 - T_2)}$$

$$= 1 - \frac{T_4 - T_1}{T_3 - T_2} \tag{3.17}$$

Considering isentropic processes 1→2 and 3→4, we have

$$\frac{T_2}{T_1} = \left(\frac{V_1}{V_2}\right)^{(\gamma-1)} \tag{3.18}$$

and

$$\frac{T_3}{T_4} = \left(\frac{V_4}{V_3}\right)^{(\gamma-1)} \tag{3.19}$$

But the volume ratios V_1/V_2 and V_4/V_3 are equal to the compression ratio, r. Therefore,

$$\frac{V_1}{V_2} = \frac{V_4}{V_3} = r \tag{3.20}$$

therefore,

$$\frac{T_2}{T_1} = \frac{T_3}{T_4} \tag{3.21}$$

From Eqn.3.21, it can be easily shown that

$$\frac{T_4}{T_3} = \frac{T_1}{T_2} = \frac{T_4 - T_1}{T_3 - T_2} \tag{3.22}$$

$$\eta_{Otto} \quad = \quad 1 - \frac{T_1}{T_2} \qquad (3.23)$$

$$= \quad 1 - \frac{1}{\left(\dfrac{V_1}{V_2}\right)^{(\gamma-1)}} \qquad (3.24)$$

$$= \quad 1 - \frac{1}{r^{(\gamma-1)}} \qquad (3.25)$$

Note that the thermal efficiency of Otto cycle is a function of compression ratio r and the ratio of specific heats, γ. As γ is assumed to be a constant for any working fluid, the efficiency is increased by increasing the compression ratio. Further, the efficiency is independent of heat supplied and pressure ratio. The use of gases with higher γ values would increase efficiency of Otto cycle. Fig.3.6 shows the effect of γ and r on the efficiency.

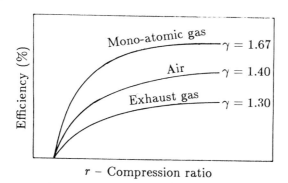

Fig.3.6 Effect of r and γ on Efficiency for Otto Cycle

3.5.2 Work Output

The net work output for an Otto cycle can be expressed as

$$W \quad = \quad \frac{p_3 V_3 - p_4 V_4}{\gamma - 1} - \frac{p_2 V_2 - p_1 V_1}{\gamma - 1} \qquad (3.26)$$

Also

$$\frac{p_2}{p_1} \quad = \quad \frac{p_3}{p_4} \quad = \quad r^\gamma$$

$$\frac{p_3}{p_2} \quad = \quad \frac{p_4}{p_1} \quad = \quad r_p \quad (say) \qquad (3.27)$$

$$V_1 \quad = \quad r V_2 \quad and \quad V_4 \quad = \quad r V_3$$

therefore,

$$W = \frac{p_1 V_1}{\gamma - 1} \left(\frac{p_3 V_3}{p_1 V_1} - \frac{p_4 V_4}{p_1 V_1} - \frac{p_2 V_2}{p_1 V_1} + 1 \right) \qquad (3.28)$$

$$= \frac{p_1 V_1}{\gamma - 1} \left(\frac{r_p r^\gamma}{r} - r_p - \frac{r^\gamma}{r} + 1 \right)$$

$$= \frac{p_1 V_1}{\gamma - 1} \left(r_p r^{\gamma - 1} - r_p - r^{\gamma - 1} + 1 \right)$$

$$= \frac{p_1 V_1}{\gamma - 1} (r_p - 1)(r^{\gamma - 1} - 1) \qquad (3.29)$$

3.5.3 Mean Effective Pressure

The mean effective pressure of the cycle is given by

$$p_m = \frac{\text{Work output}}{\text{Swept volume}} \qquad (3.30)$$

$$\text{Swept volume} = V_1 - V_2 = V_2(r - 1)$$

$$p_m = \frac{\frac{1}{\gamma - 1} p_1 V_1 (r_p - 1)(r^{(\gamma - 1)} - 1)}{V_2(r - 1)}$$

$$= \frac{p_1 r (r_p - 1)(r^{(\gamma - 1)} - 1)}{(\gamma - 1)(r - 1)} \qquad (3.31)$$

Thus, it can be seen that the work output is directly proportional to pressure ratio, r_p. The mean effective pressure which is an indication of the internal work output increases with a pressure ratio at a fixed value of compression ratio and ratio of specific heats. For an Otto cycle, an increase in the compression ratio leads to an increase in the mean effective pressure as well as the thermal efficiency.

3.6 THE DIESEL CYCLE

In actual spark-ignition engines, the upper limit of compression ratio is limited by the self-ignition temperature of the fuel. This limitation on the compression ratio can be circumvented if air and fuel are compressed separately and brought together at the time of combustion. In such an arrangement fuel can be injected into the cylinder which contains compressed air at a higher temperature than the self-ignition temperature of the fuel. Hence the fuel ignites on its own accord and requires no special device like an ignition system in a spark-ignition engine. Such engines work on heavy liquid fuels. These engines are called compression-ignition engines and they work on a ideal cycle known as Diesel cycle. The difference between Otto and Diesel cycles is in the process of heat addition. In Otto cycle the heat addition takes place at constant volume whereas in the Diesel cycle it is at constant pressure. For this reason, the Diesel cycle is often referred to as the constant-pressure cycle. It is better to avoid this term as it creates confusion with Joules cycle. The Diesel cycle is shown on p-V and T-s diagrams in Fig.3.7(a) and 3.7(b) respectively.

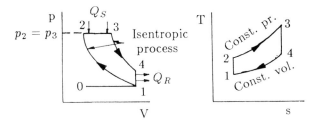

Fig.3.7 Diesel Cycle

To analyze the diesel cycle the suction and exhaust strokes, represented by $0 \rightarrow 1$ and $1 \rightarrow 0$, are neglected as in the case of the Otto cycle. Here, the volume ratio $\frac{V_1}{V_2}$ is the compression ratio, r. The volume ratio $\frac{V_3}{V_2}$ is called the cut-off ratio, r_c.

3.6.1 Thermal Efficiency

The thermal efficiency of the Diesel cycle is given by

$$\eta_{Diesel} = \frac{Q_S - Q_R}{Q_S}$$

$$= \frac{mC_p(T_3 - T_2) - mC_v(T_4 - T_1)}{mC_p(T_3 - T_2)} \qquad (3.32)$$

$$= 1 - \frac{C_v(T_4 - T_1)}{C_p(T_3 - T_2)}$$

$$= 1 - \frac{1}{\gamma}\left(\frac{T_4 - T_1}{T_3 - T_2}\right) \qquad (3.33)$$

Considering the process $1 \to 2$

$$T_2 = T_1\left(\frac{V_1}{V_2}\right)^{(\gamma-1)} = T_1 r^{(\gamma-1)} \tag{3.34}$$

Considering the constant pressure process $2 \to 3$, we have

$$\frac{V_2}{T_2} = \frac{V_3}{T_3}$$

$$\frac{T_3}{T_2} = \frac{V_3}{V_2} = r_c \quad \text{(say)}$$

$$T_3 = T_2 r_c \tag{3.35}$$

From Eqns.3.34 and 3.35, we have

$$T_3 = T_1 r^{(\gamma-1)} r_c \tag{3.36}$$

Considering process $3 \to 4$, we have

$$T_4 = T_3\left(\frac{V_3}{V_4}\right)^{(\gamma-1)} \tag{3.37}$$

$$= T_3\left(\frac{V_3}{V_2} \times \frac{V_2}{V_4}\right)^{(\gamma-1)}$$

$$= T_3\left(\frac{r_c}{r}\right)^{(\gamma-1)} \tag{3.38}$$

From Eqns.3.36 and 3.37, we have

$$T_4 = T_1 r^{(\gamma-1)} r_c \left(\frac{r_c}{r}\right)^{(\gamma-1)} = T_1 r_c^{\gamma}$$

$$\eta_{Diesel} = 1 - \frac{1}{\gamma}\left[\frac{T_1(r_c^{\gamma} - 1)}{T_1\left(r^{(\gamma-1)} r_c - r^{(\gamma-1)}\right)}\right]$$

$$= 1 - \frac{1}{\gamma}\left[\frac{(r_c^{\gamma} - 1)}{r^{(\gamma-1)} r_c - r^{(\gamma-1)}}\right]$$

$$= 1 - \frac{1}{r^{(\gamma-1)}}\left[\frac{r_c^{\gamma} - 1}{\gamma(r_c - 1)}\right] \tag{3.39}$$

It may be noted that the efficiency of the Diesel cycle is different from that of the Otto cycle only in the bracketed factor. This factor is always greater than unity. Hence for a given compression ratio, the Otto cycle is more efficient. In diesel engines the fuel cut-off ratio, r_c, depends on output, being maximum for maximum output. Therefore, unlike the Otto cycle the air-standard efficiency of the Diesel cycle depends on output. The higher efficiency of the Otto cycle as compared to the Diesel cycle for the same compression ratio is of no practical importance. In practice the operating compression ratios of diesel engines are much higher compared to spark-ignition engines working on Otto cycle. The normal range of compression ratio for diesel engine is 16 to 20 whereas for spark-ignition engines it is 6 to 10. Due to the higher compression ratios used in diesel engines the efficiency of a diesel engine is more than that of the gasoline engine.

3.6.2 Work Output

The net work output for a Diesel cycle is given by

$$W = p_2(V_3 - V_2) + \frac{p_3V_3 - p_4V_4}{\gamma - 1} - \frac{p_2V_2 - p_1V_1}{\gamma - 1} \tag{3.40}$$

$$= p_2V_2(r_c - 1) + \frac{p_3r_cV_2 - p_4rV_2}{\gamma - 1} - \frac{p_2V_2 - p_1rV_2}{\gamma - 1}$$

$$= V_2\left[\frac{p_2(r_c - 1)(\gamma - 1) + p_3r_c - p_4r - (p_2 - p_1r)}{\gamma - 1}\right]$$

$$= V_2\left[\frac{p_2(r_c - 1)(\gamma - 1) + p_3\left(r_c - \frac{p_4}{p_3}r\right) - p_2\left(1 - \frac{p_1}{p_2}r\right)}{\gamma - 1}\right]$$

$$= p_2V_2\left[\frac{(r_c - 1)(\gamma - 1) + \left(r_c - r_c^\gamma r^{(1-\gamma)}\right) - \left(1 - r^{(1-\gamma)}\right)}{\gamma - 1}\right]$$

$$= \frac{p_1V_1r^{(\gamma-1)}\left[\gamma(r_c - 1) - r^{(1-\gamma)}(r_c^\gamma - 1)\right]}{\gamma - 1} \tag{3.41}$$

3.6.3 Mean Effective Pressure

The expression for mean effective pressure can be shown to be

$$p_m = \frac{p_1V_1\left[r^{(\gamma-1)}\gamma(r_c - 1) - (r_c^\gamma - 1)\right]}{(\gamma - 1)V_1\left(\frac{r-1}{r}\right)} \tag{3.42}$$

$$= \frac{p_1\left[\gamma r^\gamma(r_c - 1) - r(r_c^\gamma - 1)\right]}{(\gamma - 1)(r - 1)} \tag{3.43}$$

3.7 THE DUAL CYCLE

In the Otto cycle, combustion is assumed at constant volume while in Diesel cycle combustion is at constant pressure. In practice they are far from real. Since, some time interval is required for the chemical reactions during combustion process, the combustion cannot take place at constant volume. Similarly, due to rapid uncontrolled combustion in diesel engines, combustion does not occur at constant pressure. The Dual cycle, also called a mixed cycle or limited pressure cycle, is a compromise between Otto and Diesel cycles. Figures 3.8(a) and 3.8(b) show the Dual cycle on *p-V* and *T-s* diagrams respectively.

In a Dual cycle a part of the heat is first supplied to the system at constant volume and then the remaining part at constant pressure.

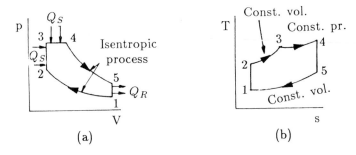

(a)   (b)

Fig.3.8 Dual Cycle

3.7.1 Thermal Efficiency

The efficiency of the cycle may be written as

$$\eta_{Dual} = \frac{Q_S - Q_R}{Q_S} \tag{3.44}$$

$$= \frac{mC_v(T_3 - T_2) + mC_p(T_4 - T_3) - mC_v(T_5 - T_1)}{mC_v(T_3 - T_2) + mC_p(T_4 - T_3)}$$

$$= 1 - \frac{T_5 - T_1}{(T_3 - T_2) + \gamma(T_4 - T_3)} \tag{3.45}$$

Now,

$$T_2 = T_1\left(\frac{V_1}{V_2}\right)^{(\gamma-1)} = T_1 r^{(\gamma-1)} \tag{3.46}$$

$$T_3 = T_2\left(\frac{p_3}{p_2}\right) = T_1 r_p r^{(\gamma-1)} \tag{3.47}$$

where r_p is the pressure ratio in the constant volume heat addition process and is equal to $\frac{p_3}{p_2}$

Cut-off ratio r_c is given by $\left(\frac{V_4}{V_3}\right)$

$$T_4 = T_3\frac{V_4}{V_3} = T_3 r_c$$

Substituting for T_3 from Eqn.3.47

$$T_4 = T_1 r_c r_p r^{(\gamma-1)} \tag{3.48}$$

and

$$T_5 = T_4\left(\frac{V_4}{V_5}\right)^{(\gamma-1)} \tag{3.49}$$

$$= T_1 r_p r_c r^{(\gamma-1)}\left(\frac{V_4}{V_5}\right)^{(\gamma-1)} \tag{3.50}$$

Now

$$\frac{V_4}{V_5} = \frac{V_4}{V_1} = \frac{V_4}{V_3}\times\frac{V_3}{V_1} \tag{3.51}$$

$$= \frac{V_4}{V_3}\times\frac{V_2}{V_1} \quad (\text{since } V_2 = V_3)$$

therefore,

$$\frac{V_4}{V_5} = \frac{r_c}{r} \tag{3.52}$$

where $\frac{V_4}{V_5}$ is the expansion ratio. Now,

$$T_5 = T_1 r_p r_c r^{\gamma-1}\left(\frac{r_c}{r}\right)^{\gamma-1}$$

$$= T_1 r_p r_c^{\gamma} \tag{3.53}$$

Substituting for T_2, T_3, T_4 and T_5 into Eqn.3.45 and simplifying

$$\eta = 1 - \frac{1}{r^{(\gamma-1)}}\left[\frac{r_p r_c^{\gamma} - 1}{(r_p - 1) + r_p\gamma(r_c - 1)}\right] \tag{3.54}$$

It can be seen from the above equation that a value of $r_p > 1$ results in an increased efficiency for a given value of r_c and γ. Thus the efficiency of Dual cycle lies between that of the Otto cycle and the Diesel cycle having same compression ratio.

With $r_c = 1$, it becomes an Otto cycle, and with $r_p = 1$, it becomes a Diesel cycle.

3.7.2 Work Output

The net work output of the cycle is given by

$$W = p_3(V_4 - V_3) + \frac{p_4V_4 - p_5V_5}{\gamma - 1} - \frac{p_2V_2 - p_1V_1}{\gamma - 1}$$

$$= \frac{p_1V_1}{\gamma - 1}\left[(\gamma - 1)\left(\frac{p_4V_4}{p_1V_1} - \frac{p_3V_3}{p_1V_1}\right) + \frac{p_4V_4}{p_1V_1} - \frac{p_5V_5}{p_1V_1} - \frac{p_2V_2}{p_1V_1} + 1\right]$$

$$= \frac{p_1V_1}{\gamma - 1}\left[(\gamma - 1)\left(r_cr_pr^{\gamma-1} - r_pr^{\gamma-1}\right) + r_cr_pr^{\gamma-1} - r_pr_c^\gamma - r^{\gamma-1} + 1\right]$$

$$= \frac{p_1V_1}{\gamma - 1}\left[\gamma r_cr_pr^{\gamma-1} - \gamma r_pr^{\gamma-1} + r_pr^{\gamma-1} - r_pr_c^\gamma - r^{\gamma-1} + 1\right]$$

$$= \frac{p_1V_1}{\gamma - 1}\left[\gamma r_pr^{\gamma-1}(r_c - 1) + r^{\gamma-1}(r_p - 1) - (r_pr_c^\gamma - 1)\right] \qquad (3.55)$$

3.7.3 Mean Effective Pressure

The mean effective pressure is given by

$$p_m = \frac{\text{Work output}}{\text{Swept volume}} = \frac{W}{V_s}$$

$$= \frac{1}{V_1 - V_2}\frac{p_1V_1}{\gamma - 1}\left[\gamma r_pr^{\gamma-1}(r_c - 1) + r^{\gamma-1}(r_p - 1) - (r_pr_c^\gamma - 1)\right]$$

$$= \frac{1}{\left(1 - \frac{V_2}{V_1}\right)}\frac{p_1}{(\gamma - 1)}\left[\gamma r_pr^{\gamma-1}(r_c - 1) + r^{\gamma-1}(r_p - 1) - (r_pr_c^\gamma - 1)\right]$$

$$= p_1\frac{\left[\gamma r_pr^\gamma(r_c - 1) + r^\gamma(r_p - 1) - r(r_pr_c^\gamma - 1)\right]}{(\gamma - 1)(r - 1)} \qquad (3.56)$$

3.8 COMPARISON OF THE OTTO, DIESEL AND DUAL CYCLES

The important variable factors which are used as the basis for comparison of the cycles are compression ratio, peak pressure, heat addition, heat rejection and the net work. In order to compare the performance of the Otto, Diesel and Dual combustion cycles some of the variable factors must be fixed. In this section, a comparison of these three cycles is made for the same compression ratio, same heat addition, constant maximum pressure and temperature, same heat rejection and net work output. This analysis will show which cycle is more efficient for a given set of operating conditions.

3.8.1 Same Compression Ratio and Heat Addition

The Otto cycle $1 \to 2 \to 3 \to 4 \to 1$, the Diesel cycle $1 \to 2 \to 3' \to 4' \to 1$ and the Dual cycle $1 \to 2 \to 2'' \to 3'' \to 4'' \to 1$ are shown in p-V and T-s diagrams in Fig.3.9(a) and 3.9(b) respectively for the same compression ratio and heat input.

From the T-s diagram, it can be seen that Area 5236 = Area 523'6' = Area 522''3''6'' as this area represents the heat input which is the same for all cycles.

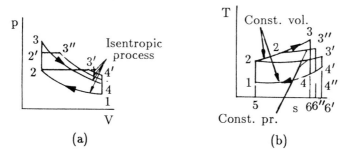

Fig.3.9 Same Compression Ratio and Heat Addition

All the cycles start from the same initial state point 1 and the air is compressed from state 1 to 2 as the compression ratio is same. It is seen from the T-s diagram for the same heat input, the heat rejection in Otto cycle (area 5146) is minimum and heat rejection in Diesel cycle (514'6') is maximum. *Consequently Otto cycle has the highest work output and efficiency. Diesel cycle has the least efficiency and Dual cycle having the efficiency between the two.*

One more observation can be made i.e., Otto cycle allows the working medium to expand more whereas Diesel cycle is least in this respect. The reason is heat is added before expansion in the case of former (Otto cycle) and the last portion of heat supplied to the fluid has a relatively short expansion in case of the latter (Diesel cycle).

3.8.2 Same Compression Ratio and Heat Rejection

Efficiency of Otto cycle is given by [Figs.3.10(a) and 3.10(b)]

$$\eta_{Otto} = 1 - \frac{Q_R}{Q_S}$$

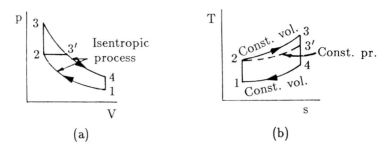

Fig.3.10 Same Compression Ratio and Heat Rejection

where Q_S is the heat supplied in the Otto cycle and is equal to the area under the curve 2→3 on the T-s diagram [Fig.3.10(b)]. The efficiency of the Diesel cycle is given by

$$\eta_{Diesel} \quad = \quad 1 - \frac{Q_R}{Q'_S}$$

where Q'_S is heat supplied in the Diesel cycle and is equal to the area under the curve 2→3′ on the T-s diagram [Fig.3.10(b)].

From the T-s diagram in Fig.3.10 it is clear that $Q_S > Q'_S$ i.e., heat supplied in the Otto cycle is more than that of the Diesel cycle. Hence, *it is evident that, the efficiency of the Otto cycle is greater than the efficiency of the Diesel cycle for a given compression ratio and heat rejection.*

3.8.3 Same Peak Pressure, Peak Temperature and Heat Rejection

Figures 3.11(a) and 3.11(b) show the Otto cycle 1→2→3→4 and Diesel cycle 1→2′→3→4 on p-V and T-s coordinates, where the peak pressure and temperature and the amount of heat rejected are the same.

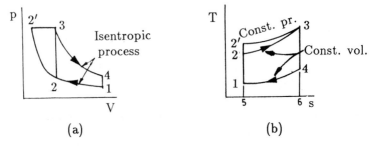

Fig.3.11 Same Peak Pressure and Temperature

The efficiency of the Otto cycle $1\rightarrow2\rightarrow3\rightarrow4$ is given by

$$\eta_{Otto} = 1 - \frac{Q_R}{Q_S}$$

where Q_S in the area under the curve $2\rightarrow3$ in Fig.3.11(b).
The efficiency of the Diesel cycle, $1\rightarrow2\rightarrow3'\rightarrow3\rightarrow4$ is

$$\eta_{Diesel} = 1 - \frac{Q_R}{Q'_S}$$

where Q'_S is the area under the curve $2'\rightarrow3$ in Fig.3.11(b).

It is evident from Fig.3.11 that $Q'_S > Q_S$. Therefore, the Diesel cycle efficiency is greater than the Otto cycle efficiency when both engines are built to withstand the same thermal and mechanical stresses.

3.8.4 Same Maximum Pressure and Heat Input

For same maximum pressure and same heat input the Otto cycle (12341) and Diesel cycle (12′3′4′1) are shown on p-V and T-s diagrams in Figs.3.12(a) and 3.12(b) respectively.

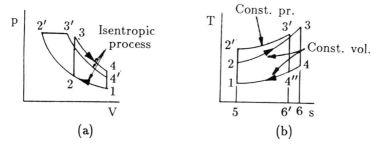

(a)　　　　　　　　　(b)

Fig.3.12 Same Maximum Pressure and Heat Input

It is evident from the figure that the heat rejection for Otto cycle (area 1564 on T-s diagram) is more than the heat rejected in Diesel cycle (156′4′). Hence Diesel cycle is more efficient than Otto cycle for the condition of same maximum pressure and heat input. One can make a note that with these conditions the Diesel cycle has higher compression ratio $\frac{V_1}{V_{2'}}$ than that of Otto cycle $\frac{V_1}{V_2}$. One should also note that the cycle which is having higher efficiency allows maximum expansion. The Dual cycle efficiency will be between these two.

3.8.5 Same Maximum Pressure and Work Output

The efficiency, η, can be written as

$$\eta \;=\; \frac{\text{Work done}}{\text{Heat supplied}} \;=\; \frac{\text{Work done}}{\text{Work done} + \text{Heat rejected}}$$

Refer to T-s diagram in Fig.3.12(b). For same work output the area 1234 (work output of Otto cycle) and area 12'3'4' (work output of Diesel cycle) are same. To achieve this, the entropy at 3 should be greater than entropy at 3'. It is clear that the heat rejection for Otto cycle is more than that of Diesel cycle. Hence, for these conditions the Diesel cycle is more efficient than the Otto cycle. The efficiency of Dual cycle lies between the two cycles.

3.9 THE LENOIR CYCLE

The Lenoir cycle consists of the following processes [see Fig.3.13(a)]. Constant volume heat addition (1→2); isentropic expansion (2→3); constant pressure heat rejection (3→1). The Lenoir cycle is used for pulse jet engines.

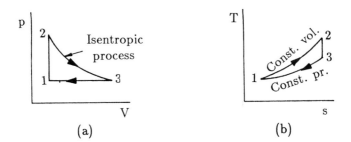

Fig.3.13 Lenoir Cycle

$$\eta_{Lenoir} \;=\; \frac{Q_S - Q_F}{Q_S}$$

$$Q_S \;=\; mC_v(T_2 - T_1) \tag{3.57}$$

$$Q_R \;=\; mC_p(T_3 - T_1) \tag{3.58}$$

$$\eta_{Lenoir} \;=\; \frac{mC_v(T_2 - T_1) - mC_p(T_3 - T_1)}{mC_v(T_2 - T_1)} \tag{3.59}$$

$$\;=\; 1 - \gamma\left(\frac{T_3 - T_1}{T_2 - T_1}\right)$$

Taking $p_2/p_1 = r_p$, we have $T_2 = T_1 r_p$ and

$$\frac{T_3}{T_2} = \left(\frac{p_3}{p_2}\right)^{\left(\frac{\gamma-1}{\gamma}\right)} \tag{3.60}$$

$$T_3 = T_2\left(\frac{1}{r_p}\right)^{\left(\frac{\gamma-1}{\gamma}\right)} = T_1 r_p \left(\frac{1}{r_p}\right)^{\left(\frac{\gamma-1}{\gamma}\right)} = T_1 r_p^{(1/\gamma)}$$

$$\eta = 1 - \gamma\left(\frac{T_1 r_p^{(1/\gamma)} - T_1}{T_1 r_p - T_1}\right) \tag{3.61}$$

$$= 1 - \gamma\left(\frac{r_p^{(1/\gamma)} - 1}{r_p - 1}\right) \tag{3.62}$$

Thus the efficiency of the Lenoir cycle depends upon the pressure ratio as well as the ratio of specific heats, viz., γ.

3.10 THE ATKINSON CYCLE

Atkinson cycle is an ideal cycle for Otto engine exhausting to a gas turbine. In this cycle the isentropic expansion $(3\rightarrow4)$ of an Otto cycle (1234) is further allowed to proceed to the lowest cycle pressure so as to increase the work output. With this modification the cycle is known as Atkinson cycle. The cycle is shown on p-V and T-s diagrams in Figs.3.14(a) and 3.14(b) respectively.

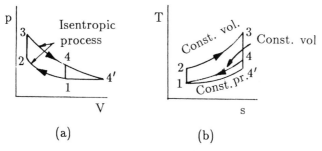

Fig.3.14 Atkinson Cycle

$$\eta_{Atkinson} = \frac{Q_S - Q_R}{Q_S} \tag{3.63}$$

$$= \frac{mC_v(T_3 - T_2) - mC_p(T_{4'} - T_1)}{mC_v(T_3 - T_2)} \tag{3.64}$$

$$= 1 - \gamma \left(\frac{T_{4'} - T_1}{T_3 - T_2} \right) \tag{3.65}$$

the compression ratio, $r = \frac{V_1}{V_2}$ and the expansion ratio $e = \frac{V_{4'}}{V_3}$. Now,

$$\frac{T_2}{T_1} = \left(\frac{V_1}{V_2} \right)^{(\gamma - 1)} = r^{(\gamma - 1)} \tag{3.66}$$

Therefore,

$$T_2 = T_1 r^{(\gamma - 1)} \tag{3.67}$$

$$\frac{T_3}{T_2} = \frac{p_3}{p_2} = \left(\frac{p_3}{p_{4'}} \times \frac{p_{4'}}{p_2} \right)$$

$$= \left(\frac{p_3}{p_{4'}} \times \frac{p_1}{p_2} \right) \tag{3.68}$$

$$\frac{p_3}{p_{4'}} = \left(\frac{V_{4'}}{V_3} \right)^{\gamma} = e^{\gamma} \tag{3.69}$$

$$\frac{p_1}{p_2} = \left(\frac{V_2}{V_1} \right)^{\gamma} = \frac{1}{r^{\gamma}} \tag{3.70}$$

Substituting Eqns.3.69 and 3.70 in Eqn.3.68,

$$\frac{T_3}{T_2} = \frac{e^{\gamma}}{r^{\gamma}} \tag{3.71}$$

$$T_3 = T_2 \frac{e^{\gamma}}{r^{\gamma}} = T_1 r^{(\gamma - 1)} \frac{e^{\gamma}}{r^{\gamma}} = T_1 \frac{e^{\gamma}}{r} \tag{3.72}$$

$$\frac{T_{4'}}{T_3} = \left(\frac{V_3}{V_{4'}} \right)^{(\gamma - 1)} = \frac{1}{e^{(\gamma - 1)}}$$

$$T_{4'} = T_3 \frac{1}{e^{(\gamma - 1)}} \tag{3.73}$$

$$= T_1 \left(\frac{e^{\gamma}}{r} \right) \left(\frac{1}{e^{(\gamma - 1)}} \right)$$

$$T_{4'} = T_1 \frac{e}{r} \tag{3.74}$$

Substituting the values of T_2, T_3, $T_{4'}$ in the Eqn.3.65,

$$\eta_{Atkinson} = 1 - \gamma\left[\frac{T_1 e/r - T_1}{T_1 e^{\gamma}/r - T_1 r^{(\gamma-1)}}\right]$$

$$= 1 - \gamma\left[\frac{e - r}{e^{\gamma} - r^{\gamma}}\right] \tag{3.75}$$

3.11 THE BRAYTON CYCLE

The Brayton cycle is a theoretical cycle for gas turbines. This cycle consists of two reversible adiabatic or isentropic processes and two constant pressure processes.

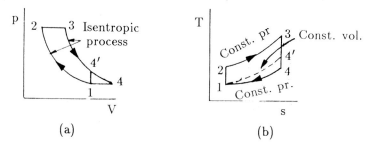

(a)   (b)

Fig.3.15 Brayton Cycle

Figure 3.15 shows the Brayton cycle on p-V and T-s coordinates. The cycle is similar to the Diesel cycle in compression and heat addition. The isentropic expansion of the Diesel cycle is further extended followed by constant pressure heat rejection.

$$\eta_{Brayton} = \frac{Q_S - Q_R}{Q_S}$$

$$= \frac{mC_p(T_3 - T_2) - mC_p(T_4 - T_1)}{mC_p(T_3 - T_2)} \tag{3.76}$$

$$= 1 - \frac{T_4 - T_1}{T_3 - T_2}$$

If r is compression ratio i.e., (V_1/V_2) and r_p is the pressure ratio i.e., (p_2/p_1) then,

$$\frac{T_3}{T_4} = \left(\frac{p_3}{p_4}\right)^{\left(\frac{\gamma-1}{\gamma}\right)} = \left(\frac{p_2}{p_1}\right)^{\left(\frac{\gamma-1}{\gamma}\right)} \tag{3.77}$$

$$= \left(\frac{V_1}{V_2}\right)^{(\gamma-1)} = r^{(\gamma-1)} \tag{3.78}$$

$$T_4 = \frac{T_3}{r^{(\gamma-1)}} \tag{3.79}$$

$$T_1 = \frac{T_2}{r^{(\gamma-1)}} \tag{3.80}$$

$$\eta_{Brayton} = 1 - \frac{(T_3/r^{(\gamma-1)}) - (T_2/r^{(\gamma-1)})}{T_3 - T_2}$$

$$= 1 - \frac{1}{r^{(\gamma-1)}} \tag{3.81}$$

$$r = \frac{V_1}{V_2} = \left(\frac{p_2}{p_1}\right)^{\frac{1}{\gamma}} = r_p^{\frac{1}{\gamma}} \tag{3.82}$$

$$= 1 - \frac{1}{\left(r_p^{\frac{1}{\gamma}}\right)^{\gamma-1}} = 1 - \frac{1}{r_p^{\left(\frac{\gamma-1}{\gamma}\right)}} \tag{3.83}$$

From Eqn.3.83, it is seen that the efficiency of the Brayton cycle depends only on the pressure ratio and the ratio of specific heat, γ.

$$\text{Network output} = \text{Expansion work} - \text{Compression work}$$

$$= C_p(T_3 - T_4) - C_p(T_2 - T_1)$$

$$= C_p T_1 \left(\frac{T_3}{T_1} - \frac{T_4}{T_1} - \frac{T_2}{T_1} + 1\right)$$

$$\frac{W}{C_p T_1} = \frac{T_3}{T_1} - \frac{T_4}{T_3}\frac{T_3}{T_1} - \frac{T_2}{T_1} + 1 \tag{3.84}$$

It can be easily seen from the Eqn.3.84 (work output) that, the work output of the cycle depends on initial temperature, T_1, the ratio of the maximum to minimum temperature, $\frac{T_3}{T_1}$, pressure ratio, r_p and γ which are used in the calculation of $\frac{T_2}{T_1}$. Therefore, for the same pressure ratio and initial conditions work output depends on the maximum temperature of the cycle.

Worked out Examples

OTTO CYCLE

3.1. An engine working on Otto cycle has the following conditions : Pressure at the beginning of compression is 1 bar and pressure at the end of compression is 11 bar. Calculate the compression ratio and air-standard efficiency of the engine. Assume $\gamma = 1.4$.

SOLUTION:

$$r \;=\; \frac{V_1}{V_2} \;=\; \left(\frac{p_2}{p_1}\right)^{\left(\frac{1}{\gamma}\right)} \;=\; 11^{\frac{1}{1.4}}$$

$$= \; 5.54 \qquad \underset{\displaystyle =}{\text{Ans}}$$

$$Air - standard\ efficiency \;=\; 1 - \frac{1}{r^{\gamma-1}}$$

$$= \; 1 - \left(\frac{1}{5.54}\right)^{0.4} \;=\; 0.496$$

$$= \; 49.6\% \qquad \underset{\displaystyle =}{\text{Ans}}$$

3.2. In an engine working on ideal Otto cycle the temperatures at the beginning and end of compression are 50 °C and 373 °C. Find the compression ratio and the air-standard efficiency of the engine.

SOLUTION:

$$r \;=\; \frac{V_1}{V_2} \;=\; \left(\frac{T_1}{T_2}\right)^{\frac{1}{\gamma-1}} \;=\; \left(\frac{646}{323}\right)^{\frac{1}{0.4}}$$

$$= \; 5.66 \qquad \underset{\displaystyle =}{\text{Ans}}$$

$$\eta_{Otto} \;=\; 1 - \frac{1}{r^{\gamma-1}} \;=\; 1 - \frac{T_1}{T_2}$$

$$= \; 1 - \frac{323}{646} \;=\; 0.5 \;=\; 50\% \qquad \underset{\displaystyle =}{\text{Ans}}$$

3.3. In an Otto cycle air at 17 °C and 1 bar is compressed adiabatically until the pressure is 15 bar. Heat is added at constant volume

until the pressure rises to 40 bar. Calculate the air-standard efficiency, the compression ratio and the mean effective pressure for
the cycle. Assume $C_v = 0.717 \text{ kJ/kg K}$ and $R = 8.314 \text{ kJ/kmol K}$.

SOLUTION:

Consider the process 1 − 2

$$p_1 V_1^\gamma = p_2 V_2^\gamma$$

$$\frac{V_1}{V_2} = r = \left(\frac{p_2}{p_1}\right)^{\frac{1}{\gamma}}$$

$$= \left(\frac{15}{1}\right)^{\frac{1}{1.4}} = 6.91$$

$$\eta = 1 - \left(\frac{1}{r}\right)^{\gamma-1}$$

$$= 1 - \left(\frac{1}{6.91}\right)^{0.4} = 0.539$$

$$= 53.9\%$$

Ans

$$T_2 = \frac{p_2 V_2}{p_1 V_1} T_1$$

$$= \frac{15}{1} \times \frac{1}{6.91} \times 290 = 629.5 \text{ K}$$

Consider the process 2 − 3

$$T_3 = \frac{p_3 T_2}{p_2} = \frac{40}{15} \times 629.5 = 1678.7$$

Heat supplied $= C_v(T_3 - T_2)$

$$= 0.717 \times (1678.7 - 629.5) = 752.3 \text{ kJ/kg}$$

Work done $= \eta \times q_s$

$$= 0.539 \times 752.3 = 405.5 \text{ kJ/kg}$$

$$p_m = \frac{\text{Work done}}{\text{Swept volume}}$$

$$v_1 \ = \ \frac{V_1}{m} \ = \ M\frac{RT_1}{p_1}$$

$$= \ \frac{8314 \times 290}{29 \times 1 \times 10^5} \ = \ 0.8314 \ \text{m}^3/\text{kg}$$

$$v_1 - v_2 \ = \ \frac{5.91}{6.91} \times 0.8314 \ = \ 0.711 \ \text{m}^3/\text{kg}$$

$$p_m \ = \ \frac{405.5}{0.711} \times 10^3 \ = \ 5.68 \times 10^5 \ \text{N}/\text{m}^2$$

$$= \ \textbf{5.68 bar} \qquad \underset{\Longleftarrow}{\text{Ans}}$$

3.4. Fuel supplied to an SI engine has a calorific value 42000 kJ/kg. The pressure in the cylinder at 30% and 70% of the compression stroke are 1.3 bar and 2.6 bar respectively. Assuming that the compression follows the law $pV^{1.3}$ = constant. Find the compression ratio. If the relative efficiency of the engine compared with the air-standard efficiency is 50%. Calculate the fuel consumption in kg/kW h.

SOLUTION:

$$V_2 \ = \ 1$$

$$V_{1'} \ = \ 1 + 0.7(r-1) \ = \ 0.7r + 0.3$$

$$V_{2'} \ = \ 1 + 0.3(r-1) \ = \ 0.3r + 0.7$$

$$\frac{V_{1'}}{V_{2'}} \ = \ \left(\frac{p_2}{p_1}\right)^{\frac{1}{n}} \ = \ \left(\frac{2.6}{1.3}\right)^{\frac{1}{1.3}} \ = \ 1.7$$

$$\frac{0.7r + 0.3}{0.3r + 0.7} \ = \ 1.7$$

$$r \ = \ \textbf{4.68} \qquad \underset{\Longleftarrow}{\text{Ans}}$$

$$Air - standard \ efficiency \ = \ 1 - \frac{1}{r^{\gamma-1}}$$

$$= \ 1 - \frac{1}{4.68^{0.4}} \ = \ 0.46 \ = \ 46\%$$

$$\text{Relative efficiency} = \frac{\text{Indicated thermal efficiency}}{\text{Air} - \text{standard efficiency}}$$

$$\text{Ind. th. efficiency} = 0.5 \times 0.46$$

$$= 0.23$$

$$\eta_{th} = \frac{ip}{CV \times \dot{m}}$$

where $\dot{m}$ is in kg/s

$$\frac{\dot{m}}{ip} = \frac{1}{42000 \times 0.23}$$

$$= 1.035 \times 10^{-4} \text{ kg/kW s}$$

$$= 1.035 \times 10^{-4} \times 3600 \text{ kg/kW h}$$

$$isfc = \mathbf{0.373} \text{ kg/kW h} \qquad \underset{=}{\underline{\text{Ans}}}$$

3.5. A gas engine working on the Otto cycle has a cylinder of diameter 200 mm and stroke 250 mm. The clearance volume is 1570 cc. Find the air-standard efficiency. Assume $C_p = 1.004$ kJ/kg K and $C_v = 0.717$ kJ/kg K for air.

SOLUTION:

$$\text{Stroke volume, } V_s = \frac{\pi}{4}d^2 L = \frac{\pi}{4} \times 20^2 \times 25$$

$$= 7853.98 \text{ cc}$$

$$\text{Compression ratio, } r = 1 + \frac{V_s}{V_c} = 1 + \frac{7853.98}{1570}$$

$$= 6.00$$

$$\gamma = \frac{C_p}{C_v} = \frac{1.004}{0.717} = 1.4$$

$$\text{Air} - \text{standard efficiency} = 1 - \frac{1}{r^{\gamma - 1}}$$

$$= 1 - \frac{1}{6^{(0.4)}} = 0.512$$

$$= \mathbf{51.2\%} \qquad \underset{=}{\underline{\text{Ans}}}$$

3.6. In a S.I. engine working on the ideal Otto cycle, the compression ratio is 5.5. The pressure and temperature at the beginning of compression are 1 bar and 27 °C respectively. The peak pressure is 30 bar. Determine the pressure and temperatures at the salient points, the air-standard efficiency and the mean effective pressure. Assume ratio of specific heats to be 1.4 for air.

SOLUTION:

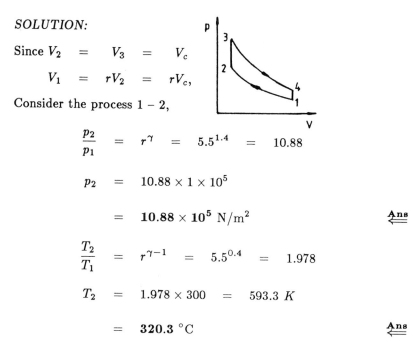

Since $V_2 = V_3 = V_c$

$\quad\quad V_1 = rV_2 = rV_c,$

Consider the process 1 – 2,

$$\frac{p_2}{p_1} = r^\gamma = 5.5^{1.4} = 10.88$$

$$p_2 = 10.88 \times 1 \times 10^5$$

$$= \mathbf{10.88 \times 10^5 \ N/m^2} \qquad \underline{\underline{\text{Ans}}}$$

$$\frac{T_2}{T_1} = r^{\gamma-1} = 5.5^{0.4} = 1.978$$

$$T_2 = 1.978 \times 300 = 593.3 \ K$$

$$= \mathbf{320.3 \ °C} \qquad \underline{\underline{\text{Ans}}}$$

Consider the process 2 – 3,

$$p_3 = 30 \times 10^5 \ N/m^2$$

$$\frac{T_3}{T_2} = \frac{p_3}{p_2} = \frac{30}{10.88} = 2.757$$

$$T_3 = 2.757 \times 593.3 = 1635.73 \ K$$

$$= \mathbf{1362.73 \ °C} \qquad \underline{\underline{\text{Ans}}}$$

Consider the process 3 – 4,

$$\frac{p_3}{p_4} = \left(\frac{V_4}{V_3}\right)^\gamma = \left(\frac{V_1}{V_2}\right)^\gamma$$

$$= r^\gamma = 5.5^{1.4} = 10.88$$

$$p_4 = \frac{p_3}{10.88}$$

$$= 2.76 \times 10^5 \text{ N/m}^2 \qquad \overset{\text{Ans}}{\Longleftarrow}$$

$$\frac{T_3}{T_4} = r^{\gamma-1} = 5.5^{0.4} = 1.978$$

$$T_4 = \frac{T_3}{1.978} = \frac{1635.73}{1.978} = 826.96 \text{ K}$$

$$= 553.96 \text{ }^\circ\text{C} \qquad \overset{\text{Ans}}{\Longleftarrow}$$

$$\eta_{Otto} = 1 - \frac{1}{r^{(\gamma-1)}} = 1 - \frac{1}{5.5^{0.4}} = 0.4943$$

$$= 49.43\% \qquad \overset{\text{Ans}}{\Longleftarrow}$$

$$p_m = \frac{Indicated\ work/cycle}{V_s}$$

$$= \frac{Area\ of\ p - V\ diagram\ 1234}{V_s}$$

$$Area\ 1234 = Area\ under\ 3\text{-}4 - Area\ under\ 2\text{-}1$$

$$= \frac{p_3 V_3 - p_4 V_4}{\gamma - 1} - \frac{p_2 V_2 - p_1 V_1}{\gamma - 1}$$

$$= \frac{30 \times 10^5 \times V_c - 2.76 \times 10^5 \times 5.5 V_c}{0.4}$$

$$- \frac{10.88 \times 10^5 \times V_c - 1 \times 10^5 \times 5.5 V_c}{0.4}$$

$$= 23.63 \times 10^5 \times V_c = p_m \times V_s$$

$$p_m = \frac{23.63 \times 10^5 \times V_c}{V_s}$$

$$= \frac{23.63 \times 10^5 \times V_c}{4.5 \times V_c} = 5.25 \times 10^5 \text{ N/m}^2$$

$$= 5.25 \text{ bar} \qquad \overset{\text{Ans}}{\Longleftarrow}$$

3.7. A gas engine operating on the ideal Otto cycle has a compression ratio of 6:1. The pressure and temperature at the commencement of compression are 1 bar and 27 °C. The heat added during the constant volume combustion process is 1170 kJ/kg. Determine the peak pressure and temperature, work output per kg of air and air-standard efficiency. Assume $C_v = 0.717$ kJ/kg K and $\gamma = 1.4$ for air.

SOLUTION:

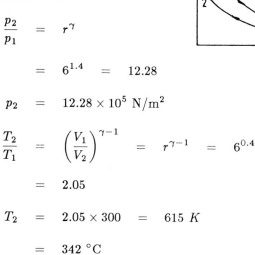

Consider the process 1 – 2

$$\frac{p_2}{p_1} = r^\gamma$$

$$= 6^{1.4} = 12.28$$

$$p_2 = 12.28 \times 10^5 \text{ N/m}^2$$

$$\frac{T_2}{T_1} = \left(\frac{V_1}{V_2}\right)^{\gamma-1} = r^{\gamma-1} = 6^{0.4}$$

$$= 2.05$$

$$T_2 = 2.05 \times 300 = 615 \; K$$

$$= 342 \; °C$$

Consider the process 2 – 3

For unit mass flow

$$q_s = q_{2-3} = C_v(T_3 - T_2)$$

$$= 1170 \text{ kJ/kg}$$

$$T_3 - T_2 = \frac{1170}{0.717} = 1631.8$$

$$T_3 = 1631.8 + 615 = 2246.8 \text{ K}$$

$$= \mathbf{1973.8} \; °C \qquad\qquad \underset{\Longleftarrow}{\mathbf{A\,ns}}$$

$$\frac{p_3}{p_2} = \frac{T_3}{T_2} = \frac{2246.8}{615} = 3.65$$

Peak pressure, $p_3 = 3.65 \times 12.28 \times 10^5$

$$= \quad 44.82 \times 10^5 \text{ N/m}^2$$

$$= \quad 44.82 \text{ bar} \qquad \underset{\Longleftarrow}{\text{Ans}}$$

$$\text{Work output} \quad = \quad \text{Area of } p - V \text{ diagram}$$

$$= \quad \text{Area under } (3 - 4) - \text{Area under } (2 - 1)$$

$$= \quad \frac{p_3 V_3 - p_4 V_4}{\gamma - 1} - \frac{p_2 V_2 - p_1 V_1}{\gamma - 1}$$

$$= \quad \frac{mR}{\gamma - 1}[(T_3 - T_4) - (T_2 - T_1)]$$

$$R \quad = \quad C_p - C_v \quad = \quad 1.004 - 0.717$$

$$= \quad 0.287 \text{ kJ/kg K}$$

$$\frac{T_3}{T_4} \quad = \quad \left(\frac{V_3}{V_4}\right)^{\gamma - 1} \quad = \quad r^{(\gamma - 1)}$$

$$= \quad 6^{0.4} \quad = \quad 2.048$$

$$T_4 \quad = \quad \frac{T_3}{2.048} \quad = \quad \frac{2246.8}{2.048}$$

$$= \quad 1097.1 \text{ K}$$

$$\text{Work output/kg} \quad = \quad \frac{0.287}{0.4} \times [(2246.8 - 1097.1) - (615 - 300)]$$

$$= \quad 598.9 \text{ kJ} \qquad \underset{\Longleftarrow}{\text{Ans}}$$

$$\eta_{Otto} \quad = \quad 1 - \frac{1}{r^{(\gamma - 1)}} \quad = \quad 1 - \frac{1}{6^{0.4}} \quad = \quad 0.5116$$

$$= \quad 51.16\% \qquad \underset{\Longleftarrow}{\text{Ans}}$$

3.8. A spark-ignition engine working on ideal Otto cycle has the compression ratio 6. The initial pressure and temperature of air are 1 bar and 37 °C. The maximum pressure in the cycle is 30 bar. For unit mass flow, calculate (i) p, V and T at various salients points of the cycle and (ii) the ratio of heat supplied to the heat rejected. Assume $\gamma = 1.4$ and $R = 8.314$ kJ/kmol K.

SOLUTION:

Consider point 1,

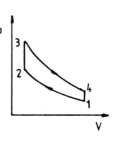

$$n = \frac{m}{M} = \frac{1}{29}$$

$$V_1 = \frac{n\,R\,T_1}{p_1}$$

$$= \frac{1 \times 8314 \times 310}{29 \times 10^5}$$

$$= \textbf{0.889 m}^3 \qquad \underset{\Longleftarrow}{\textbf{Ans}}$$

Consider point 2,

$$p_2 = p_1\, r^\gamma = 10^5 \times 6^{1.4}$$

$$= 12.3 \times 10^5 \text{ N/m}^2$$

$$= \textbf{12.3 bar} \qquad \underset{\Longleftarrow}{\textbf{Ans}}$$

$$V_2 = \frac{V_1}{6} = \frac{0.889}{6}$$

$$= \textbf{0.148 m}^3 \qquad \underset{\Longleftarrow}{\textbf{Ans}}$$

$$T_2 = \frac{p_2 V_2}{p_1 V_1}\, T_1$$

$$= \frac{12.3 \times 10^5 \times 0.148 \times 310}{1 \times 10^5 \times 0.889} = 634.8 \text{ K}$$

$$= \textbf{361.8 }^\circ\textbf{C} \qquad \underset{\Longleftarrow}{\textbf{Ans}}$$

Consider point 3,

$$V_3 = V_2 = \textbf{0.148 m}^3 \qquad \underset{\Longleftarrow}{\textbf{Ans}}$$

$$p_3 = 30 \times 10^5 \text{ N/m}^2$$

$$= \textbf{30 bar} \qquad \underset{\Longleftarrow}{\textbf{Ans}}$$

$$\frac{p_3}{T_3} = \frac{p_2}{T_2}$$

$$T_3 = \frac{30 \times 10^5}{12.3 \times 10^5} \times 634.8 = 1548 \text{ K}$$

$$= 1275 \,°\text{C} \qquad \underleftarrow{\text{Ans}}$$

Consider point 4,

$$p_3 V_3^\gamma = p_4 V_4^\gamma$$

$$p_4 = p_3 \left(\frac{V_3}{V_4}\right)^\gamma = 30 \times 10^5 \left(\frac{1}{6}\right)^{1.4}$$

$$= 2.44 \times 10^5 \text{ N/m}^2$$

$$= 2.44 \text{ bar} \qquad \underleftarrow{\text{Ans}}$$

$$V_4 = V_1 = 0.889 \text{ m}^3 \qquad \underleftarrow{\text{Ans}}$$

$$T_4 = T_1 \frac{p_4}{p_1}$$

$$= 310 \times \frac{2.44 \times 10^5}{1 \times 10^5} = 756.4 \text{ K}$$

$$= 483.4 \,°\text{C} \qquad \underleftarrow{\text{Ans}}$$

$$C_v = \frac{R}{M(\gamma - 1)} = \frac{8.314}{29 \times 0.4}$$

$$= 0.717 \text{ kJ/kg K}$$

For unit mass,

$$\text{Heat supplied} = C_v(T_3 - T_2)$$

$$= 0.717 \times (1548 - 635.5) = 654.3 \text{ kJ}$$

$$\text{Heat rejected} = C_v(T_4 - T_1)$$

$$= 0.717 \times (756.4 - 310) = 320.1 \text{ kJ}$$

$$\frac{\text{Heat supplied}}{\text{Heat rejected}} = \frac{654.3}{320.1}$$

$$= 2.04 \qquad \underleftarrow{\text{Ans}}$$

3.9. In an Otto engine, pressure and temperature at the beginning of compression are 1 bar and 37 °C respectively. Calculate the theoretical thermal efficiency of this cycle if the pressure at the end of the adiabatic compression is 15 bar. Peak temperature during the cycle is 2000 K. Calculate (i) the heat supplied per kg of air (ii) the work done per kg of air and (iii) the pressure at the end of adiabatic expansion. Take $C_v = 0.717$ kJ/kg K and $\gamma = 1.4$.

SOLUTION:

Consider the process 1 – 2

$$T_2 = T_1 \left(\frac{p_2}{p_1}\right)^{\frac{\gamma-1}{\gamma}} = 310 \times (15)^{(0.4/1.4)}$$

$$= 672 \text{ K}$$

$$\eta = 1 - \frac{T_1}{T_2} = 1 - \frac{310}{672} = 0.539$$

$$= 53.9\%$$

Consider the process 2 – 3

For unit mass flow

$$\text{Heat supplied, } q_s = C_v(T_3 - T_2) = 0.717 \times (2000 - 672)$$

$$= 952.2 \text{ kJ/kg} \qquad \overset{\textbf{Ans}}{\Longleftarrow}$$

$$\eta = \frac{\text{Work done per kg of air}}{\text{Heat supplied per kg of air}} = \frac{w}{q_s}$$

$$w = \eta q_s = 0.539 \times 952.2$$

$$= 513.2 \text{ kJ/kg} \qquad \overset{\textbf{Ans}}{\Longleftarrow}$$

$$p_3 = p_2\left(\frac{T_3}{T_2}\right) = 15 \times \frac{2000}{672}$$

$$= 44.64 \text{ bar}$$

Consider the process 3 – 4

$$\frac{p_3}{p_4} = \left(\frac{V_4}{V_3}\right)^{\gamma}$$

$$= \left(\frac{V_1}{V_2}\right)^\gamma = \frac{p_2}{p_1}$$

$$p_4 = p_3\left(\frac{p_1}{p_2}\right) = \frac{44.64 \times 1}{15}$$

$$= \textbf{2.98} \text{ bar} \qquad \underset{\Longleftarrow}{\textbf{Ans}}$$

3.10. Compare the efficiencies of ideal Atkinson cycle and Otto cycle for a compression ratio is 5.5. The pressure and temperature of air at the beginning of compression stroke are 1 bar and 27 °C respectively. The peak pressure is 25 bar for both cycles. Assume $\gamma = 1.4$ for air.

SOLUTION:

$$\eta_{Otto} = 1 - \frac{1}{r^{\gamma-1}} = 1 - \frac{1}{5.5^{0.4}} = 49.43\%$$

(Refer Fig.3.14)

$$\eta_{Atkinson} = 1 - \frac{\gamma(e-r)}{e^\gamma - r^\gamma}$$

$$r = 5.5$$

$$e = \frac{V_{4'}}{V_3} = \left(\frac{p_3}{p_{4'}}\right)^{\frac{1}{\gamma}} = \left(\frac{25}{1}\right)^{\frac{1}{1.4}} = 9.966$$

$$e^\gamma = 9.966^{1.4} = 25$$

$$r^\gamma = 5.5^{1.4} = 10.88$$

$$\eta_{Atkinson} = 1 - \frac{1.4 \times (9.966 - 5.5)}{25 - 10.88} = 55.72\%$$

$$\frac{\eta_{Atkinson}}{\eta_{Otto}} = \frac{55.72}{49.43}$$

$$= \textbf{1.127} \qquad \underset{\Longleftarrow}{\textbf{Ans}}$$

DIESEL CYCLE

3.11. A Diesel engine has a compression ratio of 20 and cut-off takes place at 5% of the stroke. Find the air-standard efficiency. Assume $\gamma = 1.4$.

SOLUTION:

$$r = \frac{V_1}{V_2} = 20$$

$$V_1 = 20V_2$$

$$V_s = 20V_2 - V_2 = 19V_2$$

$$V_3 = 0.05V_s + V_2$$

$$= 0.05 \times 19V_2 + V_2 = 1.95V_2$$

$$r_c = \frac{V_3}{V_2} = \frac{1.95V_2}{V_2} = 1.95$$

$$\eta = 1 - \frac{1}{r^{\gamma-1}} \frac{r_c^{\gamma} - 1}{\gamma(r_c - 1)}$$

$$= 1 - \frac{1}{20^{0.4}} \times \left[\frac{1.95^{1.4} - 1}{1.4 \times (1.95 - 1)} \right] = 0.649$$

$$= 64.9\% \qquad\qquad \underset{=}{\text{Ans}}$$

3.12. Determine the ideal efficiency of the diesel engine having a cylinder with bore 250 mm, stroke 375 mm and a clearance volume of 1500 cc, with fuel cut-off occurring at 5% of the stroke. Assume $\gamma = 1.4$ for air.

SOLUTION:

$$V_s = \frac{\pi}{4} d^2 L = \frac{\pi}{4} \times 25^2 \times 37.5$$

$$= 18407.8 \text{ cc}$$

$$r = 1 + \frac{V_s}{V_c} = 1 + \frac{18407.8}{1500} = 13.27$$

$$\eta = 1 - \frac{1}{r^{\gamma-1}} \frac{r_c^{\gamma} - 1}{\gamma(r_c - 1)}$$

$$r_c = \frac{V_3}{V_2}$$

$$\text{Cut-off volume} \quad = \quad V_3 - V_2 \quad = \quad 0.05V_s$$

$$= \quad 0.05 \times 12.27V_c$$

$$V_2 \quad = \quad V_c$$

$$V_3 \quad = \quad 1.6135V_c$$

$$r_c \quad = \quad \frac{V_3}{V_2} \quad = \quad 1.6135$$

$$\eta \quad = \quad 1 - \frac{1}{13.27^{0.4}} \times \frac{1.6135^{1.4} - 1}{1.4 \times (1.6135 - 1)}$$

$$= \quad 0.6052$$

$$= \quad \mathbf{60.52\%} \qquad \underset{\Longleftarrow}{\text{Ans}}$$

3.13. In an engine working on Diesel cycle inlet pressure and temperature are 1 bar and 17 °C respectively. Pressure at the end of
adiabatic compression is 35 bar. The ratio of expansion i.e. after
constant pressure heat addition is 5. Calculate the heat addition,
heat rejection and the efficiency of the cycle. Assume $\gamma = 1.4$,
$C_p = 1.004$ kJ/kg K and $C_v = 0.717$ kJ/kg K.

SOLUTION:

Consider the process 1 - 2

$$\frac{V_1}{V_2} \quad = \quad r \quad = \quad \left(\frac{p_2}{p_1}\right)^{\frac{1}{\gamma}}$$

$$= \quad \left(\frac{35}{1}\right)^{\frac{1}{1.4}} \quad = \quad 12.674$$

$$\text{Cut-off ratio} \quad = \quad \frac{V_3}{V_2} \quad = \quad \frac{V_3}{V_1} \times \frac{V_1}{V_2}$$

$$= \quad \frac{\text{Compression ratio}}{\text{Expansion ratio}}$$

$$= \quad \frac{12.674}{5} \quad = \quad 2.535$$

$$\frac{T_2}{T_1} = \left(\frac{p_2}{p_1}\right)^{\frac{\gamma-1}{\gamma}}$$

$$= \left(\frac{35}{1}\right)^{0.286} = 2.76$$

$$T_2 = 2.76 \times 290 = 801.7 \text{ K}$$

Consider the process 2 – 3

$$T_3 = T_2 \frac{V_3}{V_2} = 801.7 \times \frac{V_3}{V_2}$$

$$= 801.7 \times 2.535 = 2032.3 \text{ K}$$

Consider the process 3 – 4

$$T_4 = T_3 \left(\frac{V_3}{V_4}\right)^{\gamma-1} = 2032.3 \times \left(\frac{1}{5}\right)^{0.4}$$

$$= 1067.6 \text{ K}$$

$$\text{Heat added} = C_p(T_3 - T_2) = 1.004 \times (2032.3 - 801.7)$$

$$= \textbf{1235.5 kJ/kg} \qquad \underleftarrow{\textbf{Ans}}$$

$$\text{Heat rejected} = C_v(T_4 - T_1) = 0.717 \times (1067.6 - 290)$$

$$= \textbf{557.5 kJ/kg} \qquad \underleftarrow{\textbf{Ans}}$$

$$\text{Efficiency} = \frac{\text{Heat supplied} - \text{Heat rejected}}{\text{Heat supplied}}$$

$$= \frac{1235.5 - 557.5}{1235.5} = 0.549$$

$$= \textbf{54.9\%} \qquad \underleftarrow{\textbf{Ans}}$$

3.14. A Diesel engine is working with a compression ratio of 15 and expansion ratio of 10. Calculate the air-standard efficiency of the cycle. Assume $\gamma = 1.4$.

SOLUTION:

$$r = \frac{V_1}{V_2} = 15$$

$$r_e = \frac{V_4}{V_3} = 10$$

$$\eta = 1 - \frac{1}{\gamma} \frac{1}{r^{\gamma-1}} \left[\frac{\left(\frac{r}{r_e}\right)^{\gamma} - 1}{\left(\frac{r}{r_e}\right) - 1} \right]$$

$$= 1 - \frac{1}{1.4} \times \frac{1}{15^{0.4}} \times \left[\frac{\left(\frac{15}{10}\right)^{1.4} - 1}{\left(\frac{15}{10}\right) - 1} \right] = 0.63$$

$$= 63\% \qquad \qquad \text{Ans}$$

3.15. A Diesel engine works on Diesel cycle with a compression ratio of 15 and cut-off ratio of 1.75. Calculate the air-standard efficiency assuming $\gamma = 1.4$.

SOLUTION:

$$\eta = 1 - \frac{1}{r^{\gamma-1}} \frac{1}{\gamma} \left(\frac{r_c^{\gamma} - 1}{r - 1} \right)$$

$$= 1 - \frac{1}{15^{0.4}} \times \frac{1}{1.4} \times \left(\frac{1.75^{1.4} - 1}{1.75 - 1} \right) = 0.617$$

$$= 61.7\% \qquad \qquad \text{Ans}$$

3.16. A Diesel cycle operates at a pressure of 1 bar at the beginning of compression and the volume is compressed to $\frac{1}{16}$ of the initial volume. Heat is supplied until the volume is twice that of the clearance volume. Calculate the mean effective pressure of the cycle. Take $\gamma = 1.4$.

SOLUTION:

$$V_1 = 16V_2$$

$$V_3 = 2V_2$$

$$\text{Swept volume} = V_1 - V_2$$

$$= (r - 1)V_2 = 15V_2$$

$$V_2 = \frac{V_s}{15}$$

Consider the process 1 – 2

$$p_2 = p_1 \left(\frac{V_1}{V_2}\right)^\gamma = 1 \times 16^{1.4} = 48.5 \text{ bar}$$

$$p_2 = p_3 = 48.5 \text{ bar}$$

$$p_4 = p_3 \left(\frac{V_3}{V_4}\right)^{1.4} = 48.5 \times \left(\frac{2}{16}\right)^{1.4}$$

$$= \frac{48.5}{8^{1.4}}$$

$$= 2.64 \text{ bar} \qquad (Since, \ V_4 = V_1)$$

Mean effective pressure, p_m, is given by

$$p_m = \frac{1}{V_s} \left[p_2(V_3 - V_2) + \frac{p_3V_3 - p_4V_4}{\gamma - 1} - \frac{p_2V_2 - p_1V_1}{\gamma - 1} \right]$$

$$= \frac{V_2}{V_s} \left[p_3 \left(\frac{V_3}{V_2} - 1\right) + \frac{\left(p_3 \frac{V_3}{V_2} - p_4 \frac{V_4}{V_2}\right)}{\gamma - 1} - \frac{\left(p_2 - p_1 \frac{V_1}{V_2}\right)}{\gamma - 1} \right]$$

$$= \frac{1}{15} \times \left[48.5 \times (2 - 1) + \frac{48.5 \times 2 - 2.64 \times 16}{1.4 - 1} - \frac{48.5 - 1 \times 16}{1.4 - 1} \right]$$

$$= \textbf{6.95 bar} \qquad\qquad \underset{\underline{\underline{\Leftarrow}}}{\text{Ans}}$$

3.17. In an engine working on the Diesel cycle the ratios of the weights of air and fuel supplied is 50 : 1. The temperature of air at the beginning of the compression is 60 °C and the compression ratio used is 14 : 1. What is the ideal efficiency of the engine. Calorific value of fuel used is 42000 kJ/kg. Assume $C_p = 1.004$ kJ/kg K and $C_v = 0.717$ kJ/kg K for air.

SOLUTION:

$$\eta = 1 - \frac{1}{r^{\gamma-1}} \frac{r_c^\gamma - 1}{\gamma(r_c - 1)}$$

$$\gamma = \frac{C_p}{C_v} = \frac{1.004}{0.717} = 1.4$$

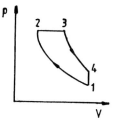

$$\text{Cut-off ratio, } r_c \quad = \quad \frac{V_3}{V_2} \quad = \quad \frac{T_3}{T_2}$$

Consider the process 1 – 2

$$\frac{T_2}{T_1} \quad = \quad \left(\frac{V_1}{V_2}\right)^{(\gamma-1)}$$

$$= \quad r^{(\gamma-1)} \quad = \quad 14^{0.4} \quad = \quad 2.874$$

$$T_2 \quad = \quad 2.874 \times 333 \quad = \quad 957.04 \text{ K}$$

Consider the process 2 – 3

$$\text{Heat added/kg of air} \quad = \quad C_p\,(T_3 - T_2) \quad = \quad F/A \times CV$$

$$T_3 - T_2 \quad = \quad \frac{F/A \times CV}{C_p} \quad = \quad \frac{42000}{50 \times 1.004}$$

$$= \quad 836.6$$

$$T_3 \quad = \quad 1793.64 \text{ } K$$

$$r_c \quad = \quad \frac{T_3}{T_2} \quad = \quad \frac{1793.64}{957.04} \quad = \quad 1.874$$

$$\eta \quad = \quad 1 - \frac{1}{1.4 \times 14^{0.4}} \times \left(\frac{1.874^{1.4} - 1}{0.874}\right)$$

$$= \quad 0.60$$

$$= \quad 60\% \qquad\qquad \underset{\Leftarrow}{\text{Ans}}$$

3.18. In an ideal Diesel cycle, the pressure and temperature are 1.03 bar and 27 °C respectively. The maximum pressure in the cycle is 47 bar and the heat supplied during the cycle is 545 kJ/kg. Determine (i) the compression ratio (ii) the temperature at the end of compression (iii) the temperature at the end of constant pressure combustion and (iv) the air-standard efficiency. Assume $\gamma = 1.4$ and $C_p = 1.004$ kJ/kg K for air.

SOLUTION:

$$p_2 \quad = \quad p_3 \quad = \quad 47 \times 10^5 \text{ N/m}^2$$

$$\frac{p_2}{p_1} = \left(\frac{V_1}{V_2}\right)^{\gamma} = r^{\gamma}$$

$$r = \left(\frac{p_2}{p_1}\right)^{\left(\frac{1}{\gamma}\right)} = \left(\frac{47}{1.03}\right)^{\left(\frac{1}{1.4}\right)}$$

$$= \mathbf{15.32} \qquad \underleftarrow{\textbf{Ans}}$$

$$\frac{T_2}{T_1} = \left(\frac{V_1}{V_2}\right)^{(\gamma-1)} = r^{(\gamma-1)}$$

$$= 15.32^{0.4} = 2.979$$

$$T_2 = 2.979 \times 300 = 893.7 \text{ K}$$

$$= \mathbf{620.7 \,°C} \qquad \underleftarrow{\textbf{Ans}}$$

$$\textit{Heat supplied}/kg = C_p\,(T_3 - T_2) = 545$$

$$T_3 - T_2 = \frac{545}{1.004} = 542.8$$

$$T_3 = 542.8 + 893.7 = 1436.5 \text{ K}$$

$$= \mathbf{1163.5 \,°C} \qquad \underleftarrow{\textbf{Ans}}$$

$$\eta = 1 - \frac{1}{r^{\gamma-1}}\,\frac{r_c^{\gamma} - 1}{\gamma(r_c - 1)}$$

$$r_c = \frac{V_3}{V_2} = \frac{T_3}{T_2}$$

$$= \frac{1436.5}{893.7} = 1.61$$

$$\eta_{Diesel} = 1 - \left[\frac{1}{1.4 \times 15.32^{0.4}} \times \left(\frac{1.61^{1.4} - 1}{0.61}\right)\right]$$

$$= 0.6275$$

$$= \mathbf{62.75\%} \qquad \underleftarrow{\textbf{Ans}}$$

3.19. A diesel engine operating on the air-standard Diesel cycle has six cylinders of 100 mm bore and 120 mm stroke. The engine speed is 1800 rpm. At the beginning of compression the pressure and temperature of air are 1.03 bar and 35 °C. If the clearance volume is 1/8th of the stroke volume, calculate (i) the pressure and temperature at the salient points of the cycle (ii) the compression ratio (iii) the efficiency of the cycle and (iv) the power output if the air is heated to 1500 °C Assume C_p and C_v of air to be 1.004 and 0.717 kJ/kg K respectively.

SOLUTION:

$$r = 1 + \frac{V_s}{V_c} = 1 + 8$$

$$= 9 \qquad \text{Ans}$$

Consider the process 1 – 2,

$$\frac{p_2}{p_1} = r^\gamma = 9^{1.4} = 21.67$$

$$p_2 = 21.67 \times 1.03 \times 10^5 = 22.32 \times 10^5 \text{ N/m}^2$$

$$= \textbf{22.32 bar} \qquad \text{Ans}$$

$$\frac{T_2}{T_1} = r^{(\gamma-1)} = 9^{0.4} = 2.408$$

$$T_2 = 308 \times 2.408 = 741.6 \text{ K}$$

$$= \textbf{468.6 °C} \qquad \text{Ans}$$

Consider the process 2 – 3,

$$p_3 = p_2 = 22.32 \times 10^5 \text{ N/m}^2$$

$$= \textbf{22.32 bar} \qquad \text{Ans}$$

$$T_3 = 1773 \text{ K}$$

$$= \textbf{1500 °C} \qquad \text{Ans}$$

Consider the process 3 – 4,

$$\frac{T_3}{T_4} = r_e^{(\gamma-1)}$$

$$r_c = \frac{T_3}{T_2} = \frac{1773}{741.6} = 2.39$$

$$r_e = \frac{r}{r_c} = \frac{9}{2.391} = 3.764$$

$$\frac{T_3}{T_4} = 1.7$$

$$T_4 = \frac{T_3}{1.7} = \frac{1773}{1.7} = 1042.9 \text{ K}$$

$$= \textbf{769.9} \text{ °C} \qquad \underset{\rightleftharpoons}{\textbf{Ans}}$$

$$\frac{p_3}{p_4} = r_e^\gamma = 3.764^{1.4} = 6.396$$

$$p_4 = \frac{p_3}{6.396} = \frac{22.32 \times 10^5}{6.396}$$

$$= 3.49 \times 10^5 \text{ N/m}^2$$

$$= \textbf{3.49} \text{ bar} \qquad \underset{\rightleftharpoons}{\textbf{Ans}}$$

$$\eta_{Cycle} = \frac{\text{Work output}}{\text{Heat added}} = 1 - \frac{\text{Heat rejected}}{\text{Heat added}}$$

$$= 1 - \frac{q_{4-1}}{q_{2-3}}$$

$$q_{4-1} = C_v (T_4 - T_1)$$

$$= 0.717 \times (1042.9 - 308) = 526.9 \text{ kJ/kg}$$

$$q_{2-3} = C_p (T_3 - T_2)$$

$$= 1.004 \times (1773 - 741.6)$$

$$= 1035.5 \text{ kJ/kg}$$

$$\eta_{Cycle} = 1 - \frac{526.9}{1035.5} = 0.4912$$

$$= \textbf{49.12\%} \qquad \underset{\rightleftharpoons}{\textbf{Ans}}$$

$$\text{Work output} \quad = \quad q_{2-3} - q_{4-1}$$

$$= \quad 1035.5 - 526.9 \quad = \quad 508.6 \text{ kJ/kg}$$

$$\text{Power output} \quad = \quad \text{Work output} \times \dot{m}_a$$

$$\dot{m}_a \quad = \quad \frac{p_1 V_1}{R T_1} \times \frac{N}{2}$$

$$R \quad = \quad C_p - C_v \quad = \quad 0.287 \text{ kJ/kg K}$$

$$V_1 \quad = \quad V_s + V_c \quad = \quad \frac{9}{8} V_s$$

$$V_s \quad = \quad 6 \times \frac{\pi}{4} d^2 \ L \quad = \quad 6 \times \frac{\pi}{4} \times 10^2 \times 12$$

$$= \quad 5654.8 \text{ cc} \quad = \quad 5.65 \times 10^{-3} \text{ m}^3$$

$$V_1 \quad = \quad 5.65 \times 10^{-3} \times \frac{9}{8} \quad = \quad 6.36 \times 10^{-3} \text{ m}^3$$

$$\dot{m}_a \quad = \quad \frac{1.03 \times 10^5 \times 6.36 \times 10^{-3} \times 30}{287 \times 308 \times 2}$$

$$= \quad 0.111 \text{ kg/s}$$

$$\text{Power output} \quad = \quad 508.6 \times 0.111$$

$$= \quad \textbf{56.45 kW} \qquad \underset{\Longleftarrow}{\text{Ans}}$$

3.20. The mean effective pressure of an ideal Diesel cycle is 8 bar. If the initial pressure is 1.03 bar and the compression ratio is 12, determine the cut-off ratio and the air-standard efficiency. Assume ratio of specific heats for air to be 1.4.

SOLUTION:

$$\text{Work output} \quad = \quad p_m \times V_s$$

$$= \quad \text{Area 1234}$$

$$= \quad \text{Area under } 2\text{-}3 + \text{Area under } 3\text{-}4$$

$$- \text{Area under } 2\text{-}1$$

$$= p_2(V_3 - V_2) + \frac{p_3 V_3 - p_4 V_4}{\gamma - 1} - \frac{p_2 V_2 - p_1 V_1}{\gamma - 1}$$

$$r = \frac{V_1}{V_2} = 1 + \frac{V_s}{V_c} = 12$$

$$V_s = 11 V_c \quad ; \quad V_2 = V_c$$

$$V_1 = V_4 = 12 V_2 = 12 V_c$$

$$V_3 = r_c V_2 = r_c V_c$$

$$\frac{p_2}{p_1} = r^\gamma = 12^{1.4} = 32.42$$

$$p_2 = 32.42 \times 1.03 \times 10^5$$

$$= 33.39 \times 10^5 \text{ N/m}^2 = p_3$$

$$\frac{p_3}{p_4} = \left(\frac{r}{r_c}\right)^{1.4} = \frac{12^{1.4}}{r_c^{1.4}} = \frac{32.42}{r_c^{1.4}}$$

$$p_4 = \frac{33.39}{32.42} \times r_c^{1.4} \times 10^5 = 1.03 r_c^{1.4} \times 10^5$$

$$\text{Area } 1234 = 33.39 \left(r_c V_c - V_c\right)$$

$$+ \frac{33.39 \times r_c V_c - 1.03 r_c^{1.4} \times 12 V_c}{0.4}$$

$$- \frac{33.39 \times V_c - 1.03 \times 12 V_c}{0.4} \times 10^5$$

$$p_m \times V_s = 8 \times 11 V_c \times 10^5$$

$$\text{Area } 1234 = p_m \times V_s$$

Substituting and simplifying

$$0.672 r_c - 0.178 r_c^{1.4} = 1$$

Solving by iteration,

$$r_c = \mathbf{2.38}$$

Ans

$$\eta = 1 - \frac{1}{r^{\gamma-1}} \frac{r_c^\gamma - 1}{\gamma(r_c - 1)}$$

$$= 1 - \frac{1}{1.4 \times 12^{0.4}} \times \left(\frac{2.38^{1.4} - 1}{1.38}\right) = 0.5466$$

$$= 54.66\% \qquad \qquad \text{Ans}$$

DUAL CYCLE

3.21. An air-standard Dual cycle has a compression ratio of 10. The pressure and temperature at the beginning of compression are 1 bar and 27 °C. The maximum pressure reached is 42 bar and the maximum temperature is 1500 °C. Determine (i) the temperature at the end of constant volume heat addition (ii) cut-off ratio (iii) work done per kg of air and (iv) the cycle efficiency. Assume $C_p = 1.004$ kJ/kg K and $C_v = 0.717$ kJ/kg K for air.

SOLUTION:

$$\frac{V_s}{V_c} = r - 1 = 9$$

$$V_s = 9V_c$$

$$\gamma = \frac{C_p}{C_v} = \frac{1.004}{0.717} = 1.4$$

Consider the process 1 - 2

$$\frac{T_2}{T_1} = r^{(\gamma-1)} = 10^{0.4} = 2.512$$

$$T_2 = 2.512 \times 300 = 753.6 \text{ K}$$

$$\frac{p_2}{p_1} = r^\gamma = 10^{1.4} = 25.12$$

$$p_2 = 25.12 \times 10^5 \text{ N/m}^2$$

Consider the process 3 - 4

$$\frac{T_3}{T_2} = \frac{p_3}{p_2} = \frac{42}{25.12} = 1.672$$

$$T_3 = 1.672 \times 753.6 = 1260 \text{ K}$$

$$= 987 \text{ °C} \qquad \qquad \text{Ans}$$

$$r_c = \frac{T_4}{T_3} = \frac{1773}{1260}$$

$$= \mathbf{1.407} \qquad \underset{\Longleftarrow}{\mathbf{Ans}}$$

$$\text{Work done/kg} = \text{Heat supplied} - \text{Heat rejected}$$

$$\text{Heat supplied/kg} = C_v (T_3 - T_2) + C_p (T_4 - T_3)$$

$$= 0.717 \times (1260 - 753.6)$$

$$+ 1.004 \times (1773 - 1260)$$

$$= \mathbf{878.1 \ kJ} \qquad \underset{\Longleftarrow}{\mathbf{Ans}}$$

Consider the process 4 – 5

$$\frac{T_4}{T_5} = \left(\frac{V_5}{V_4}\right)^{(\gamma-1)} = \left(\frac{r}{r_c}\right)^{(\gamma-1)}$$

$$= \left(\frac{10}{1.407}\right)^{0.4} = 2.191$$

$$T_5 = \frac{T_4}{2.191}$$

$$= \mathbf{809.2 \ K} \qquad \underset{\Longleftarrow}{\mathbf{Ans}}$$

$$\text{Heat rejected/kg} = C_v (T_5 - T_1)$$

$$= 0.717 \times (809.2 - 300) = 365.1 \ kJ$$

$$\text{Work output/kg} = 878.1 - 365.1$$

$$= \mathbf{513 \ kJ} \qquad \underset{\Longleftarrow}{\mathbf{Ans}}$$

$$\eta_{Dual} = \frac{\text{Work output}}{\text{Heat added}}$$

$$= \frac{513}{878.1} = 0.5842$$

$$= \mathbf{58.42\%} \qquad \underset{\Longleftarrow}{\mathbf{Ans}}$$

3.22. For an engine working on the ideal Dual cycle, the compression ratio is 10 and the maximum pressure is limited to 70 bar. If the heat supplied is 1680 kJ/kg, find the pressures and temperatures at the various salient points of the cycle and the cycle efficiency. The pressure and temperature of air at the commencement of compression are 1 bar and 100 °C respectively. Assume $C_p = 1.004$ kJ/kg K and $C_v = 0.717$ kJ/kg K for air.

SOLUTION:

$$\frac{V_s}{V_c} = r - 1 = 9$$

$$\gamma = \frac{C_p}{C_v} = \frac{1.004}{0.717} = 1.4$$

Consider the process 1 – 2

$$\frac{p_2}{p_1} = r^\gamma = 10^{1.4}$$

$$= 25.12$$

$$p_2 = 25.12 \times 10^5 \text{ N/m}^2$$

$$= \textbf{25.12 bar} \qquad \underset{\Longleftarrow}{\textbf{Ans}}$$

$$\frac{T_2}{T_1} = r^{(\gamma - 1)}$$

$$= 10^{0.4} = 2.512$$

$$T_2 = 2.512 \times 373 = 936.9 \text{ K}$$

$$= \textbf{663.9 °C} \qquad \underset{\Longleftarrow}{\textbf{Ans}}$$

Consider the process 2 – 3 and 3 – 4

$$\frac{T_3}{T_2} = \frac{p_3}{p_2}$$

$$= \frac{70}{25.12} = 2.787$$

$$T_3 = 2.787 \times 936.9 = 2611.1 \text{ K}$$

$$= \textbf{2338 °C} \qquad \underset{\Longleftarrow}{\textbf{Ans}}$$

Heat added during constant volume combustion

$$= C_v (T_3 - T_2)$$

$$= 0.717 \times (2611.1 - 936.9)$$

$$= 1200.4 \text{ kJ/kg}$$

Total heat added $= 1680 \text{ kJ/kg}$

Hence, heat added during constant pressure combustion

$$= 1680 - 1200.4 \quad = \quad 479.6 \text{ kJ/kg}$$

$$= C_p (T_4 - T_3)$$

$$T_4 - T_3 = \frac{479.6}{1.004} = 477.7 \text{ K}$$

$$T_4 = 477.7 + 2611.1 = 3088.8 \text{ K}$$

$$= \mathbf{2815.8 \ °C} \qquad \underset{\Longleftarrow}{\text{Ans}}$$

Cut-off ratio, $r_c = \dfrac{V_4}{V_3} = \dfrac{T_4}{T_3}$

$$= \frac{3088.8}{2611.1} = 1.183$$

Consider the process 4 – 5

$$\frac{T_4}{T_5} = \left(\frac{r}{r_c}\right)^{(\gamma - 1)}$$

$$= 8.453^{0.4} = 2.35$$

$$T_5 = \frac{T_4}{2.35} = \frac{3088.8}{2.35} = 1314.4 \text{ K}$$

$$= \mathbf{1041.4 \ °C} \qquad \underset{\Longleftarrow}{\text{Ans}}$$

$$\frac{p_4}{p_5} = \left(\frac{r}{r_c}\right)^{\gamma} = 19.85$$

$$p_5 = \frac{p_4}{19.85} = \frac{70 \times 10^5}{19.85}$$

$$= \quad 3.53 \times 10^5 \text{ N/m}^2$$

$$= \quad \textbf{3.53 bar} \qquad\qquad \underset{\Longleftarrow}{\textbf{Ans}}$$

$$\text{Heat rejected} \quad = \quad C_v(T_5 - T_1)$$

$$= \quad 0.717 \times (1314.4 - 373) \quad = \quad 674.98 \text{ kJ/kg}$$

$$\eta \quad = \quad \frac{1680 - 674.98}{1680}$$

$$= \quad \textbf{59.82\%} \qquad\qquad \underset{\Longleftarrow}{\textbf{Ans}}$$

3.23. An oil engine works on the Dual cycle, the heat liberated at constant pressure being twice that liberated at constant volume. The compression ratio of the engine is 8 and the expansion ratio is 5.3. But the compression and expansion processes follow the law $pV^{1.3} = C$. The pressure and temperature at the beginning of compression are 1 bar and 27 °C respectively. Assuming $C_p = 1.004$ kJ/kg K and $C_v = 0.717$ kJ/kg K for air, find the air-standard efficiency and the mean effective pressure.

SOLUTION:

$$\gamma \quad = \quad \frac{C_p}{C_v} \quad = \quad \frac{1.004}{0.717} \quad = \quad 1.4$$

$$\frac{V_s}{V_c} \quad = \quad r - 1 \quad = \quad 7$$

$$V_s \quad = \quad 7V_c$$

$$r_e \quad = \quad \frac{r}{r_c} \quad = \quad 5.3$$

$$r_c \quad = \quad \frac{8}{5.3} \quad = \quad 1.509$$

Mean effective pressure

$$= \quad \frac{\text{Area } 12345}{V_s}$$

$$\text{Area } 12345 \quad = \quad \text{Area under } 3\text{-}4$$

$$+ \quad \text{Area under } 4\text{-}5 - \text{Area under } 2\text{-}1$$

$$= \quad p_3 \left(V_3 - V_4 \right) + \quad \frac{p_4 V_4 - p_5 V_5}{n-1} - \frac{p_2 V_2 - p_1 V_1}{n-1}$$

$$V_2 \quad = \quad V_3 \quad = \quad V_c$$

$$V_1 \quad = \quad V_5 \quad = \quad r V_c \quad = \quad 8 V_c$$

$$V_4 \quad = \quad r_c V_3 \quad = \quad 1.509 \, V_c$$

$$\frac{T_2}{T_1} \quad = \quad r^{(n-1)} \quad = \quad 8^{0.3} \quad = \quad 1.866$$

$$T_2 \quad = \quad 1.866 \times 300 \quad = \quad 559.82 \text{ K}$$

$$\frac{p_2}{p_1} \quad = \quad r^n \quad = \quad 8^{1.3} \quad = \quad 14.93$$

$$p_2 \quad = \quad 14.93 \times p_1 \quad = \quad 14.93 \times 10^5 \text{ N/m}^2$$

Heat released during constant pressure combustion

$$= \quad 2 \times \text{Heat released during constant volume combustion}$$

$$C_p \left(T_4 - T_3 \right) \quad = \quad 2 C_v \left(T_3 - T_2 \right)$$

$$1.004 \times \left(T_4 - T_3 \right) \quad = \quad 2 \times 0.717 \times \left(T_3 - T_2 \right)$$

$$T_4 - T_3 \quad = \quad 1.428 \times \left(T_3 - T_2 \right)$$

$$\frac{T_4}{T_3} \quad = \quad \frac{V_4}{V_3} \quad = \quad r_c \quad = \quad 1.509$$

$$T_4 \quad = \quad 1.509 \, T_3$$

Therefore,

$$1.509 T_3 - T_3 \quad = \quad 1.428 \times \left(T_3 - 559.82 \right)$$

$$T_3 \quad = \quad 869.88 \text{ K}$$

$$T_4 \quad = \quad 1312.65 \text{ K}$$

$$\frac{p_3}{p_2} \quad = \quad \frac{T_3}{T_2} \quad = \quad \frac{869.88}{559.82} \quad = \quad 1.554$$

$$p_3 \quad = \quad 1.554 \times 14.93 \times 10^5$$

$$= \quad 23.20 \times 10^5 \text{ N/m}^2 \quad = \quad p_4$$

$$\frac{T_4}{T_5} \quad = \quad r_e^{(n-1)} \quad = \quad 5.3^{0.3} \quad = \quad 1.649$$

$$T_5 \quad = \quad \frac{1312.65}{1.649} \quad = \quad 796.03 \text{ K}$$

$$\frac{p_4}{p_5} \quad = \quad r_e^n \quad = \quad 5.3^{1.3} \quad = \quad 8.741$$

$$p_5 \quad = \quad \frac{p_4}{8.741} \quad = \quad \frac{23.2 \times 10^5}{8.741}$$

$$= \quad 2.654 \times 10^5 \text{ N/m}^2$$

$$\text{Area } 12345 \quad = \quad \left[\frac{23.2 \times 1.509 V_c - 2.654 \times 8 V_c}{0.3} \right.$$

$$+ \quad 23.2 \times (1.509 V_c - V_c)$$

$$\left. - \quad \frac{14.93 \times V_c - 1 \times 8 V_c}{0.3} \right] \times 10^5$$

$$= \quad 34.63 \times V_c \times 10^5 \text{ N/m}^2$$

$$V_c \quad = \quad \frac{V_1}{8}$$

$$\text{Area } 12345 \quad = \quad p_m \times V_s \quad = \quad p_m \times 7 \times V_c$$

Therefore,

$$p_m \quad = \quad \frac{34.63}{7} \quad = \quad 4.95 \times 10^5 \text{ N/m}^2$$

$$= \quad \textbf{4.95 bar} \qquad\qquad \underleftarrow{\textbf{Ans}}$$

$$\eta \quad = \quad \frac{w}{q_s}$$

$$v_1 \quad = \quad \frac{mRT_1}{p_1} \quad = \quad \frac{1 \times 287 \times 300}{1 \times 10^5}$$

$$= \quad 0.861 \text{ m}^3/\text{kg}$$

$$w = 34.63 \times 10^5 \times \frac{v_1}{8}$$

$$= 34.63 \times 10^5 \times \frac{0.861}{8}$$

$$= 3.727 \times 10^5 \text{ J/kg} = 372.7 \text{ kJ/kg}$$

$$q_s = q(T_3 - T_2) + C_p(T_4 - T_3)$$

$$= 667.1 \text{ kJ/kg}$$

$$\eta = \frac{372.7}{667.1} \times 100$$

$$= 55.9\% \qquad \underset{\Longleftarrow}{\text{Ans}}$$

COMPARISON OF CYCLES

3.24. A four-cylinder, four-stroke, spark-ignition engine has a displacement volume of 300 cc per cylinder. The compression ratio of the engine is 10 and operates at a speed of 3000 rev/min. The engine is required to develop an output of 40 kW at this speed. Calculate the cycle efficiency, the necessary rate of heat addition, the mean effective pressure and the maximum temperature of the cycle. Assume that the engine operates on the Otto cycle and that the pressure and temperature at the inlet conditions are 1 bar and 27 °C respectively.

If the above engine is a compression-ignition engine operating on the Diesel cycle and receiving heat at the same rate, calculate efficiency, the maximum temperature of the cycle, the cycle efficiency, the power output and the mean effective pressure. Take $C_v = 0.717$ kJ/kg K and $\gamma = 1.4$.

SOLUTION:
Consider the Otto cycle, Fig.3.9(a)

$$\eta = 1 - \frac{1}{r^{\gamma-1}}$$

$$= 1 - \frac{1}{10^{0.4}} = 0.602$$

$$= 60.2\% \qquad \underset{\Longleftarrow}{\text{Ans}}$$

$$\eta = \frac{\text{Power output}}{\text{Heat supplied}}$$

$$\text{Heat supplied} = \frac{40}{0.602} = 66.5 \text{ kW}$$

$$= \textbf{66.5 kJ/s} \qquad \underset{\Leftarrow}{\textbf{Ans}}$$

$$\text{Number of cycles/sec} = \frac{3000}{2 \times 60} = 25$$

Net work output per cycle from each cylinder

$$= \frac{40}{4 \times 25} = 0.4 \text{ kJ}$$

$$p_m = \frac{W}{V_s} = \frac{0.4 \times 1000}{300 \times 10^{-6}}$$

$$= \textbf{13.3} \times \textbf{10}^5 \textbf{ N/m}^2 \qquad \underset{\Leftarrow}{\textbf{Ans}}$$

$$T_2 = T_1 \left(\frac{V_1}{V_2}\right)^{\gamma-1}$$

$$= 300 \times 10^{0.4} = 753.6 \text{ K}$$

Heat supplied/cylinder/cycle (Q_{2-3})

$$= \frac{66.5}{4 \times 25} = 0.665 \text{ kJ}$$

Now,

$$Q_{2-3} = mC_v (T_3 - T_2)$$

$$v_1 = \frac{RT_1}{p_1} = \frac{287 \times 300}{1 \times 10^5}$$

$$= 0.861 \text{ m}^3/\text{kg}$$

This initial volume of air in the cylinder is

$$V_1 = V_2 + V_s = \left(\frac{V_1}{10}\right) + V_s$$

$$0.9V_1 = V_s$$

$$V_1 = \frac{V_s}{0.9} = \frac{300 \times 10^{-6}}{0.9}$$

$$= 333 \times 10^{-6} \text{ m}^3$$

$$m = \frac{V_1}{v_1} = \frac{333 \times 10^{-6}}{0.861}$$

$$= 0.387 \times 10^{-3} \text{ kg}$$

The temperature rise resulting from heat addition is

$$T_3 - T_2 = \frac{Q_{2-3}}{m \ C_v}$$

$$= \frac{0.665}{0.387 \times 10^{-3} \times 0.717}$$

$$= 2396.6 \text{ K}$$

$$T_3 = T_2 + 2396.6 = 753.6 + 2396.6$$

$$= 3150.2 \text{ K}$$

$$= \mathbf{2847} \ ^\circ\text{C} \qquad \underset{\Longleftarrow}{\text{Ans}}$$

Now let us consider the Diesel cycle

T_2 *is the same as in the previous case, i.e.*

$$T_2 = 753.6$$

Heat supplied per cycle per cylinder is also same, i.e.

$$Q_{2-3'} = 0.665 \text{ kJ}$$

$$Q_{2-3'} = m \ C_p (T_{3'} - T_2)$$

$$T_{3'} - T_2 = \frac{0.665}{0.387 \times 10^{-3}} \times \frac{1}{1.004}$$

$$= 1711.5 \text{ K}$$

$$T_{3'} = 1711.5 + 753.6 = 2465.1 \text{ K}$$

$$= \mathbf{2162.1} \ ^\circ\text{C} \qquad \underset{\Longleftarrow}{\text{Ans}}$$

$$\text{Cut-off ratio, } r_c \quad = \quad \frac{V_3}{V_2} \quad = \quad \frac{T_{3'}}{T_2}$$

$$= \quad \frac{2465.1}{753.6} \quad = \quad 3.27$$

$$\text{Air} - \text{standard efficiency} \quad = \quad 1 - \frac{1}{r^\gamma - 1} \left[\frac{r_c^\gamma - 1}{\gamma \left(r_c - 1 \right)} \right]$$

$$= \quad 1 - \frac{1}{10^{0.4}} \times \left[\frac{3.27^{1.4} - 1}{1.4 \times (3.27 - 1)} \right]$$

$$= \quad 1 - 0.398 \times 1.338 \quad = \quad 0.467$$

$$= \quad \textbf{46.7\%} \qquad \overset{\textbf{Ans}}{\Longleftarrow}$$

$$\text{Power output} \quad = \quad \eta \times \text{total rate of heat added}$$

$$= \quad 0.467 \times 66.5 \quad = \quad 31.1 \text{ kW}$$

$$\text{Power output/cylinder} \quad = \quad \frac{31.1}{4} \quad = \quad 7.76 \text{ kW}$$

$$\text{Work done/cylinder/cycle}$$

$$= \quad \frac{7.76}{25} \quad = \quad 0.3104 \text{ kJ}$$

$$p_m \quad = \quad \frac{W}{V_s} \quad = \quad \frac{0.3104 \times 1000}{300 \times 10^{-6}}$$

$$= \quad \textbf{10.37} \times \textbf{10}^5 \text{ N/m}^2 \qquad \overset{\textbf{Ans}}{\Longleftarrow}$$

As discussed in the text, this problem illustrates that for the same compression ratio and heat input Otto cycle is more efficient.

3.25. The compression ratio of an engine is 10 and the temperature and pressure at the start of compression is 37 °C and 1 bar. The compression and expansion processes are both isentropic and the heat is rejected at exhaust at constant volume. The amount of heat added during the cycle is 2730 kJ/kg. Determine the mean effective pressure and thermal efficiency of the cycle if (i) the maximum pressure is limited to 70 bar and heat is added at both constant volume and constant pressure and (ii) if all the heat is added at constant volume. In this case how much additional work per kg of charge would be obtained if it were possible to expand isentropically the exhaust gases to their original pressure of 1 bar. Assume that the charge has the same physical properties as that of air.

SOLUTION:

$$v_1 = \frac{V_1}{m} = \frac{RT_1}{p_1}$$

$$= \frac{287 \times 310}{1 \times 10^5} = 0.89 \text{ m}^3/\text{kg}$$

Consider the process 1-2

$$\frac{T_2}{T_1} = \left(\frac{V_1}{V_2}\right)^{\gamma-1} = 310 \times 10^{0.4}$$

$$= 778.7 \text{ K}$$

$$p_2 = p_1 \left(\frac{V_1}{V_2}\right)^{\gamma} = 25.12 \text{ bar}$$

Consider the limited pressure cycle (123451)

$$T_3 = T_2 \frac{p_3}{p_2} = 778.7 \times \frac{70}{25.12} = 2170 \text{ K}$$

Heat supplied at constant volume

$$= 0.717 \times (2170 - 778.7)$$

$$= 997.56 \text{ kJ/kg}$$

Heat supplied at constant pressure

$$= 2730 - 997.56 = 1732.44 \text{ kJ/kg}$$

Now,

$$1732.4 = 1.004 \times (T_4 - 2170)$$

$$T_4 = 2170 + \frac{1732.4}{1.004} = 3895.5 \text{ K}$$

$$v_4 = v_3 \frac{T_4}{T_3} = 0.89 \times \frac{3895.5}{2170}$$

$$= 0.16 \text{ m}^3/\text{kg}$$

$$T_5 \;=\; T_4 \left(\frac{v_4}{v_5}\right)^{\gamma-1}$$

$$=\; 3895.5 \times \left(\frac{0.16}{0.89}\right)^{0.4} \;=\; 1961 \text{ K}$$

$$\text{Heat rejected} \;=\; C_v \left(T_5 - T_1\right)$$

$$=\; 0.717 \times \left(1961 - 310\right) \;=\; 1184 \text{ kJ/kg}$$

$$\text{Work done} \;=\; \text{Heat supplied} - \text{Heat rejected}$$

$$=\; 2730 - 1184 \;=\; 1546 \text{ kJ/kg}$$

$$\eta \;=\; \frac{w}{q} \;=\; \frac{1546}{2730} \;=\; 0.57$$

$$=\; \mathbf{57\%} \qquad \overset{\text{Ans}}{\Longleftarrow}$$

$$p_m \;=\; \frac{w}{v_1 - v_2} \;=\; \frac{w}{v_1 \left(1 - \frac{1}{r}\right)}$$

$$=\; \frac{1546 \times 10^3}{0.89 \times \left(1 - \frac{1}{\cdot 10}\right)}$$

$$=\; 19.32 \times 10^5 \text{ N/m}^2 \;=\; \mathbf{19.32} \text{ bar} \qquad \overset{\text{Ans}}{\Longleftarrow}$$

Consider the constant-volume cycle (123'4'1)

$$q_{2-3'} \;=\; 2730 \text{ kJ/kg}$$

$$2730 \;=\; 0.717 \times \left(T_{3'} - 778\right)$$

$$T_{3'} \;=\; 4586 \text{ K}$$

$$p_{3'} \;=\; T_{3'} \left(\frac{p_2}{T_2}\right)$$

$$=\; 4586 \times \frac{25.12}{778.7} \;=\; 147.9 \text{ bar}$$

$$\frac{T_{3'}}{T_{4'}} \;=\; \left(\frac{V_{4'}}{V_{3'}}\right)^{\gamma-1}$$

$$T_{4'} = 4586 \times \left(\frac{1}{10}\right)^{0.4}$$

$$= 1852.7$$

Heat rejected $= 0.717 \times (1825.7 - 310)$

$$= 1087 \text{ kJ/kg}$$

$$w = 2730 - 1087 = 1643 \text{ kJ}$$

$$\eta = \frac{1643}{2730} = 0.602$$

$$= \mathbf{60.2\%} \qquad \underset{\underline{\underline{}}}{\text{Ans}}$$

$$p_m = \frac{1643 \times 10^3}{\left(1 - \frac{1}{10}\right) \times 0.89}$$

$$= 20.53 \times 10^5 \text{ N/m}^2$$

$$= \mathbf{20.53 \text{ bar}} \qquad \underset{\underline{\underline{}}}{\text{Ans}}$$

If the gases were expanded isentropically to their original pressure of
1 bar, then the temperature T_6 at the end of expansion would be

$$T_6 = T_{3'}\left(\frac{p_6}{p_{3'}}\right)^{\frac{\gamma-1}{\gamma}}$$

$$= 4586 \times \left(\frac{1}{147.9}\right)^{\frac{0.4}{1.4}}$$

$$= 1100.12$$

Heat rejected at constant pressure

$$= 1.004 \times (1100.12 - 310)$$

$$= 793.3 \text{ kJ/kg}$$

Work increase $= 1087 - 793.3$

$$= \mathbf{293.7 \text{ kJ/kg}} \qquad \underset{\underline{\underline{}}}{\text{Ans}}$$

Questions :-

3.1 What is the simplest way by which an IC engine cycle can be analysed? Do IC engines operate on a thermodynamic cycle?

3.2 What is the use of air-standard cycle analysis?

3.3 Mention the various assumption made in air-standard cycle analysis.

3.4 What is Carnot cycle and what is its importance? How is this cycle reversible?

3.5 Draw the Carnot cycle on p-V and T-s diagrams. Derive an expression for its efficiency. Comment on the significance of this result as it related to source and sink temperature.

3.6 Define mean effective pressure and comment its application in internal combustion engines.

3.7 Name the cycles which have the same efficiency of Carnot cycle. Are these cycles reversible in the sense the Carnot cycle is?

3.8 Draw the Stirling cycle on p-V and T-s diagrams and show how the cycle is reversible?

3.9 Derive the expression for Stirling cycle efficiency and show that the expression is same as that of Carnot cycle.

3.10 If you include the efficiency of the heat exchanger show how the expression is modified.

3.11 Draw the p-V and T-s diagram of Ericsson cycle and show how it is made reversible.

3.12 Compare Carnot, Stirling and Ericsson cycles operating between the same source and sink temperatures and with equal changes in specific volume.

3.13 Draw the Otto cycle on p-V and T-s diagrams mark the various processes.

3.14 Derive an expression for the efficiency of Otto cycle and comment on the effect of compression ratio on the efficiency with respect of ratio of specific heats by means of a suitable graph.

3.15 Obtain an expression for mean effective pressure of an Otto cycle.

3.16 What is the basic difference between an Otto cycle and Diesel cycle? Derive the expression for the efficiency and mean effective pressure of the Diesel cycle.

3.17 Show that the efficiency of the Diesel cycle is lower than that of Otto cycle for the same compression ratio. Comment why the higher efficiency of the Otto cycle compared to Diesel cycle for the same compression ratio is only of a academic interest and not practical importance.

3.18 Compare the Otto cycle for the same peak pressure and temperature. Illustrate the cycles on *p-V* and *T-s* diagrams.

3.19 Draw the *p-V* and *T-s* diagrams of a Dual cycle. Why this cycle is also called limited pressure or mixed cycle?

3.20 Derive the expressions for the efficiency and mean effective pressure of a Dual cycle.

3.21 Compare Otto, Diesel and Dual cycles for the
    (i) same compression ratio and heat input
   (ii) same maximum pressure and heat input
  (iii) same maximum pressure and temperature
  (iv) same maximum pressure and work output

3.22 Sketch the Lenoir cycle on *p-V* and *T-s* diagrams and obtain an expression for its air-standard efficiency.

3.23 Compare the Otto cycle and Atkinson cycle. Derive the expression for the efficiency of Atkinson cycle.

3.24 Derive an expression for the air-standard efficiency of the Joule cycle in terms of
    (i) compression ratio
   (ii) pressure ratio.

3.25 Where do the following cycles have applications
    (i) Otto cycle
   (ii) Diesel cycle
  (iii) Dual cycle
  (iv) Stirling cycle
   (v) Ericsson cycle
  (vi) Atkinson cycle
 (vii) Lenoir cycle
(viii) Joule cycle

Exercise:-

3.1. Assume working substance for a Carnot cycle to be air with $C_p = 1$ kJ/kg K and $C_v = 0.717$ kJ/kg K. Temperature at which heat is added is 2000 K and temperature at which heat is rejected is 300 K. The amount of heat added per kg of the working substance is 840 kJ/kg. Calculate for the cycle (i) the maximum pressure developed (ii) the compression ratio assuming adiabatic compression and (iii) the efficiency of the cycle. The pressure at the beginning of isothermal compression is 1 bar.

Ans: (i) 3304.9 bar (ii) 114.5 (iii) 85%

3.2. An engine operates on Otto cycle between pressures 1 bar and 30 bar. The ratio of pressure at constant volume is 4. The temperature at the end of compression is 200 °C and the law of compression and expansion is $PV^{1.3} =$ constant. If the engine now operates on Carnot cycle for the same range of temperature, find the efficiency of the cycle. *Ans:* 84.302 %

3.3. An Otto cycle engine having a clearance volume of 250 cc has a compression ratio of 8. The ratio of pressure rise at constant volume is 4. If the initial pressure is 1 bar, find the work done per cycle and the theoretical mean effective pressure. Take $\gamma = 1.4$. *Ans:* (i) 1946.1 J/cycle (ii) 11.12 bar

3.4. Find the *mep* for the ideal air-standard Otto cycle having a maximum pressure of 40 bar and minimum pressure of 1 bar. The compression ratio is 5:1. Take $\gamma = 1.4$. *Ans:* 9.043 bar

3.5. The pressure and temperature of a Diesel cycle at the start are 1 bar and 20 °C respectively and the compression ratio is 14. The pressure at the end of expansion is 2.5 bar. Find the percentage of working stroke at which heat is supplied and heat supplied per kg of air. Assume $\gamma = 1.4$ and $C_p = 1.004$ kJ/kg K.

Ans: (i) 7.11% (ii) 782.06 kJ

3.6. An oil engine works on Diesel cycle, the compression ratio being 15. The temperature at the start of compression is 17 °C and 700 kJ of heat is supplied at constant pressure per kg of air and it attains a temperature of 417 °C at the end of adiabatic expansion. Find the air-standard efficiency of the cycle. What would be the theoretical work done per kg of air. Take $C_v = 0.717$ kJ/kg K and $\gamma = 1.4$. *Ans:* (i) 59.03% (ii) 413.14 kJ

3.7. An internal combustion engine works on Diesel cycle with a compression ratio of 8 and expansion ratio of 5. Calculate the air-standard efficiency. Assume $\gamma = 1.41$. *Ans: 52.6 %*

3.8. A Diesel engine works on Diesel cycle with a compression ratio of 16 and cut-off ratio of 1.8. Calculate the thermal efficiency assuming $\gamma = 1.4$. *Ans: 62.38 %*

3.9. An internal combustion engine works on Diesel cycle with a compression ratio of 14 and cut-off takes place at 10 % of the stroke. Find the ratio of cut-off and the air-standard efficiency. *Ans:* (i) 2.3 (ii) 57.8 %

3.10. An ideal Diesel cycle operates on a pressure of 1 bar and a temperature of 27 °C at the beginning of compression and a pressure of 2 bar at the end of adiabatic expansion. Calculate the amount of heat required to be supplied per kg of air if the ideal thermal efficiency is taken as 60 %. Take $C_v = 0.717$ kJ/kg K. *Ans: 537.75 kJ/kg*

3.11. The pressure and temperature of a Diesel cycle at the start are 1 bar and 17 °C. The pressure at the end of compression is 40 bar and that at the end of expansion is 2 bar. Find the air-standard efficiency. Assume $\gamma = 1.4$. *Ans: 61.4 %*

3.12. A Diesel cycle operates at a pressure of 1 bar at the beginning of compression and the volume is compressed to $\frac{1}{15}$ of the initial volume. Heat is then supplied until the volume is twice that of the clearance volume. Determine the mean effective pressure. Take $\gamma = 1.4$. *Ans: 6.68 bar*

3.13. A semi-diesel engine works on dual combustion cycle. The pressure and temperature at the beginning of the compression is 1 bar and 27 °C respectively and the compression ratio being 12. If the maximum pressure is 50 bar and heat received at constant pressure is for $\frac{1}{30}$th of the stroke, find the work done per kg of air and the thermal efficiency. Take $C_v = 0.717$ and $C_p = 1.004$. *Ans: 61.5 %*

3.14. A compression-ignition engine has a compression ratio of 10 and $\frac{2}{3}$ of heat of combustion is liberated at constant volume and the remainder at constant pressure. The pressure and temperature at the beginning are 1 bar and 27 °C and the maximum pressure is 40 bar. Find the temperatures at the end of compression and expansion, if it follows the law $pV^{1.35}$ = constant, and $\gamma = 1.4$.   *Ans*: (i) 398.6 °C (ii) 379.8 °C

3.15. A compression-ignition engine works on dual combustion cycle. The pressure and temperature at the beginning of compression are 1 bar and 27 °C respectively and the pressure at the end of compression is 25 bar. If 420 kJ of heat is supplied per kg of air during constant volume heating and the pressure at the end of adiabatic expansion is found to be 3 bar, find the ideal thermal efficiency. Assume $C_p = 1.004$ kJ/kg K and $C_v = 0.717$ kJ/kg K.

Ans: 41.25 %

3.16. The cycle of an internal combustion engine with isochoric heat supply is performed with the compression ratio equal to 8. Find heat supplied to the cycle and the useful work, if the removed heat is 500 kJ/kg and the working fluid is air.

Ans: (i) 1148.633 kJ/kg (ii) 648.633 kJ/kg

3.17. The initial parameters (at the beginning of compression) of the cycle of an internal combustion engine with isobaric heat supply are 0.1 MPa and 80 °C. The compression ratio is 16 and the heat supplied is 850 kJ/kg. Calculate the parameters at the characteristic points of the cycle and the thermal efficiency, if the working fluid is air.

Ans: (i) p_2 = 48.5 bar (ii) p_3 = 48.5 bar (iii) p_4 = 2.26 bar
(iv) T_2 = 1070.1 K (v) T_3 = 1916 K   (vi) T_4 = 798.3 K
(vii) η_{th} = 62.4%

3.18. The pressure ratio $\lambda = 1.5$ in the process of isochoric heat supply for the cycle of an internal combustion engine with a mixed supply of heat = 1034 kJ/kg and the compression ratio = 13. Find the thermal efficiency and temperature at the characteristic points of the cycle if the initial parameters are 0.09 MPa and 70 °C and the working substance is air.

Ans: (i) η_{th} = 62.1% (ii) T_2 = 957 K (iii) T_3 = 1435 K
(iv) T_4 = 2125 K (v) T_5 = 890 K

3.19. The parameters of the initial state of one kilogram of air in the cycle of an internal combustion engine are 0.095 MPa and 65 °C. The compression ratio is 11. Compare the values of the thermal efficiency for isobaric and isochoric heat supply in amounts of 800 kJ, assuming that $k = 1.4$.

Ans: η_{t_p} = 55.67 %, η_{t_v} = 61.67 %

3.20. Find the thermal efficiency of the cycle of an internal combustion engine with a mixed heat supply, if the minimum temperature of the cycle is 85 °C and the maximum temperature is 1700 K. The compression ratio is 15 and the pressure ratio in the process of heat supply is 1.3. The working fluid is air.

Ans: $\eta_{th} = 65.24$ %

3.21. The pressure ratio during the compression in the cycle of an internal combustion engine with isochoric heat supply is equal to 18. Find the compression ratio, supplied and removed heat, work and efficiency, if during heat removal the temperature drops from 600 to 100 °C and the working fluid is air. Assume $\gamma = 1.4$ and $C_v = 0.717$ kJ/kg K.

Ans: (i) r $= 7.8881$ (ii) $q_1 = 815.24$ kJ/kg
(iii) $q_2 = 357$ kJ/kg (iv) $w = 458.24$ kJ/kg
(v) $\eta_{th} = 56.2\%$

3.22. An oil engine working on the dual combustion cycle has a cylinder diameter of 20 cm and stroke of 40 cm. The compression ratio is 13.5 and the explosion ratio 1.42. Cut-off occurs at 5.1% of the stroke. Find the air-standard efficiency. Take $\gamma = 1.4$.

Ans: 61.66 %

3.23. A compression-ignition engine working on Dual cycle takes in two-fifth of its total heat supply at constant volume and the remaining at constant pressure. Calculate :

(i) The pressure and temperature at the five cardinal points of the cycle.

(ii) The ideal thermal efficiency of the cycle.

Given : compression ratio = 13.1, Maximum pressure in the cycle = 45 bar, air intake at 1 bar and 15 °C, $C_p = 1.004$ kJ/kg K and $C_v = 0.717$ kJ/kg K.

Ans: (i) $p_1 = 1$ bar (ii) $p_2 = 36.6$ bar (iii) $p_3 = 45$ bar
(iv) $p_4 = 45$ bar (v) $p_5 = 1.582$ bar (vi) $T_1 = 288$ K
(vii) $T_2 = 805.9$ K (viii) $T_3 = 989.3$ K (ix) $T_4 = 1185.7$ K
(x) $T_5 = 455.5$ K (xi) $\eta_{th} = 63.3\%$

3.24. An oil engine working on the dual combustion cycle has a cylinder diameter of 25 cm and stroke 35 cm. The clearance volume is 1500 cc and cut-off takes place at 5% of the stroke. The explosion ratio 1.4. Find the air-standard efficiency of the engine. Assume $\gamma = 1.4$ for air.

Ans: 60.74 %

3.25. A gas turbine unit works on an air-standard Brayton cycle. The pressure ratio across the compression is 6. Air enters the compressor at 1 bar and 27 °C. The maximum temperature of the cycle is 850 °C. Calculate the specific output and the efficiency of the cycle. What will be the power developed by the unit for a mass flow rate of 10 kg/s. Would you recommend this cycle for a reciprocating engine? For air $\gamma = 1.4$ and $C_p = 1.005$ kJ/kg K.

Ans: (i) 248 kJ/kg (ii) 2480 kW

       (iii) The volume of the cylinder will be too large due to high specific volume at state 4, therefore this cycle is not recommended for reciprocating engine.

4

FUEL–AIR CYCLES
AND
THEIR ANALYSIS

4.1 INTRODUCTION

In the previous chapter, a detailed discussion of air-standard cycles, particularly for IC engines has been given. The analysis was based on highly simplifying assumptions. Because of this, the estimated engine performance by air-standard cycle analysis is on the higher side compared to the actual performance. For example, the actual indicated thermal efficiency of an SI engine, say with a compression ratio of 8:1, is of the order of 28% whereas the air-standard efficiency is 56.5%. This large deviation may to some extent be attributed to progressive burning of the fuel, incomplete combustion and valve operation etc. However, the main reasons for this may be attributed to the over simplified assumptions made in the analysis.

In an actual engine, the working fluid is a mixture of air, fuel vapour and residual gases from the previous cycle. Further, the specific heats of the working fluid are not constant but increase with temperature. Finally, the products of combustion are subjected to a certain dissociation at high temperatures. If the actual physical properties of the gases in the cylinder before and after the combustion are taken into account, a reasonably close values to the actual pressures and temperatures existing within the engine cylinder can be estimated. The mean effective pressures and efficiencies, calculated by this analysis, in the case of well designed engines are higher only by a few per cent from the actual values obtained by tests. The analysis based on the actual properties of the working medium viz., fuel and air is called the fuel-air cycle analysis and even this analysis has simplifying assumptions. However, they are more justifiable and close to the actual conditions than those used in the air-standard cycle analysis.

4.2 FUEL–AIR CYCLES AND THEIR SIGNIFICANCE

By air-standard cycle analysis, it is understood how the efficiency is improved by increasing the compression ratio. However, analysis cannot bring out the effect of air-fuel ratio on the thermal efficiency because the working medium was assumed to be air. In this chapter, the presence of fuel in the cylinder is taken into account and accordingly the working medium will be a mixture of fuel and air. By fuel-air cycle analysis it will be possible to bring out the effect of fuel-air ratio on thermal efficiency and also study how the peak pressures and temperatures during the cycle vary with respect to fuel-air ratio. In general, influence of many of the engine operating variables on the pressures and temperatures within the engine cylinder may be better understood by the examination of the fuel-air cycles. The fuel-air cycle analysis takes into account the following :

(i) *The actual composition of the cylinder gases :* The cylinder gases contains fuel, air, water vapour and residual gas. The fuel-air ratio changes during the operation of the engine which changes the relative amounts of CO_2, water vapour, etc.

(ii) *The variation in the specific heat with temperature :* Specific heats increase with temperature except for mono-atomic gases. Therefore, the value of γ also changes with temperature.

(iii) *The effect of dissociation :* The fuel and air do not completely combine chemically at high temperatures (above 1600 K) and this leads to the presence of CO, H_2, H and O_2 at equilibrium conditions.

(iv) *The variation in the number of molecules :* The number of molecules present after combustion depend upon fuel-air ratio and upon the pressure and temperature after the combustion.

Besides taking the above factors into consideration, the following assumptions are commonly made :

(i) There is no chemical change in either fuel or air prior to combustion.

(ii) Subsequent to combustion, the charge is always in chemical equilibrium.

(iii) There is no heat exchange between the gases and the cylinder walls in any process, i.e. they are adiabatic. Also the compression and expansion processes are frictionless.

(iv) In case of reciprocating engines it is assumed that fluid motion can be ignored inside the cylinder.

With particular reference to constant-volume fuel-air cycle, it is also assumed that

(v) The fuel is completely vaporized and perfectly mixed with the air, and

(vi) The burning takes place instantaneously at top dead centre (at constant volume).

As already mentioned, the air-standard cycle analysis shows the general effect of only compression ratio on engine efficiency whereas the fuel-air cycle analysis gives the effect of variation of fuel-air ratio, inlet pressure and temperature on the engine performance. It will be noticed that compression ratio and fuel-air ratio are very important parameters of the engine while inlet conditions are not so important.

The actual efficiency of a good engine is about 85 per cent of the estimated fuel-air cycle efficiency. Thus a very good estimate of the power to be expected from the actual engine can be made from fuel-air cycle analysis. Also, peak pressures and exhaust temperatures which affect the engine structure and design can be very closely estimated. Thus the effect of many variables on the performance of an engine can be understood better by fuel-air cycle analysis.

4.3 COMPOSITION OF CYLINDER GASES

The air-fuel ratio changes during the engine operation. This change in air-fuel ratio affects the composition of the gases before combustion as well as after combustion particularly the percentage of carbon dioxide, carbon monoxide, water vapour etc in the exhaust gases.

In four-stroke engines, fresh charge as it enters the engine cylinder, comes into contact with the burnt gases left in the clearance space of the previous cycle. The amount of exhaust gases in clearance space varies with speed and load on the engine. Fuel-air cycle analysis takes into account this fact and the results are computed for preparing the combustion charts. However, with the availability of fast digital computers, nowadays it is possible to analyze the effect of cylinder gas composition on the performance of the engine by means of suitable numerical techniques. The computer analysis can produce fast and accurate results. Thus, fuel-air cycle analysis can be done more easily through computers rather than through manual calculations.

4.4 VARIABLE SPECIFIC HEATS

All gases, except mono-atomic gases, show an increase in specific heat with temperature. The increase in specific heat does not follow any particular law. However, over the temperature range generally encountered for gases in heat engines (300 K to 2000 K) the specific heat curve is nearly a straight line which may be approximately expressed in the form

$$
\left.\begin{array}{rcl}
C_p & = & a_1 + k_1 T \\[2mm]
C_v & = & b_1 + k_1 T
\end{array}\right\} \tag{4.1}
$$

where a_1, b_1 and k_1 are constants. Now,

$$
R \;\; = \;\; C_p - C_v \;\; = \;\; a_1 - b_1 \tag{4.2}
$$

where R is the characteristic gas constant.

Above 1500 K the specific heat increases much more rapidly and may be expressed in the form

$$
\left.\begin{array}{rcl}
C_p & = & a_1 + k_1 T + k_2 T^2 \\[2mm]
C_v & = & b_1 + k_1 T + k_2 T^2
\end{array}\right\} \tag{4.3}
$$

In Eqn.4.3 if the term T^2 is neglected it becomes same as Eqn.4.1. Many expressions are available even upto sixth order of T (i.e. T^6) for the calculation of C_p and C_v.

The physical explanation for increase in specific heat is that as the temperature is raised, larger fractions of the heat would be required to produce motion of the atoms within the molecules. Since temperature is the result of motion of the molecules, as a whole, the energy which goes into moving the atoms does not contribute to proportional temperature rise. Hence more heat is required to raise the temperature of unit mass through one degree at higher levels. This heat by definition is the specific heat. For air, the values are

C_p = 1.005 kJ/kg K at 300 K C_v = 0.717 kJ/kg K at 300 K
C_p = 1.343 kJ/kg K at 2000 K C_v = 1.055 kJ/kg K at 2000 K

Since the difference between C_p and C_v is constant, the value of γ decreases with increase in temperature. Thus, if the variation of specific heats is taken into account during the compression stroke, the final temperature and pressure would be lower than if constant values of specific heat are used. This point is illustrated in Fig.4.1.

With variable specific heats, the temperature at the end of compression will be 2', instead of 2. The magnitude of drop in temperature

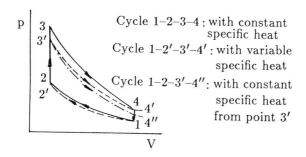

p | 3
3'

Cycle 1–2–3–4 : with constant
               specific heat
Cycle 1–2'–3'–4' : with variable
               specific heat
Cycle 1–2–3'–4": with constant
               specific heat
               from point 3'

Fig.4.1 Loss of Power due to Variation of Specific Heat

is proportional to the drop in the value of ratio of specific heats. For the process $1 \rightarrow 2$, with constant specific heats

$$T_2 = T_1 \left(\frac{v_1}{v_2} \right)^{\gamma - 1} \tag{4.4}$$

with variable specific heats,

$$T_{2'} = T_1 \left(\frac{v_1}{v_2} \right)^{k - 1} \tag{4.5}$$

where $k = \frac{C_p}{C_v}$.

For given values of T_1, v_1 and v_2, the magnitude of $T_{2'}$ depends on k. Constant volume combustion, from point $2'$ will give a temperature $T_{3'}$ instead of T_3. This is due to the fact that the rise in the value of C_v because of variable specific heat, which reduces the temperature as already explained.

The process, $2' \rightarrow 3'$ is heat addition with the variation in specific heat. From $3'$, if expansion takes place at constant specific heats, this would result in the process $3' \rightarrow 4''$ whereas actual expansion due to variable specific heat will result in $3' \rightarrow 4'$ and $4'$ is higher than $4''$. The magnitude in the difference between $4'$ and $4''$ is proportional to the reduction in the value of γ.

Consider the process $3'4''$

$$T_{4''} = T_{3'} \left(\frac{v_3}{v_4} \right)^{\gamma - 1} \tag{4.6}$$

for the process $3'4'$

$$T_{4'} = T_{3'} \left(\frac{v_3}{v_4} \right)^{k - 1} \tag{4.7}$$

Reduction in the value of k due to variable specific heat results in increase of temperature from $T_{4''}$ to $T_{4'}$.

4.5 DISSOCIATION

Dissociation process can be considered as the disintegration of combustion products at high temperature. Dissociation can also be looked as the reverse process to combustion. During dissociation the heat is absorbed whereas during combustion the heat is liberated. In IC engines, mainly dissociation of CO_2 into CO and O_2 occurs, whereas there is a very little dissociation of H_2O.

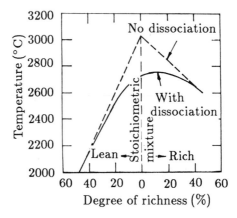

Fig.4.2 Effect of Dissociation on Temperature

The dissociation of CO_2 into CO and O_2 starts commencing around 1000 °C and the reaction equation can be written as

$$CO_2 \quad \rightleftharpoons \quad 2CO + O_2 + Heat$$

Similarly, the dissociation of H_2O occurs at temperatures above 1300 °C and is written as

$$H_2O \quad \rightleftharpoons \quad 2H_2 + O_2 + Heat$$

The presence of CO and O_2 in the gases tends to prevent dissociation of CO_2; this is noticeable in a rich fuel mixture, which, by producing more CO, suppresses dissociation of CO_2. On the other hand, there is no dissociation in the burnt gases of a lean fuel-air mixture. This is mainly due to the fact that the temperature produced is too low for this phenomenon to occur. Hence, the maximum extent of dissociation occurs in the burnt gases of the chemically correct fuel-air mixture when the temperatures are expected to be high but decreases with the leaner and richer mixtures.

In case of internal combustion engines heat transfer to the cooling medium causes a reduction in the maximum temperature and pressure. As the temperature falls during the expansion stroke the separated constituents recombine; the heat absorbed during dissociation is thus again released, but it is too late in the stroke to recover entirely the lost power. A portion of this heat is carried away by the exhaust gases.

Figure 4.2 shows a typical curve that indicates the reduction in the temperature of the exhaust gas mixtures due to dissociation with respect to air-fuel ratio. With no dissociation maximum temperature is attained at chemically correct air-fuel ratio. With dissociation maximum temperature is obtained when mixture is slightly rich. Dissociation reduces the maximum temperature by about 300 °C even at the chemically correct air-fuel ratio. In the Fig.4.2, lean mixtures and rich mixtures are marked clearly.

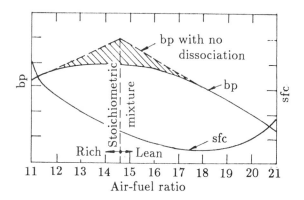

Fig.4.3 Effect of Dissociation on Power

The effect of dissociation on output power is shown in Fig.4.3 for a typical four-stroke spark-ignition engine operating at constant speed. If there is no dissociation, the brake power output is maximum when the mixture ratio is stoichiometric. The shaded area between the brake power graphs shows the loss of power due to dissociation. When the mixture is quite lean there is no dissociation. As the air-fuel ratio decreases i.e., as the mixture becomes rich the maximum temperature rises and dissociation commences. The maximum dissociation occurs at chemically correct mixture strength. As the mixture becomes richer, dissociation effect tends to decline due to incomplete combustion.

Dissociation effects are not so pronounced in a CI engine as in an SI engine. This is mainly due to

(i) the presence of a heterogeneous mixture and

(ii) excess air to ensure complete combustion.

Both these factors tend to reduce the peak gas temperature attained in the CI engine.

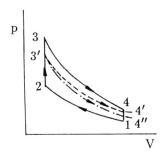

Fig.4.4 Effect of Dissociation shown on a *p-V* Diagram

Figure 4.4 shows the effect of dissociation on *p-V* diagram of Otto cycle. Because of lower maximum temperature due to dissociation the maximum pressure is also reduced and the state after combustion will be represented by 3' instead of 3. If there was no reassociation due to fall of temperature during expansion the expansion process would be represented by 3'→4'' but due to reassociation the expansion follows the path 3'→4'. By comparing with the ideal expansion 3→4, it is observed that the effect of dissociation is to lower the temperature and consequently the pressure at the beginning of the expansion stroke. This causes a loss of power and also efficiency. Though during recombining the heat is given back it is too late to contribute a convincing positive increase in the output of the engine.

4.6 EFFECT OF NUMBER OF MOLES

As already mentioned the number of molecules present in the cylinder after combustion depends upon the fuel-air ratio, type and extend of reaction in the cylinder. According to the gas law

$$pV = N \bar{R} T$$

the pressure depends on the number of molecules or moles present. This has direct effect on the amount of work the cylinder gases can impart on the piston.

4.7 COMPARISON OF AIR–STANDARD AND FUEL–AIR CYCLES

In this section reasons for difference between air-standard cycles and fuel-air cycles is discussed. The magnitude of difference between the two cycles can be attributed to the following factors :

(i) character of the cycle (due to assumptions)
(ii) relative fuel-air ratio (actual $F/A \div$ stoichiometric F/A)
(iii) chemical composition of the fuel

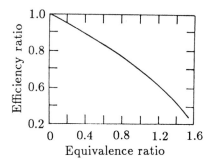

Fig.4.5 Variation of Efficiency Ratio with Mixture Strength

Figure 4.5 shows variation of efficiency with mixture strength of fuel-air cycle relative to that of air cycle showing the gain in efficiency as the mixture becomes leaner. It is seen from Fig.4.5 that the efficiency ratio (fuel-air cycle efficiency/air-standard cycle efficiency) increases as the mixture becomes leaner and leaner tending towards the air-standard cycle efficiency. It is to be noted that this, trend exists at all compression ratios.

At very low fuel-air ratio the mixture would tend to behave like a perfect gas with constant specific heat. Cycles with lean to very lean mixtures tend towards air-standard cycles. In such cycles the pressure and temperature rises. Some of the chemical reactions involved tend to be more complete as the pressure increases. These considerations apply to constant-volume as well as constant-pressure cycles.

The simple air-standard analysis cannot predict the variation of thermal efficiency with mixture strength since air is assumed to be the working medium. However, fuel-air analysis suggests that the thermal efficiency will deteriorate as the mixture supplied to an engine is

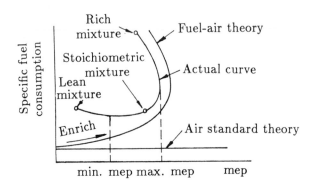

Fig.4.6 Specific Fuel Consumption vs Mean Effective Pressure at Constant Speed and Constant Throttle Setting

enriched. This is explained by the increasing losses due to variable specific heats and dissociation as the mixture strength approaches chemically correct values. This is because, the gas temperature goes up after combustion as the mixture strength approaches chemically correct values. Enrichment beyond the chemically correct ratio will lead to incomplete combustion and loss in thermal efficiency. Hence, as the mixture is made leaner the drop in thermal efficiency will increase further. However, beyond a certain leaning, the combustion becomes erratic with loss of efficiency. Thus the maximum efficiency is within the lean zone very near the stoichiometric ratio. This gives rise to combustion loop, as shown in Fig.4.6 which can be plotted for different mixture strengths for an engine running at constant speed and at a constant throttle setting. This loop gives an idea about the effect of mixture strength on the specific fuel consumption.

4.8 EFFECT OF OPERATING VARIABLES

The effect of the common engine operating variables on the pressure and temperature within the engine cylinder is better understood by fuel-air cycle analysis. The details are discussed in the following sections.

4.8.1 Compression Ratio

The fuel-air cycle efficiency increases with the compression ratio in the same manner as the air-standard cycle efficiency, principally for the same reason (more scope of expansion work). This is shown in Fig.4.7.

The variation of indicated thermal efficiency with respect to the relative air-fuel ratio for various compression ratios is given in Fig.4.8.

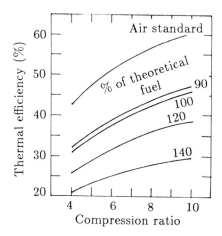

Fig.4.7 Effect of Compression Ratio and
Mixture Strength on Efficiency

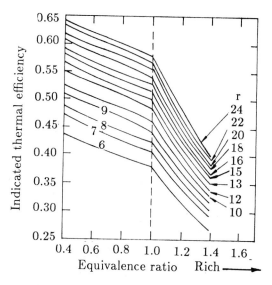

Fig.4.8 Effect of Mixture Strength on Thermal Efficiency
for various Compression Ratios

The relative fuel-air ratio, F_R, is defined as ratio of actual fuel-air
ratio to chemically correct fuel-air ratio on mass basis. The maximum
pressure and maximum temperature increase with compression ratio

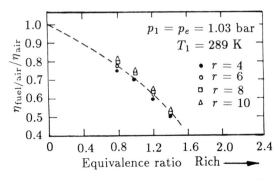

Fig.4.9 Variation of Efficiency with Mixture Strength
for a Constant Volume Fuel-Air Cycle

since the temperature, T_2, and pressure, p_2, at the end of compression are higher. However, it may be observed from Fig.4.9 that the ratio of fuel-air cycle efficiency to air-standard efficiency is independent of the compression ratio for a given fuel-air ratio for a constant-volume fuel-air cycle.

4.8.2 Fuel–Air Ratio

(i) *Efficiency :* As the mixture is made lean (less fuel) the temperature rise due to combustion will be lowered as a result of reduced energy input per unit mass of mixture. This will result in lower specific heat. Further, it will lower the losses due to dissociation and variation in specific heat. The efficiency is therefore, higher and, in fact, approaches the air-cycle efficiency as the fuel-air ratio is reduced as shown in Fig.4.10.

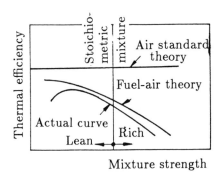

Fig.4.10 Effect of Mixture Strength on Thermal Efficiency

(ii) *Maximum Power :* Fuel-air ratio affects the maximum power output of the engine. The variation is as shown in Fig.4.11. As the mixture becomes richer, after a certain point the efficiency falls rapidly as can be seen from the experimental curve. This is because in addition to higher specific heats and chemical equilibrium losses, there is insufficient air which will result in formation of CO and H_2 during combustion, which represents a direct wastage of fuel. However, fuel-air cycle analysis cannot exactly imitate the experimental curve due to various assumptions made.

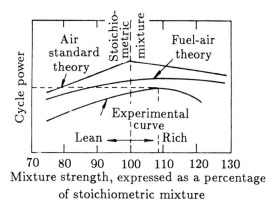

Fig.4.11 Effect of Fuel-Air Ratio on Power

(iii) *Maximum temperature :* At a given compression ratio the temperature after combustion reachers a maximum when the mixture is slightly rich, i.e., around 6% or so $(F/A = 0.072$ or $A/F = 14 : 1)$ as shown in Fig.4.12. At chemically correct ratio there is still some oxygen present at the point 3 (in the p-V diagram, refer Fig.4.1) because of chemical equilibrium effects a rich mixture will cause more fuel to combine with oxygen at that point thereby raising the temperature T_3. However, at richer mixtures increased formation of CO counters this effect.

(iv) *Maximum Pressure :* The pressure of a gas in a given space depends upon its temperature and the number of molecules The curve of p_3, therefore follows T_3, but because of the increasing number of molecules p_3 does not start to decrease until the mixture is some what richer than that for maximum T_3 (at $F/A = 0.083$ or A/F 12 : 1), i.e. about 20 per cent rich (Fig.4.12).

(v) *Exhaust Temperature :* The exhaust gas temperature, T_4 is maximum at the chemically correct mixture as shown in Fig.4.13. At this point the fuel and oxygen are completely used up, as the effect of chemical equilibrium is not significant. At lean mixtures, because of less fuel, T_3 is less and hence T_4 is less. At rich mixtures less sensible energy is developed and hence T_4 is less. That

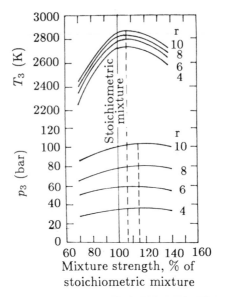

Fig.4.12 Effect of Fuel-Air Ratio on T_3 and p_3

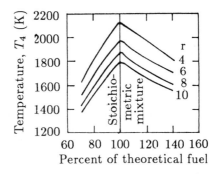

Fig.4.13 Effect of Fuel-Air Ratio on
the Exhaust Gas Temperature

is, T_4 varies with fuel-air ratio in the same manner as T_3 except that maximum T_4 is at the chemically correct fuel-air ratio in place of slightly rich fuel-air ratio (6%) as in case of T_3.

However, the behaviour of T_4 with compression ratio is different from that of T_3 as shown in Fig.4.13. Unlike T_3, the exhaust gas temperature, T_4 is lower at high compression ratios, because the increased expansion causes the gas to do more work on the piston leaving less heat to be rejected at the end of the stroke. The same effect is present in the case of air-cycle analysis also.

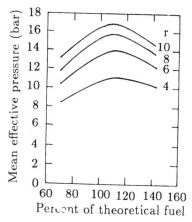

Fig.4.14 Effect of Fuel-Air Ratio on *mep*

(vi) *Mean Effective Pressure (mep)* : The mean effective pressure increases with compression ratio. It follows the trend of p_3 and p_4 and hence it is maximum at a fuel-air ratio slightly richer than the chemically correct ratio as shown in Fig.4.14. Table 4.1 shows a summary of conditions which give maximum pressure and temperature in a constant-volume cycle assuming fuel-air cycle approximations.

Table 4.1 *Condition for Maximum Temperature and Pressure in a Constant Volume Fuel-Air Cycle*

Variable	Maximum at	Reason
1. Temperature, T_3 (See Fig.4.12)	6% rich $(F/A=0.072;$ $A/F14:1, F_r = 1.06$	Because of chemical equilibrium some O_2 still present at chemically correct fuel-air ratio. More fuel can be burnt, limit is reached at 6% rich. If richer than 6% CO formation.
2. Pressure, p_3 (See Fig.4.12)	20% rich $(F/A = 0.083;$ A/F 12:1)	$pV = N\overline{R}T$ p depends on T and N
3. Temperature, T_4 (See Fig.4.13)	Chemically correct fuel-air ratio	No effect of chemical equilibrium due to low temperature, less heat at lean mixture, and incomplete combustion at rich mixture.
4. Mean effective pressure (see Fig.4.14)	6% rich $F/A=0.0745$ $A/F=13.5,$ $F_r = 1.05$ to 1.1	Mean effective pressure follows the trend of p_3 and p_4.

Worked out Examples

4.1. What will be the effect on the efficiency of an Otto cycle having a compression ratio of 8, if C_v increases by 1.6%?

SOLUTION:

$$\eta_{Otto} = 1 - \frac{1}{r^{\gamma-1}}$$

$$C_p - C_v = R$$

$$\frac{C_p}{C_v} = \gamma$$

$$\gamma - 1 = \frac{R}{C_v}$$

$$\eta = 1 - \left(\frac{1}{r}\right)^{R/C_v}$$

$$1 - \eta = r^{-R/C_v}$$

$$\ln(1 - \eta) = \frac{-R}{C_v}\ln r$$

Differentiating

$$-\frac{1}{1 - \eta}d\eta = \frac{R}{C_v^2}\ln r\, dC_v$$

$$d\eta = -\frac{(1 - \eta)R\ln r}{C_v^2}dC_v$$

$$\frac{d\eta}{\eta} = -\frac{(1 - \eta)(\gamma - 1)\ln r}{\eta}\frac{dC_v}{C_v}$$

Now,

$$\eta = 1 - \left(\frac{1}{8}\right)^{0.4} = 0.565 = 56.5\%$$

$$\frac{d\eta}{\eta} = -\frac{(1 - 0.565) \times (1.4 - 1) \times \ln 8}{0.565} \times \frac{1.6}{100}$$

$$= -1.025\% \qquad \underline{\underline{\text{Ans}}}$$

4.2. What will be the effect on the efficiency of a diesel cycle having a compression ratio of 20 and a cut-off ratio is 5% of the swept volume, if the C_v increases by 1%. Take $C_v = 0.717$ and $R = 0.287$ kJ/kg K.

SOLUTION:

$$\eta_{Diesel} = 1 - \left(\frac{1}{r}\right)^{\gamma-1}\left(\frac{1}{\gamma}\frac{r_c^\gamma - 1}{r_c - 1}\right)$$

$$1 - \eta = \frac{1}{\gamma}\frac{r_c^\gamma - 1}{r^{\gamma-1}(r_c - 1)}$$

Taking logarithm

$$\ln(1 - \eta) = -\ln\gamma + \ln(r_c^\gamma - 1) - \ln(r_c - 1) - (\gamma - 1)\ln r$$

$$\gamma - 1 = \frac{R}{C_v}$$

$$\gamma = \left(1 + \frac{R}{C_v}\right)$$

Substituting this in the above equation

$$\ln(1 - \eta) = -\ln\left(\frac{R}{C_v} + 1\right) + \ln\left(r_c^{\left(\frac{R}{C_v}+1\right)} - 1\right)$$

$$- \ln(r_c - 1) - \frac{R}{C_v}\ln r$$

Differentiating we get,

$$-\frac{d\eta}{\eta} = \frac{\frac{R}{C_v^2}dC_v}{\frac{R}{C_v} + 1} - \frac{\frac{R}{C_v^2}\left(r_c^{\left(\frac{R}{C_v}+1\right)}\right)\ln r_c dC_v}{r_c^{\left(\frac{R}{C_v}+1\right)} - 1} + \frac{R}{C_v^2}\ln r\, dC_v$$

$$\frac{d\eta}{\eta} = -\frac{dC_v}{C_v}\frac{R}{C_v}\left(\frac{1 - \eta}{\eta}\right)$$

$$\times \left(\frac{1}{\frac{R}{C_v} + 1} + \ln r - \frac{r_c^{\frac{R}{C_v}+1}\ln(r_c)}{r_c^{\frac{R}{C_v}+1} - 1}\right)$$

$$\frac{d\eta}{\eta} = -\frac{dC_v}{C_v}\left(\frac{1 - \eta}{\eta}\right)(\gamma - 1)\left[\frac{1}{\gamma} + \ln r - \frac{r_c^\gamma\ln(r_c)}{r_c^\gamma - 1}\right]$$

$$\gamma = 1.4$$

$$\frac{V_1}{V_2} = r = 20$$

$$V_1 = 20V_2$$

$$V_s = 20V_2 - V_2 = 19V_2$$

$$V_3 = 0.05V_s + V_2$$

$$= (0.05 \times 19V_2) + V_2 = 1.95V_2$$

$$r_c = \frac{V_3}{V_2} = \frac{1.95V_2}{V_2} = 1.95$$

$$\gamma = 1.4$$

$$\eta = 1 - \frac{1}{\gamma} \frac{1}{r^{\gamma-1}} \frac{r_c^\gamma - 1}{r_c - 1}$$

$$= 1 - \frac{1}{1.4} \times \left(\frac{1}{20}\right)^{0.4} \times \frac{1.95^{1.4} - 1}{1.95 - 1} = 0.649$$

$$\frac{d\eta}{\eta} = -0.01 \times \frac{1 - 0.649}{0.649}$$

$$\times 0.4 \times \left[\frac{1}{1.4} + \ln(20) - \frac{1.95^{1.4} \times \ln(1.95)}{1.95^{1.4} - 1}\right]$$

$$= 0.565\% \qquad \underset{\Longleftarrow}{\text{Ans}}$$

4.3. A petrol engine having a compression ratio of 6 uses a fuel with calorific value of 42 MJ/kg. The air-fuel ratio is 15:1. Pressure and temperature at the start of the suction stroke is 1 bar and 57 °C respectively. Determine the maximum pressure in the cylinder if the index of compression is 1.3 and the specific heat at constant volume is given by $C_v = 0.678 + 0.00013\,T$, where T is in Kelvin. Compare this value with that obtained when $C_v = 0.717$ kJ/kg K.

SOLUTION:

Consider the process 1-2

$$p_2 V_2^n = p_1 V_1^n$$

$$p_2 = p_1 \left(\frac{V_1}{V_2}\right)^n$$

$$= 1 \times 6^{1.3} = 10.27 \text{ bar}$$

$$T_2 = T_1 \left(\frac{p_2 V_2}{p_1 V_1}\right)$$

$$= \quad 330 \times \left(\frac{10.27}{1} \times \frac{1}{6} \right) \quad = \quad 565 \text{ K}$$

Average temperature during combustion

$$= \quad \frac{T_3 + T_2}{2}$$

$$C_{v_{mean}} \quad = \quad 0.678 + 0.00013 \times \left(\frac{T_3 + T_2}{2} \right)$$

Assuming unit quantity of air

$$Q_{2-3}/\text{kg of air} \quad = \quad \frac{42}{15} \quad = \quad 2.8 \text{ MJ}$$

$$\text{Mass of charge/kg of air} \quad = \quad 1 + \frac{1}{15} \quad = \quad \frac{16}{15} \text{ kg}$$

$$Q_{2-3} \quad = \quad C_{v_{mean}} \dot{m}(T_3 - T_2)$$

$$2.8 \times 10^3 \quad = \quad \left[0.678 + 13 \times 10^{-5} \times \left(\frac{T_3 + 565}{2} \right) \right]$$

$$\times \frac{16}{15} \times (T_3 - 565)$$

Solving for T_3 we get

$$T_3 \quad = \quad 3375 \text{ K}$$

$$p_3 \quad = \quad p_2 \frac{T_3}{T_2} \quad = \quad 10.27 \times \frac{3375}{565}$$

$$= \quad \textbf{61.22 bar} \qquad \underset{\Leftarrow}{\textbf{Ans}}$$

For constant specific heat

$$2.8 \times 10^3 \quad = \quad 0.717 \times \frac{16}{15} \times (T_3 - 565)$$

On solving we get

$$T_3 \quad = \quad 4226$$

$$p_3 \quad = \quad 10.27 \times \frac{4226}{565}$$

$$= \quad \textbf{76.81 bar} \qquad \underset{\Leftarrow}{\textbf{Ans}}$$

4.4. The air-fuel ratio of a Diesel engine is 29:1. If the compression ratio is 16:1 and the temperature at the end of compression is 900 K, find at what percentage of stroke is the combustion completed.

Assume that the combustion begins at the top dead centre and takes place at constant pressure. Take calorific value of the fuel as 42000 kJ/kg, $R = 0.287$ kJ/kg K and $C_v = 0.709 + 0.000028\ T$ kJ/kg K.

SOLUTION:

$$C_p = C_v + R$$

$$= (0.709 + 0.000028\ T) + 0.287$$

$$= 0.996 + 0.000028\ T$$

$$dQ = mC_p dT$$

For unit mass,

$$Q = \int_2^3 C_p dT$$

For 1 kg of fuel the total charge is 30 kg

$$Q = \frac{CV}{30} = \frac{42000}{30} = 1400$$

$$= \int_2^3 (0.996 + 0.000028\ T)dT$$

$$= 0.996(T_3 - T_2) + \frac{0.000028}{2}(T_3^2 - T_2^2)$$

$$= 0.996 \times (T_3 - 900)$$

$$+ 0.000014 \times (T_3^2 - 900^2)$$

On solving,

$$T_3 = 2256.8\ K$$

$$V_3 = \frac{T_3}{T_2}V_2 = \frac{2256.8}{900}V_2 = 2.508V_2$$

$$Stroke\ volume = V_1 - V_2 = V_2\left(\frac{V_1}{V_2} - 1\right)$$

$$= V_2(r - 1) = 15V_2$$

$$\%\ of\ stroke\ volume = \frac{2.508V_2}{15V_2} \times 100$$

$$= \mathbf{16.72\%} \qquad \underset{\rightleftharpoons}{\textbf{Ans}}$$

4.5. An oil engine, working on the dual combustion cycle, has a compression ratio of 15:1. The heat supplied per kg of air is 2000 kJ, half of which is supplied at constant volume and the other half at constant pressure. If the temperature and pressure at the beginning of compression are 100 °C and 1 bar respectively, find (i) the maximum pressure in the cycle and (ii) the percentage of stroke when cut-off occurs. Assume $\gamma = 1.4$, $R = 0.287$ kJ/kg K and $C_v = 0.709 + 0.000028T$ kJ/kg K.

SOLUTION:

$$p_1 V_1^{\gamma} = p_2 V_2^{\gamma}$$

$$p_2 = p_2 \left(\frac{V_1}{V_2}\right)^{\gamma}$$

$$= 1 \times 10^5 \times 15^{1.4} = 44.3 \times 10^5 \ \text{N/m}^2$$

$$T_1 V_1^{(\gamma-1)} = T_2 V_2^{\gamma-1}$$

$$T_2 = T_1 r^{(\gamma-1)}$$

$$= 373 \times 13^{0.4} = 1040.6 \ \text{K}$$

For unit mass :

Consider the process 2-3,

$$Q_{2-3} = \frac{1}{2} \times 2000 = 1000 \ \text{kJ}$$

$$Q_{2-3} = m \int_2^3 (0.709 + 0.000028T) \, dT$$

$$1000 = 0.709 \times (T_3 - 1040.6) +$$

$$\frac{0.000028}{2} \times (T_3^2 - 1040.6^2)$$

On solving,

$$T_3 = 2362.2 \ \text{K}$$

$$p_3 = p_2 \left(\frac{T_3}{T_2}\right) = 44.3 \times \left(\frac{2362.2}{1040.6}\right) \times 10^5$$

$$= 100.5 \times 10^5 \ \text{N/m}^2 \qquad \underset{\Longleftarrow}{\text{Ans}}$$

Consider the process 3-4,

$$Q_{3-4} = \frac{1}{2} \times 1000 = 1000 \ \text{kJ}$$

$$C_p = C_v + R = 0.996 + 0.000028\,T$$

$$Q_{3-4} = m \int_3^4 dT$$

$$1000 = \int_3^4 (0.996 + 0.000028)\,dT$$

$$= 0.996 \times (T_4 - 2362.2) + \frac{0.000028}{2}$$

$$\times (T_4^2 - 2362.2^2)$$

On solving,

$$T_4 = 3292.2 \text{ K}$$

$$V_4 = V_3 \left(\frac{T_4}{T_3}\right) = \frac{3292.3}{2362.2} V_3$$

$$= 1.394 V_3$$

Stroke volume, $V_s = V_1 - V_3 = V_3(r-1) = 14V_3$

Cut-off % of stroke $= \dfrac{V_4 - V_3}{V_s} \times 100$

$$= \frac{V_4 - V_3}{14V_3} \times 100$$

$$= \frac{1.394 - 1}{14} \times 100$$

$$= \mathbf{2.81\%} \qquad \text{Ans}$$

Questions :-

4.1 Mention the simplified various assumptions used in fuel-air cycle analysis.

4.2 What is the difference between air-standard cycle and fuel-air cycle analysis? Explain the significance of the fuel-air cycle.

4.3 Explain why the fuel-air cycle analysis is more suitable for analyzing through a computer rather than through hand calculations.

4.4 How do the specific heats vary with temperature? What is the physical explanation for this variation?

4.5 Explain with the help of a p-V diagram the loss due to variation of specific heats in an Otto cycle.

4.6 Show with the help of a p-V diagram for an Otto cycle, that the effect of dissociation is similar to that of variation of specific heats.

4.7 Explain by means of suitable graphs the effect of dissociation on maximum temperature and brake power. How does the presence of CO affect dissociation?

4.8 Explain the effect of change of number of molecules during combustion on maximum pressure in the Otto cycle.

4.9 Compare the air-standard cycle and fuel-air cycles based on
(i) character of the cycle
(ii) fuel-air ratio
(iii) chemical composition of the fuel

4.10 Is the effect of compression ratio on efficiency the same in fuel-air cycles also? Explain.

4.11 From the point of view of fuel-air cycle analysis how does fuel-air ratio affect efficiency, maximum power, temperature and pressure in a cycle.

4.12 How do exhaust temperature and mean effective pressure affect the engine performance? Explain.

Exercise:-

4.1. Find the percentage change in the efficiency of an Otto cycle having a compression ratio of 10, if C_v decreases by 2%.

Ans: 1.22%

4.2. Find the percentage increase in the efficiency of a Diesel cycle having a compression ratio of 16 and cut-off ratio is 10% of the swept volume, if C_v decreases by 2%. Take $C_v = 0.717$ and $\gamma = 1.4$.
Ans: 3.04%

4.3. The air-fuel ratio of a Diesel engine is 31:1. If the compression ratio is 15:1 and the temperature at the end of compression is 1000 K, find at what percentage of stroke is the combustion complete if the combustion begins at TDC and continuous at constant pressure. Calorific value of the fuel is 40000 kJ/kg. Assume the variable specific heat, $C_p = a + bT$, where $a = 1$ and $b = 0.28 \times 10^{-4}$.
Ans: 15.7%

4.4. An engine working on the Otto cycle, uses hexane as fuel. The air-fuel ratio of the mixture is 13.66:1 and the compression ratio is 8. Pressure and temperature at the beginning of compression are 1 bar and 77 °C respectively. If the calorific value of the fuel is 43000 kJ/kg and $C_v = 0.717$ kJ/kg K, find the maximum pressure and temperature without considering molecular contraction. Assume the compression follows the law $pV^{1.3}$ = constant. The stoichiometric equation is

$$C_6H_{14} + 9.5O_2 \quad \rightarrow \quad 6CO_2 + 7H_2O$$

Ans: (i) 117.3 bar (ii) 4744 K

4.5. In a constant-volume combustion fuel-air cycle with compression ratio 6, the mixture has a fuel-air ratio of 0.0785. The fuel used is C_8H_{18} having a lower calorific value of 44000 kJ/kg. Assume mixture pressure and temperature of the beginning of compression as 1 bar and 300 K respectively. Calculate the indicated thermal efficiency, indicated specific fuel consumption and the mean effective pressure of the cycle. Take the index for compression and expansion as 1.3. Assume that the fraction of residual gas at $r = 6$ is 0.03. Specific heat may be calculated at an average temperature of 2500 K and the drop in temperature rise during combustion due to dissociation at the above fuel-air ratio may be assumed as 13% of the temperature without considering dissociation. Take correction factor for dissociation as 0.87. C_p values of the various products are to be calculated using the following equations:

$$N_2 \quad = \quad 39.6 - \frac{8 \times 10^3}{T} + \frac{1.5 \times 10^6}{T^2} \quad \text{kJ/kmol K}$$

$$CO_2 \quad = \quad 67.8 - \frac{15 \times 10^3}{T} + \frac{1.8 \times 10^6}{T^2} \quad \text{kJ/kmol K}$$

$$CO \quad = \quad 39.5 - \frac{7.6 \times 10^3}{T} + \frac{1.4 \times 10^6}{T^2} \quad \text{kJ/kmol K}$$

$$H_2O \text{ (vap)} \quad = \quad 83.0 - 0.0125T + \frac{17.3 \times 10^3}{T} \quad \text{kJ/kmol K}$$

$$H_2 \quad = \quad 24.0 - 0.00435T + \frac{62.7}{T} \quad \text{kJ/kmol K}$$

Ans: (i) 27.4% (ii) 0.298 kg/kW h (iii) 12.5 bar

5

ACTUAL CYCLES
AND
THEIR ANALYSIS

5.1 INTRODUCTION

The actual cycles for IC engines differ from the fuel-air cycles and air-standard cycles in many respects. The actual cycle efficiency is much lower than the air-standard efficiency due to various losses occurring in the actual engine operation. The major losses are due to :

(i) Variation of specific heats with temperature
(ii) Dissociation of the combustion products
(iii) Progressive combustion
(iv) Incomplete combustion of fuel
(v) Heat transfer into the walls of the combustion chamber
(vi) Blowdown at the end of the exhaust process
(vii) Gas exchange process

An estimate of these losses can be made from previous experience and some simple tests on the engines and these estimates can be used in evaluating the performance of an engine.

5.2 COMPARISON OF THERMODYNAMIC AND ACTUAL CYCLES

The actual cycles for internal combustion engines differ from thermodynamic cycles in many respects. These differences are mainly due to :

(i) The working substance being a mixture of air and fuel vapour or finely atomized liquid fuel in air combined with the products of combustion left from the previous cycle.

(ii) The change in chemical composition of the working substance.

(iii) The variation of specific heats with temperature.

(iv) The change in the composition, temperature and actual amount of fresh charge because of the residual gases.

(v) The progressive combustion rather than the instantaneous combustion.

(vi) The heat transfer to and from the working medium.

(vii) The substantial exhaust blowdown loss, i.e., loss of work on the expansion stroke due to early opening of the exhaust valve.

(viii) Gas leakage, fluid friction etc., in actual engines.

Points (i) to (iv), being related to fuel-air cycles have already been dealt in detail in Chapter 4. Remaining points viz. (v) to (viii) are in fact responsible for the difference between fuel-air cycles and actual cycles.

Most of the factors listed above tend to decrease the thermal efficiency and power output of the actual engines. On the other hand, the analysis of the cycles while taking these factors into account clearly indicates that the estimated thermal efficiencies are not very different from those of the actual cycles.

Out of all the above factors, major influence is exercised by

(i) *Time loss factor* i.e. loss due to time required for mixing of fuel and air and also for combustion.

(ii) *Heat loss factor* i.e. loss of heat from gases to cylinder walls.

(iii) *Exhaust blowdown factor* i.e. loss of work on the expansion stroke due to early opening of the exhaust valve.

These major losses which are not considered in the previous two chapters are discussed in the following sections.

5.3 TIME LOSS FACTOR

In thermodynamic cycles the heat addition is assumed to be an instantaneous process whereas in an actual cycle it is over a definite period of time. The time required for the combustion is such that under all circumstances some change in volume takes place while it is in progress. The crankshaft will usually turn about 30 to 40° between the time the spark occurs and the time the charge is completely burnt or when the peak pressure in the cycle is reached.

The consequence of the finite time of combustion is that the peak pressure will not occur when the volume is minimum i.e., when the piston is at TDC; but will occur some time after TDC. The pressure, therefore, rises in the first part of the working stroke from b to c as shown in Fig.5.1. The point 3 represents the state of gases had the combustion been instantaneous and an additional amount of work equal to area shown hatched would have been done. This loss of work reduces the efficiency and is called *time loss due to progressive combustion* or merely *time losses*.

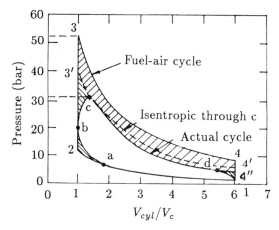

Fig.5.1 The Effect of Time Losses shown on *p-V* Diagram

The time taken for the burning depends upon the flame velocity which in turn depends upon the type of fuel and the fuel-air ratio and also on the shape and size of the combustion chamber. Further, the distance from the point of ignition to the opposite side of the combustion space also plays an important role.

In order that the peak pressure is not reached too late in the expansion stroke, the time at which the combustion starts is varied by varying the spark timing or spark advance. Figures 5.2 and 5.3 show the effect of spark timing on *p-V* diagram from a typical trial. With spark at *TDC* (Fig.5.2) the peak pressure is low due to the expansion of gases. If the spark is advanced to achieve complete combustion close to *TDC* (Fig.5.3) additional work is required to compress the burning gases. This represents a direct loss. In either case, viz., with or without spark advance the work area is less and the power output and efficiency are lowered. Therefore, a moderate or optimum spark advance (Fig.5.4) is the best compromise resulting in minimum losses on both the compression and expansion strokes. Table 5.1 compares the engine performance for various ignition timings. Figure 5.5 shows the effect of spark advance on the power output by means of the *p-V* diagram. As seen from Fig.5.6, when the ignition advance is increased there is a drastic reduction in the *imep* and the consequent loss of power. However, some times a deliberate spark retardation from

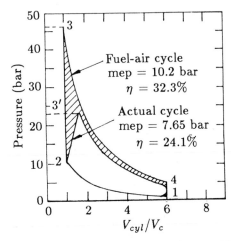

Fig.5.2 Spark at TDC, Advance $0°$

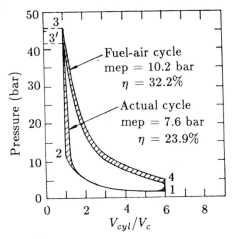

Fig.5.3 Combustion Completed at TDC , Advance $35°$

optimum may be necessary in actual practice in order to avoid knocking and to simultaneously reduce exhaust emissions of hydrocarbons and carbon monoxide.

At full throttle with the fuel-air ratio corresponding to maximum power and the optimum ignition advance the time losses may account for a drop in efficiency of about 5 per cent (fuel-air cycle efficiency is reduced by about 2 per cent). These losses are higher when the mixture is richer or leaner when the ignition advance is not optimum and also at part throttle operations the losses are higher.

Table 5.1 *Cycle Performance for Various Ignition Timings for r = 6*
(Typical Values)

Cycle	Ignition advance	Max. cycle pressure bar	*mep* bar	efficiency %	$\dfrac{\text{Actual } \eta}{\text{Fuel cycle } \eta}$
Fuel-air cycle	0°	44	10.20	32.2	1.00
Actual cycle	0°	23	7.50	24.1	0.75
"	17°	34	8.35	26.3	0.81
"	35°	41	7.60	23.9	0.74

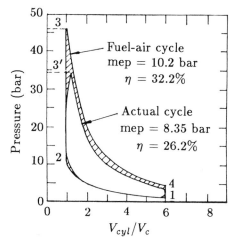

Fig.5.4 Optimum Advance 15° – 30°

It is impossible to obtain a perfect homogeneous mixture with fuel-vapour and air, since, residual gases from the previous cycle are present in the clearance volume of the cylinder. Further, only very limited time is available between the mixture preparation and ignition. Under these circumstances, it is possible that a pocket of excess oxygen is present in one part of the cylinder and a pocket of excess fuel in another part. Therefore, some fuel does not burn or burns partially to CO and the unused O_2 appears in the exhaust as shown in Fig.5.7. Energy release data show that only about 95% of the energy is released with stoichiometric fuel-air ratios. Energy release in actual engine is about 90% of fuel energy input.

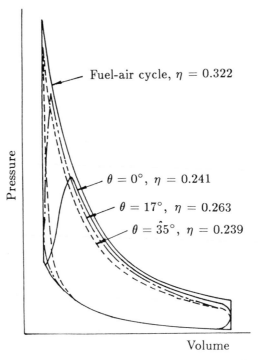

Fig.5.5 *p-V* Diagram showing Power Loss due to
Ignition Advance

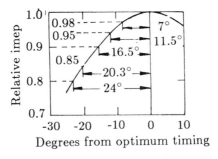

Fig.5.6 Power Loss due to Ignition Advance

It should be noted that it is necessary to use a lean mixture to
eliminate wastage of fuel, while a rich mixture is required to utilize all
the oxygen. Slightly leaner mixture would give maximum efficiency
but too lean a mixture will burn slowly increasing the time losses
or will not burn at all causing total wastage of fuel. In a rich mixture a part of the fuel will not get the necessary oxygen and will be

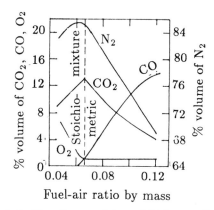

Fig.5.7 The Composition of Exhaust Gases
for Various Fuel-Air Ratios

completely lost. Also the flame speed in mixtures more than 10%
richer is low, thereby, increasing the time losses and lowering the efficiency. Even if this unused fuel and oxygen eventually combine during
the exhaust stroke and burn, the energy which is released at such a
late stage cannot be utilized.

Imperfect mixing of fuel and air may give different fuel-air ratios
on the following suction strokes or certain cylinders may get continuously leaner mixtures than others.

5.4 HEAT LOSS FACTOR

During the combustion process and the subsequent expansion stroke
the heat flows from the cylinder gases through the cylinder walls and
cylinder head into the water jacket or cooling fins. Some heat enters
the piston head and flows through the piston rings into the cylinder
wall or is carried away by the engine lubricating oil which splashes on
the underside of the piston. The heat loss along with other losses is
shown on the p-V diagram in Fig.5.8.

Heat loss during combustion will naturally have the maximum
effect on the cycle efficiency while heat loss just before the end of the
expansion stroke can have very little effect because of its contribution
to the useful work is very little. The heat lost during the combustion
does not represent a complete loss because, even under ideal conditions assumed for air-standard cycle, only a part of this heat could be
converted into work (equal to $Q \times \eta_{th}$) and the rest would be rejected
during the exhaust stroke. About 15 per cent of the total heat is lost
during combustion and expansion. Of this, however, much is lost so

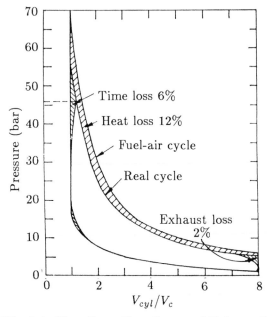

Fig.5.8 Time Loss, Heat Loss and Exhaust Loss
in Petrol Engines

late in the cycle to have contributed to useful work. If all the heat
loss is recovered only about 20% of it may appear as useful work.
Figure 5.8 shows percentage of time loss, heat loss and exhaust loss
in a Cooperative Fuel Research (CFR) engine. Losses are given as
percentage of fuel-air cycle work.

The effect of loss of heat during combustion is to reduce the
maximum temperature and therefore, the specific heats are lower. It
may be noted from the Fig.5.8 that of the various losses, heat loss
factor contributes around 12%.

5.5 EXHAUST BLOWDOWN

The cylinder pressure at the end of exhaust stroke is about 7 bar
depending on the compression ratio employed. If the exhaust valve is
opened at the bottom dead centre, the piston has to do work against
high cylinder pressures during the early part of the exhaust stroke. If
the exhaust valve is opened too early, a part of the expansion stroke
is lost. The best compromise is to open the exhaust valve 40° to
70° before BDC thereby reducing the cylinder pressure to halfway
to atmospheric before the exhaust stroke begins. This is shown in
Fig.5.9 by the roundness at the end of the diagram.

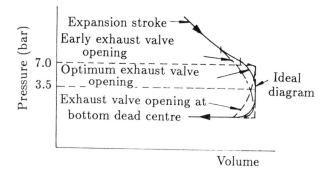

Fig.5.9 Effect of Exhaust Valve Opening Time on Blowdown

5.5.1 Loss Due to Gas Exchange Processes

The difference of work done in expelling the exhaust gases and the work done by the fresh charge during the suction stroke is called the pumping work. In other words loss due to the gas exchange process (pumping loss) is due to pumping gas from lower inlet pressure p_i to higher exhaust pressure p_e. *The pumping loss increases at part throttle because throttling reduces the suction pressure.* Pumping loss also increases with speed. The gas exchange processes affect the volumetric efficiency of the engine. The performance of the engine, to a great deal, depends on the volumetric efficiency. Hence, it is worthwhile to discuss this parameter in greater detail here.

5.5.2 Volumetric Efficiency

As already stated in section 1.8.4, volumetric efficiency is an indication of the breathing ability of the engine and is defined as the ratio of the volume of air actually inducted at ambient condition to swept volume. However, it may also be defined on mass basis as the ratio of the actual mass of air drawn into the engine during a given period of time to the theoretical mass which should have been drawn in during that same period of time, based upon the total piston displacement of the engine, and the temperature and pressure of the surrounding atmosphere.

The above definition is applicable only to the naturally aspirated engine. In the case of the supercharged engine, however, the theoretical mass of air should be calculated at the conditions of pressure and temperature prevailing in the intake manifold.

The volumetric efficiency is affected by many variables, some of the important ones are:

(i) *The density of the fresh charge :* As the fresh charge arrives in the hot cylinder, heat is transferred to it from the hot chamber walls and the hot residual exhaust gases, raising its temperature. This results in a decrease in the mass of fresh charge admitted and a reduction in volumetric efficiency. The volumetric efficiency is increased by low temperatures (provided there are no heat transfer effects) and high pressure of the fresh charge, since density is thereby increased, and more mass of charge can be inducted into a given volume.

(ii) *The exhaust gas in the clearance volume :* As the piston moves from TDC to BDC on the intake stroke, these products tend to expand and occupy a portion of the piston displacement greater than the clearance volume, thus reducing the space available to the incoming charge. In addition, these exhaust products tend to raise the temperature of the fresh charge, thereby decreasing its density and further reducing volumetric efficiency.

(iii) *The design of the intake and exhaust manifolds :* The exhaust manifold should be so designed as to enable the exhaust products to escape readily, while the intake manifold should be designed to as to bring in the maximum possible fresh charge. This implies minimum restriction is offered to the fresh charge flowing into the cylinder, as well as to the exhaust products being forced out.

(iv) *The timing of the intake and exhaust valves :* Valve timing is the regulation of the points in the cycle at which the valves are set to open and close. Since, the valves require a finite period of time to open or close for smooth operation, a slight "lead" time is necessary for proper opening and closing. The design of the valve operating cam provides for the smooth transition from one position to the other, while the cam setting determines the timing of the valve.

The effect of the *intake valve* timing on the engine air capacity is indicated by its effect on the air in'ducted per cylinder per cycle, i.e., the mass of air taken into one cylinder during one suction stroke. Figure 5.10 shows representative intake valve timing for both a low speed and high speed SI engine. In order to understand the effect of the intake valve timing on the charge inducted per cylinder per cycle, it is desirable to follow through the intake process, referring to the Fig.5.10.

While the intake valve should open, theoretically, at TDC in almost all SI engines utilize an intake valve opens a few degrees before TDC on the exhaust stroke. This is to ensure that the valve will be fully open and the fresh charge starts to flow into the cylinder as soon as the piston reaches TDC. In Fig.5.10, the intake valve starts to open 10° before TDC. It may be noted from Fig.5.10 that for a low speed engine, the intake valve closes 10° after BDC, and for a high speed engine, 60° after BDC.

As the piston descends on the intake stroke, the fresh charge is drawn in through the intake port and valve. When the piston reaches BDC and starts to ascend on the compression stroke, the inertia of the incoming fresh charge tends to cause it to continue to move into the cylinder. At low engine speeds, the charge is moving into the cylinder relatively slowly, and its inertia is relatively low. If the intake valve were to remain open much beyond BDC, the up-moving piston on the compression stroke would tend to force some of the charge, already in the cylinder back into the intake manifold, with consequent reduction in volumetric efficiency. Hence, the intake valve is closed relatively early after BDC for a slow speed engine. High speed engines, however, bring the charge in through the intake manifold at greater speeds, and the charge has greater inertia. As the piston moves up on the compression stroke, there is a "ram" effect produced by the incoming mixture which tends to pack more charge into the cylinder. In the high speed engine, therefore, the intake valve closing is delayed for a greater period of time after BDC in order to take advantage of this "ram" and induct the maximum quantity of charge.

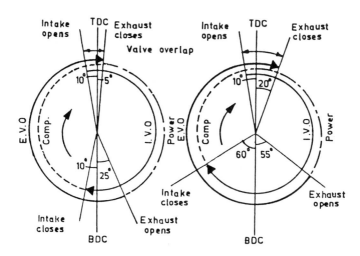

(a) Low Speed Engine   (b) High Speed Engine

Fig.5.10 Valve Timing Diagram of Four-Stroke Engines

For either a low speed or a high speed engine operating in its range of speeds, there is some point at which the charge per cylinder per cycle becomes a maximum, for a particular valve setting. If the revolutions of the low speed engine are increased beyond this point, the intake valve in effect close too soon, and the charge per cylinder per cycle is reduced. If the revolutions of the high speed engine are increased beyond this maximum, the flow may be chocked due to fluid friction. These losses can become greater than the benefit of the ram, and the charge per cylinder per cycle falls off.

The chosen intake valve setting for an engine operating over a range of speeds must necessarily be a compromise between the best setting for the low speed end of the range and the best setting for the high speed end.

The timing of the *exhaust valve* also affects the volumetric efficiency. The exhaust valve usually opens prior to the time when the piston reaches *BDC* on the expansion stroke. This reduces the work done by the expanding gases during the power stroke, but decreases the work necessary to expel the burned products during the exhaust stroke, and results in an overall gain in output.

During the exhaust stroke, the piston forces the burned gases out at high velocity. If the closing of the exhaust valve is delayed beyond *TDC*, the inertia of the exhaust gases tends to scavenge the cylinder better by carrying out a greater mass of the gas left in the clearance volume, and results in increased volumetric efficiency. Consequently, the exhaust valve is often set to close a few degrees after *TDC* on the exhaust stroke, as indicated in Fig.5.10.

It should be noted that it is quite possible for both the intake and exhaust valves to remain open, or partially open, at the same time. This is termed the *valve overlap*. This overlap, of course, must not be excessive enough to allow the burned gases to be sucked into the intake manifold, or the fresh charge to escape through the exhaust valve.

The reasons for the necessity of valve overlap and valve timings other than at *TDC* or *BDC*, has been explained above, taking into consideration only the dynamic effects of gas flow. One must realize, however, that the presence of a mechanical problem in actuating the valves has an influence in the timing of the valves.

The valve cannot be lifted instantaneously to a desired height, but must be opened gradually due to the problem of acceleration involved. If the sudden change in acceleration from positive to negative values are encountered in design of a cam. The cam follower may lose the contact with the cam and then be forced back to close contact by the valve spring, resulting in a blow against the cam. This type of action must be avoided and, hence, cam contours are so designed as to produce gradual and smooth changes in directional acceleration. As a result, the opening of the valve must commence ahead of the time at which it is fully opened. The same reasoning applies for the closing time. It can be seen, therefore, that the timing of valves depends on dynamic and mechanical considerations.

Both the intake and exhaust valves are usually timed to give the most satisfactory results for the average operating conditions of the particular engine, and the settings are determined on the prototype of the actual engine.

5.6 LOSS DUE TO RUBBING FRICTION

These losses are due to friction between the piston and the cylinder walls, friction in various bearings and also the energy spent in operating the auxiliary equipment such as cooling water pump, ignition system, fan, etc. The piston ring friction increases rapidly with engine speed. It also increases to a small extent with increase in mean effective pressure. The bearing friction and the auxiliary friction also increase with engine speed.

The efficiency of an engine is maximum at full load and decreases at part loads. It is because the percentage of direct heat loss, pumping loss and rubbing friction loss increase at part loads. The approximate losses for a gasoline engine of high compression ratio, say 8:1 using a chemically correct mixture are given in Table 5.2, as percentage of fuel energy input.

5.7 ACTUAL AND FUEL-AIR CYCLES OF CI ENGINES

In the diesel cycle the losses are less than in the Otto cycle. The main loss is due to incomplete combustion and is the cause of main difference between fuel-air cycle and actual cycle of a diesel engine. This is shown in Fig.5.11. In a fuel-air cycle the combustion is supposed to be completed at the end of the constant pressure burning whereas in actual practice *after burning* continues up to half of the expansion stroke. The ratio between the actual efficiency and the fuel-air cycle efficiency is about 0.85 in the diesel engines.

In fuel-air cycles, when allowance is made for the presence of fuel and combustion products, there is reduction in cycle efficiency. In actual cycles, allowances are also made for the losses due to phenomena such as heat transfer and finite combustion time. This reduces the cycle efficiency further. For complete analysis of actual cycles, computer models are being developed nowadays.

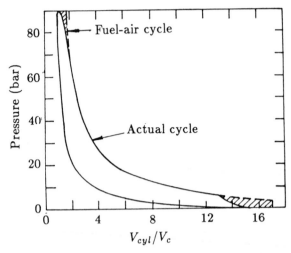

Fig.5.11 Actual Diesel Cycle Vs Equivalent Fuel Combustion
Limited Pressure Cycle for Two-Stroke Diesel Engine

Table 5.2 *Typical Losses in a Gasoline Engine for r = 8*

Sl. No.	Item	At load	
		Full load	Half load
(a)	Air-standard cycle efficiency ($\eta_{air\text{-}std}$)	56.5	56.5
1.	Losses due to variation of specific heat and chemical equilibrium, %	13.0	13.0
2.	Loss due to progressive combustion, %	4.0	4.0
3.	Loss due to incomplete combustion, %	3.0	3.0
4.	Direct heat loss, %	4.0	5.0
5.	Exhaust blowdown loss, %	0.5	0.5
6.	Pumping loss, %	0.5	1.5
7.	Rubbing friction loss, %	3.0	6.0
(b)	Fuel-air cycle efficiency $= \eta_{air\text{-}std} - (1)$	43.5	43.5
(c)	Gross indicated thermal efficiency (η_{th}) $=$ Fuel-air cycle efficiency (η_{ith}) $-(2 + 3 + 4 + 5)$	32.0	31.0
(d)	Actual brake thermal efficiency $= \eta_{ith} - (6 + 7)$	28.5	23.5

Questions :-

5.1 Why the actual cycle efficiency is much lower than the air-standard cycle efficiency? List the major losses in an actual engine.

5.2 List the main differences between actual cycles and air-standard cycles.

5.3 What are the three principal factors that influence engine performance?

5.4 Briefly explain the following:
   (i) time loss factor
   (ii) heat loss factor
   (iii) exhaust blowdown factor

5.5 Compare the actual and fuel-air cycles of a gasoline engine.

5.6 How does the composition of exhaust gases vary for various fuel-air ratios in a gasoline engine?

5.7 Discuss the effect of spark advance on the performance of an Otto cycle engine. What is meant by the optimum spark advance?

5.8 Discuss the optimum opening position of exhaust valve to reduce the exhaust blowdown loss.

5.9 What is meant by pumping loss? Discuss its effect on the engine performance?

5.10 Discuss briefly the loss due to gas exchange process.

5.11 Define volumetric efficiency and discuss the effect of various factors affecting the volumetric efficiency.

5.12 Briefly discuss the rubbing friction losses.

6

FUELS

6.1 INTRODUCTION

The study of fuels for IC engines has been carried out ever since these engines came into existence. The engine converts heat energy which is obtained from the chemical combination of the fuel with the oxygen, into mechanical energy. Since the heat energy is derived from the fuel, a fundamental knowledge of types of fuels and their characteristics is essential in order to understand the combustion phenomenon. The characteristics of the fuel used have considerable influence on the design, efficiency, output and particularly, the reliability and durability of the engine. Further, the fuel characteristics play a vital role in the atmospheric pollution caused by the engines used in automobiles.

6.2 FUELS

Internal combustion engines can be operated on different types of fuels such as liquid, gaseous and even solid fuels. Depending upon the type of fuel to be used the engine has to be designed accordingly.

6.2.1 Solid Fuels

The solid fuels find little practical application at present because of the problems in handling the fuel as well as in disposing off, the solid residue or ash after combustion. However, in the initial stages of the engine development, solid fuels such as finely powdered coal was attempted. Compared to gaseous and liquid fuels, solid fuels are quite difficult to handle and storage and feeding are quite cumbersome. Because of the complications in the design of the fuel feed systems these fuels have become unsuitable in solid form. Attempts are being made to generate gaseous or liquid fuels from charcoal for use in IC engines.

6.2.2 Gaseous Fuels

Gaseous fuels are ideal fuels and pose very few problems in using them in internal combustion engines. Being gaseous, they mix more homogeneously with air and eliminate the distribution and starting problems that are encountered with liquid fuels. Even though the gaseous fuels are the most ideal for internal combustion engines, storage and handling problems restrict their use in automobiles. Consequently, they are commonly used for stationary power plants located near the source of availability of the fuel. Some of the gaseous fuels can be liquefied under pressure for reducing the storage volume but this arrangement is very expensive as well as risky. Because of the energy crisis in the recent years considerable research efforts are being made to improve the design and performance of gas engines which became obsolete when liquid fuels began to be used.

6.2.3 Liquid Fuels

In most of the modern internal combustion engines, liquid fuels which are the derivatives of liquid petroleum are being used. The three principal commercial types of liquid fuels are benzyl, alcohol and petroleum products. However, petroleum products form the main fuels for internal combustion engines today.

6.3 CHEMICAL STRUCTURE OF PETROLEUM

Petroleum as obtained from the oil wells, is predominantly a mixture of many hydrocarbons with differing molecular structure. It also contains small amounts of sulphur, oxygen, nitrogen and impurities such as water and sand. The carbon and hydrogen atoms may be linked in different ways in a hydrocarbon molecule and this linking influences the chemical and physical properties of different hydrocarbon groups. Most petroleum fuels tend to exhibit the characteristics of that type of hydrocarbon which forms a major component of the fuel.

The carbon and hydrogen combine in different proportions and molecular structures to form a variety of hydrocarbons. The carbon to hydrogen ratio which is one of the important parameters and their nature of bonding determine the energy characteristics of the hydrocarbon fuels. Depending upon the number of carbon and hydrogen atoms the petroleum products are classified into different groups.

The differences in physical and chemical properties between the different hydrocarbon types depend on their chemical composition and affect mainly the combustion processes and hence, the proportion of fuel and air required in the engine. The basic families of hydrocarbons, their general formulae and their molecular arrangement are shown in Table 6.1.

Table 6.1 *Basic Families of Hydrocarbons*

Family of hydrocarbons	General formula	Molecular structure	Saturated/ Unsaturated	Stability
Paraffin	C_nH_{2n+2}	Chain	Saturated	Stable
Olefin	C_nH_{2n}	Chain	Unsaturated	Unstable
Naphthene	C_nH_{2n}	Ring	Saturated	Stable
Aromatic	C_nH_{2n-6}	Ring	Highly unsaturated	Most unstable

6.3.1 Paraffin Series

The normal paraffin hydrocarbons are of straight chain molecular structure. They are represented by a general chemical formula, C_nH_{2n+2}. The molecular structures of the first few members of the paraffin of hydrocarbons are shown below.

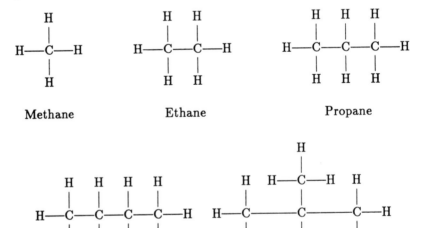

Methane   Ethane   Propane

Butane   Isobutane

In these hydrocarbons the valency of all the carbon atoms is fully utilized by single bonds with hydrogen atoms. Therefore, the paraffin hydrocarbons are saturated compounds and are characteristically very stable.

A variation of the paraffin family consists of an open chain structure with an attached branch and is usually termed a branched chain paraffin. The hydrocarbons which have the same chemical formulae but different structural formulae are known as isomers.

Isobutane shown above has the same general chemical formula and molecular weight as butane but a different molecular structure and physical characteristics. It is called an isomer of butane and is known as isobutane. Isoparaffins are also stable compounds.

6.3.2 Olefin Series

Olefins are also straight chain compounds similar to paraffins but are unsaturated because they contain one or more double bonds between carbon atoms. Their chemical formula is $C_n H_{2n}$. Mono-olefins have one double bond whereas diolefin have two in their structure.

$$H-\underset{\underset{H}{|}}{\overset{\overset{H}{|}}{C}}-\underset{\underset{H}{|}}{\overset{\overset{H}{|}}{C}}-\underset{\underset{H}{|}}{\overset{\overset{H}{|}}{C}}-\underset{\underset{H}{|}}{\overset{\overset{H}{|}}{C}}-\overset{\overset{H}{|}}{C}=\overset{\overset{H}{|}}{C}-H$$

Hexene

$$H-\overset{\overset{H}{|}}{C}=\overset{\overset{H}{|}}{C}-\overset{\overset{H}{|}}{C}=\overset{\overset{H}{|}}{C}-H$$

Butadiene

Olefins are not as stable as the single bond paraffins because of the presence of the double bonds in their structure. Consequently, these are readily oxidized in storage to form gummy deposits. Hence, olefin content in certain petroleum products is kept low by specification.

6.3.3 Naphthene Series

The naphthenes have the same chemical formula as the olefin series of hydrocarbons but have a ring structure and therefore, often they are called as cyclo-paraffins. They are saturated and tend to be stable. The naphthenes are saturated compounds whereas olefins are unsaturated. Cyclopentane is one of the compounds in the naphthene series $(C_n H_{2n})$.

Cyclopentane

6.3.4 Aromatic Series

Aromatic compounds are ring structured having a benzene molecule as their central structure and have a general chemical formula C_nH_{2n-6}. Though the presence of double bonds indicate that they are unsaturated, a peculiar nature of these double bonds causes them to be more stable than the other unsaturated compounds. Various aromatic compounds are formed by replacing one or more of the hydrogen atoms of the benzene molecules with an organic radical such as paraffins, naphthenes and olefins. By adding a methyl group (CH_3), benzene is converted to toluene $(C_6H_5CH_3)$, the base for the preparation of Trinitrotoluene (TNT) which is a highly explosive compound.

Benzene   Toluene

The above families of hydrocarbons exhibit some general characteristics due to their molecular structure which are summarized below

(i) Normal paraffins exhibit the poorest antiknock quality when used in an SI engine. But the antiknock quality improves with the increasing number of carbon atoms and the compactness of the molecular structure. The aromatics offer the best resistance to knocking in SI Engines.

(ii) For CI engines, the order is reversed i.e., the normal paraffins are the best fuels and aromatics are the least desirable.

(iii) As the number of atoms in the molecular structure increases, the boiling temperature increases. Thus fuels with fewer atoms in the molecule tend to be more volatile.

(iv) The heating value generally increases as the proportion of hydrogen atoms to carbon atoms in the molecule increases due to the higher heating value of hydrogen than carbon. Thus, paraffins have the highest heating value and the aromatics the least.

6.4 PETROLEUM REFINING PROCESS

Crude petroleum, as obtained from the oil wells contains gases (mainly methane and ethane) and certain impurities such as water, solids etc. The crude oil is separated into gasoline, kerosene, fuel oil etc. by the process of fractional distillation. This process is based on the fact that the boiling points of various hydrocarbons increase with increase in molecular weight.

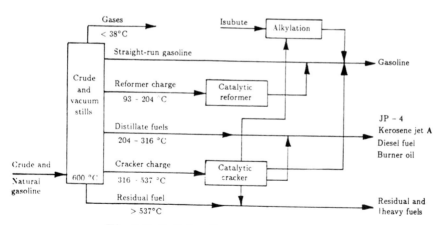

Fig.6.1 Refining Process of Petroleum

Figure 6.1 shows a simple diagram of the refining process of crude petroleum (the diagram does not show the facilities for process utilization, desulphurization etc.). In the first step, the petroleum is passed through a separator in which the gases are removed and a product known as natural gasoline is obtained. The liquid petroleum is then vapourized in a still, at temperatures of 600 °C and the vapour is admitted at the bottom of the fractionating tower. The vapour is forced to pass upwards along a labyrinth-like arrangement of plates which direct the vapour through trays of liquid fuel maintained at different temperatures. The compounds with higher boiling points get condensed out at lower levels while those with lower boiling points move up to higher levels where they get condensed in trays at appropriate temperature. Generally the top fraction is called the straight-run gasoline and the other fractions, kerosene, diesel oil, fuel oils etc. are obtained in the increasing range of boiling temperatures. Details are shown in Fig.6.1. The important products of the refining process are given in Table 6.2.

Many processes can be used to convert some of these fractions to compounds for which there is a greater demand. Some of the main refinery processes are as follows :

(i) *Cracking* consists of breaking down large and complex hydrocarbon molecules into simpler compounds with lower boiling points. Thermal cracking subjects the large hydrocarbon molecules to high temperature and pressure and they are decomposed into smaller, lower boiling point molecules. Catalytic cracking using catalysts is done at a relatively lower pressure and temperature than the thermal cracking. Due to catalysis, the naphthenes are cracked to olefins and paraffins and olefins to isoparaffins, thus forming the components needed for gasoline. Catalytic cracking gives better antiknock property for gasoline as compared to thermal cracking.

(ii) *Hydrogenation* consists of the addition of hydrogen atoms to certain hydrocarbons under high pressure and temperature to produce more desirable compounds. It is often used to convert unstable compounds to stable ones.

(iii) *Polymerization* is the process of converting olefins, the unsaturated products of cracking, into heavier and stable compounds.

(iv) *Alkylation* combines an olefin with an isoparaffin to produce a branched chain isoparaffin in the presence of a catalyst.

Example : isobutylene + isobutane $\xleftarrow{\text{Alkylation}}$ iso-octane

(v) *Isomerization* changes the relative position of the atoms within the molecule of a hydrocarbon without changing its molecular formula. For example, Isomerization is used for the conversion of n-butane into iso-butane for alkylation. Conversion of n-pentane and n-hexane into isoparaffins to improve knock rating of highly volatile gasoline is another example.

Table 6.2 *Products of Petroleum Refining Process*

S.No.	Fraction	App. boiling range, °C	Remarks
1.	Fuel gas	-160 to -44	Methane, ethane and some propane used as refinery fuel
2.	Propane	-40	L.P.G.
3.	Butane	-12 to 30	Blended with motor gasoline to increase its volatility
4.	Light Naphtha	0 to 150	Motor gasoline for catalytic reforming
5.	Heavy Naphtha	150 to 200	Catalytic reforming fuel, blended with light gas oil to form jet fuels
6.	Kerosene – Middle distillate	200 to 300	Domestic, aviation fuels
7.	Light gas oil – Middle distillate	200 to 315	Furnace fuel oil, diesel fuels
8.	Heavy gas oil	315 to 425	Feed for catalytic cracking
9.	Vacuum gas oil	425 to 600	Feed for catalytic cracking
10.	Pitch	>600	Heavy fuel oil, asphalts

(vi) *Cyclization* joins together the ends of a straight chain molecule to form a ring compound of the naphthene family.

(vii) *Aromatization* is similar to cyclization, the exception being that the product is an aromatic compound.

(viii) *Reformation* is a type of cracking process which is used to convert the low antiknock quality stocks into gasolines of higher octane rating (see section 6.6). It does not increase the total gasoline volume.

(ix) *Blending* is a process of obtaining a product of desired quality by mixing certain products in some suitable proportion.

6.5 IMPORTANT QUALITIES OF ENGINE FUELS

Fuels used in IC engines should possess certain basic qualities which are important for the smooth running of the engines. In this section, the important qualities of fuels for both SI and CI engines are reviewed.

6.5.1 SI Engine Fuels

Gasoline which is mostly used in the present day SI engines is usually a blend of several low boiling paraffins, naphthenes and aromatics in varying proportions. Some of the important qualities of gasoline are discussed below.

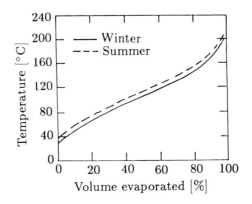

Fig.6.2 Typical Distillation Curves of Gasoline

(i) *Volatility :* Volatility is one of the main characteristic properties of gasoline which determines its suitability for use in an SI engine. Since gasoline is a mixture of different hydrocarbons, volatility depends on the fractional composition of the fuel. The usual practice of measuring the fuel volatility is the distillation of the fuel in a special device at atmospheric pressure and in the presence of its own vapour. The fraction that boils off at a definite temperature is measured. The characteristic points are the temperatures at which 10, 40, 50 and 90% of the volume evaporates as well as the temperature at which boiling of the fuel terminates. Figure 6.2 shows the fractional distillation curve of gasoline for both winter and summer grade gasoline. The method for measuring volatility has been standardized by the American Society for Testing Materials (ASTM) and the graphical representation of the result of the tests is generally referred to as the ASTM distillation curve. The more important aspects of volatility related to engine fuels are discussed in detail in conjunction with the distillation curve.

(a) *Starting and Warm up :* A certain part of the gasoline should vapourize at the room temperature for easy starting of the engine. Hence, the portion of the distillation curve between about 0 and 10% boiled off have relatively low boiling temperatures. As the engine warms up, the temperature will gradually increase to the operating temperature. Low distillation temperatures are desirable throughout the range of the distillation curve for best warm-up.

(b) *Operating Range Performance :* In order to obtain good vapourization of the gasoline, low distillation temperatures are preferable in the engine operating range. Better vapourization tends to produce both more uniform distribution of fuel to the cylinders as well as better acceleration characteristics by reducing the quantity of liquid droplets in the intake manifold.

(c) *Crankcase Dilution :* Liquid fuel in the cylinder causes loss of lubricating oil (by washing away oil from cylinder walls) which deteriorates the quality of lubrication and tends to cause damage to the engine through increased friction. The liquid gasoline may also dilute the lubricating oil and weaken the oil film between rubbing surfaces. To prevent these possibilities, the upper portion of the distillation curve should exhibit sufficiently low distillation temperatures to insure that all gasoline in the cylinder is vapourized by the time of combustion.

(d) *Vapour Lock Characteristics :* High rate of vapourization of gasoline can upset the carburettor metering or even stop the fuel flow to the engine by setting up a vapour lock in the fuel passages. This characteristic, demands the presence of relatively high boiling temperature hydrocarbons throughout the distillation range. Since this requirement is not consistent with the other requirements desired in (a), (b) and (c), a compromise must be made for the desired distillation temperatures.

(ii) *Antiknock Quality :* Abnormal burning or detonation in an SI engine combustion chamber causes a very high rate of energy release, excessive temperature and pressure inside the cylinder and

adversely affects its thermal efficiency. Therefore, the characteristics of the fuel used should be such that it resists the tendency
to produce detonation and this property is called its antiknock
property. The antiknock property of a fuel depends on the self-
ignition characteristics of its mixture and vary largely with the
chemical composition and molecular structure of the fuel. In
general, the best SI engine fuel will be that having the highest
antiknock property, since this permits the use of higher compression ratios and thus the engine thermal efficiency and the power
output can be greatly increased.

(iii) *Gum Deposits :* Reactive hydrocarbons and impurities in the
fuel have a tendency to oxidize upon storage and form liquid and
solid gummy substances. The gasoline containing hydrocarbons
of the paraffin, naphthene and aromatic families forms little gum
while cracked gasoline containing unsaturated hydrocarbons is
the worst offender. A gasoline with high gum content will cause
operating difficulties such as sticking valves and piston rings carbon deposits in the engine, gum deposits in the manifold, clog-
ging of carburettor jets and enlarging of the valve stems, cylinders
and pistons. The amount of gum increases with increased concentrations of oxygen, with rise in temperature, with exposure to
sunlight and also on contact with metals. Gasoline specifications
therefore limit both the gum content of the fuel and its tendency
to form gum during storage.

(iv) *Sulphur Content :* Hydrocarbon fuels may contain free sulphur,
hydrogen sulphide and other sulphur compounds which are objectionable for several reasons. The sulphur is a corrosive element
of the fuel that can corrode fuel lines, carburettors and injection pumps and it will unite with oxygen to form sulphur dioxide
that, in the presence of water at low temperatures, may form
sulphurous acid. Since sulphur has a low ignition temperature,
the presence of sulphur can reduce the self-ignition temperature,
then promoting knock in the SI engine.

6.5.2 CI Engine Fuels

(i) *Knock Characteristics :* Knock in the CI engine occurs because of
an ignition lag in the combustion of the fuel between the time of
injection and the time of actual burning. As the ignition lag
increases, the amount of fuel accumulated in the combustion
chamber increases and when combustion actually takes place,
abnormal amount of energy is suddenly released causing an excessive rate of pressure rise which results in an audible knock.
Hence, a good CI engine fuel should have a short ignition lag and
will ignite more readily. Furthermore, ignition lag affects the

starting, warm up, and leads to the production of exhaust smoke in CI engines. The present day measure in the cetane rating, the best fuel in general, will have a cetane rating sufficiently high to avoid objectionable knock.

(ii) *Volatility* : The fuel should be sufficiently volatile in the operating range of temperature to produce good mixing and combustion. Figure 6.3 is a representative distillation curve of a typical diesel fuel.

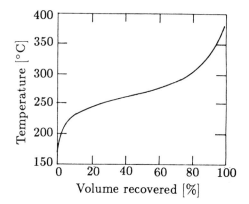

Fig.6.3 Typical Distillation Curve for Diesel

(iii) *Starting Characteristics* : The fuel should help in starting th engine easily. This requirement demands high enough volatility to form a combustible mixture readily and a high cetane rating in order that the self-ignition temperature is low.

(iv) *Smoking and Odour* : The fuel should not promote either smoke or odour in the engine exhaust. Generally, good volatility is the first prerequisite to ensure good mixing and therefore complete combustion.

(v) *Viscosity* : CI engine fuels should be able to flow through the fuel system and the strainers under the lowest operating temperatures to which the engine is subjected to.

(vi) *Corrosion and Wear* : The fuel should not cause corrosion and wear of the engine components before or after combustion. These requirements are directly related to the presence of sulphur, ash and residue in the fuel.

(vii) *Handling Ease* : The fuel should be a liquid that will readily flow under all conditions that are encountered in actual use. This requirement is measured by the pour point and the viscosity of the fuel. The fuel should also have a high flash point and a high fire point.

6.6 RATING OF FUELS

Normally fuels are rated for their antiknock qualities. The rating of fuels is done by defining two parameters called Octane number and Cetane number for gasoline and diesel oil respectively. The rating of fuels, both for SI and CI engines is discussed in this section.

6.6.1 Rating of SI Engine Fuels

Resistance to knocking is an extremely important characteristic of fuel for spark-ignition engines. These fuels differ widely in their ability to resist knock depending on their chemical composition. A satisfactory rating method for comparing the antiknock qualities of the various fuels has been established. In addition to the chemical characteristics of hydrocarbons in the fuel, other operating parameters such as fuel-air ratio, ignition timing, dilution, engine speed, shape of the combustion chamber, ambient conditions, compression ratio etc. affect the tendency to knock in the engine cylinder. Therefore, in order to determine the knock resistance characteristic of the fuel, the engine and its operating variables must be fixed at standard values.

According to a standard practice, the antiknock value of an SI engine fuel is determined by comparing its antiknock property with a mixture of two reference fuels, iso-octane (C_8H_{18}) and normal heptane (C_7H_6). Iso-octane chemically being a very good antiknock fuel, is arbitrarily assigned a rating of 100 octane number. Normal heptane (C_7H_{16}), on the other hand, has very poor antiknock qualities and is given a rating of 0 octane number. *The Octane number fuel is defined as the percentage, by volume, of iso-octane in a mixture of iso-octane and normal heptane, which exactly matches the knocking intensity of the fuel in a standard engine under a set of standard operating conditions.*

The addition of certain compounds (e.g. tetraethyl lead) to iso-octane produces fuels of greater antiknock quality (above 100 octane number). The antiknock effectiveness of tetraethyl lead, for the same quantity of lead added, decreases as the total content of lead in the fuel increases. Furthermore, each octane number at the higher range of the octane scale will produce greater antiknock effect compared to the same unit at the lower end of the scale. For instance, octane number increase from 92 to 93 produce greater antiknock effect than a similar increase from 32 to 33 octane number. Because of this non-linear variation, a new scale was derived which expresses the approximate relative engine performance and the units of this scale are known as the Performance Numbers, PN. Octane numbers, ON above 100 can be computed by

$$ON(>100) \quad = \quad 100 + \frac{28.28A}{1.0 + 0.736A + \sqrt{1.0 + 0.736A - 0.035216A^2}}$$

where A is TEL in ml/gal of fuel,
or from the performance number, PN,

$$\text{Octane Number} \quad = \quad 100 + \frac{PN - 100}{3}$$

Laboratory Method : The engine is run at specified conditions with a definite compression ratio and a definite blend of reference fuels. The intensity of knock at these standard conditions is called standard knock. The knock meter is adjusted to give a particular reading under these conditions. The test fuel is now used in the engine and air-fuel ratio is adjusted to give maximum knock intensity. The compression ratio of the engine is gradually changed until the knock meter reading is the same as in the previous run (standard knock). The compression ratio is now fixed and known blends of reference fuels are used in the engine. The blend of reference fuels which gives a knock meter reading equal to the standard value will match the knocking characteristics of the test fuel. Percentage by volume of iso-octane in the particular blend gives the octane number.

6.6.2 Rating of CI Engine Fuels

In compression-ignition engines, the knock resistance depends on chemical characteristics as well as on the operating and design conditions of the engine. Therefore, the knock rating of a diesel fuel is found by comparing the fuel under prescribed conditions of operation in a special engine with primary reference fuels. The reference fuels are normal cetane, $C_{16}H_{34}$, which is arbitrarily assigned a cetane number of 100 and alpha methyl naphthalene, $C_{11}H_{10}$, with an assigned cetane number of 0. *Cetane number of a fuel is defined as the percentage by volume of normal cetane in a mixture of normal cetane and α-methyl naphthalene which has the same ignition characteristics (ignition delay) as the test fuel when combustion is carried out in a standard engine under specified operating conditions.* Since ignition delay is the primary factor in controlling the initial autoignition in the CI engine, it is reasonable to conclude that knock should be directly related to the ignition delay of the fuel. Knock resistance property of diesel oil can be improved by adding small quantities of compounds like amyl nitrate, ethyl nitrate or ether.

Table 6.3 *Conditions for Ignition Quality Test on Diesel Fuels*

Engine speed	900 rpm
Jacket water temperature	100 °C
Inlet air temperature	65.5 °C
Injection advance	Constant at 13°*bTDC*
Ignition delay	13°

Laboratory Method : The test is carried out in a standard single cylinder engine like the CFR diesel engine or Ricardo single cylinder variable compression engine under the conditions shown in Table 6.3.

The test fuel is first used in the engine operating at the specified conditions. The fuel pump delivery is adjusted to give a particular fuel-air ratio. The injection timing is also adjusted to give an injection advance of 13 degrees. By varying the compression ratio the ignition delay can be increased or decreased until a position is found where combustion begins at *TDC*. When this position is found, the test fuel undergoes a 13 degree ignition delay. The cetane number of the unknown fuel can be estimated by noting the compression ratio for 13 degree delay and then referring to a prepared chart showing the relationship between cetane number and compression ratio. However, for accuracy two reference fuel blends differing by not more than 5 cetane numbers are selected to bracket the unknown sample. The compression ratio is varied for each reference blend to reach the standard ignition delay (13 degrees) and, by interpolation of the compression ratios, the cetane rating of the unknown fuel is determined.

Questions :-

6.1 What are the different kinds of fuels used in an IC engine?

6.2 Briefly explain the chemical structure of petroleum.

6.3 Give the general chemical formula of the following fuels:

(i) paraffin (ii) olefin (iii) diolefin (iv) naphthene (v) aromatic

Also state their molecular arrangements and mention whether they are saturated or unsaturated.

6.4 Briefly explain the petroleum refining process.

6.5 Discuss the significance of distillation curves.

6.6 Discuss the important qualities of an SI engine fuel.

6.7 Describe the important qualities of a CI engine fuel.

6.8 What is the effect of high sulphur content on the performance of SI and CI engines?

6.9 How are SI engine fuels rated?

6.10 Briefly describe the rating of CI engine fuels.

7

CARBURETION

7.1 INTRODUCTION

Spark-ignition engines normally use volatile liquid fuels. Preparation of fuel and air mixture is done outside the engine cylinder and formation of a homogeneous mixture is normally not completed in the inlet manifold. Fuel droplets which remain in suspension continue to evaporate and mix with air even during suction and compression processes. The process of mixture preparation is very important for spark-ignition engines. The purpose of carburetion is to provide a combustible mixture of fuel and air in the required quantity and quality for efficient operation of the engine under all conditions.

7.2 DEFINITION OF CARBURETION

The process of formation of a combustible fuel-air mixture by mixing the proper amount of fuel with air before admission to engine cylinder is called carburetion and the device which does this job is called a carburettor.

7.3 FACTORS AFFECTING CARBURETION

Of the various factors, the process of carburetion is influenced by

(i) the engine speed
(ii) the vapourization characteristics of the fuel
(iii) the temperature of the incoming air and
(iv) the design of the carburettor

Since modern engines are of high speed type, the time available for mixture formation is very limited. For example, an engine running at 3000 rpm has only about 10 milliseconds (ms) for mixture induction during intake stroke. When the speed becomes 6000 rpm the time available is 5 ms only. Therefore, in order to have high quality carburetion (that is mixture with high vapour content) the velocity of the air stream at the point where the fuel is injected has to be increased. This is achieved by introducing a venturi section in the path of the air and the fuel is discharged from the jet of the carburettor at the minimum cross section of the venturi (called throat).

Other factors which ensure high quality carburetion within a short period are the presence of highly volatile hydrocarbons in the fuel. Therefore, suitable evaporation characteristics of the fuel, indicated by its distillation curve, are necessary for efficient carburetion especially at high engine speeds.

The temperature and pressure of surrounding air has a large influence on efficient carburetion. Higher atmospheric air temperature increases the vaporization of fuel (percentage of fuel vapour increases with increase in mixture temperature) and produces a more homogeneous mixture. An increase in atmospheric temperature, however, leads to a decrease in power output of the engine when the air-fuel ratio is constant due to reduced mass flow into the cylinder or, in other words, reduced volumetric efficiency.

The design of the carburettor, the intake system and the combustion chamber have considerable influence on uniform distribution of mixture to the various cylinders of the engine. Proper design of carburettor elements alone ensures the supply of desired composition of the mixture under different operating conditions of the engine.

7.4 AIR–FUEL MIXTURES

An engine is generally operated at different loads and speeds. For this, proper air-fuel mixture should be supplied to the engine cylinder. Fuel and air are mixed to form three different types of mixtures.

(i) chemically correct mixture

(ii) rich mixture and

(iii) lean mixture

Chemically correct or stoichiometric mixture is one in which there is just enough air for complete combustion of the fuel. For example, to burn one kg of octane (C_8H_{18}) completely 15.12 kg of air is required. Hence chemically correct A/F ratio for C_8H_{18} is 15.12:1; usually approximated to 15:1. This chemically correct mixture will vary only slightly in numerical value between different hydrocarbon fuels. It is always computed from the chemical equation for complete combustion for a particular fuel.

A mixture which contains less air than the stoichiometric requirement is called a rich mixture (example, A/F ratio of 12:1, 10:1 etc.).

A mixture which contains more air than the stoichiometric requirement is called a lean mixture (example, A/F ratio of 17:1, 20:1 etc.).

There is, however, a limited range of A/F ratios in a homogeneous mixture, only within which combustion in an SI engine will occur. Outside this range, the ratio is either too rich or too lean to sustain flame propagation. This range of useful A/F ratio runs from approximately 19:1 (lean) to 9:1 (rich) as indicated in Fig.7.1.

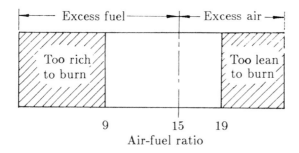

Fig.7.1 Useful Air-Fuel Mixture Range of Gasoline

The carburettor should provide an A/F ratio in accordance with engine operating requirements and this ratio must be within the combustible range.

7.5 MIXTURE REQUIREMENTS AT DIFFERENT LOADS AND SPEEDS

The air-fuel ratio at which an engine operates has a considerable influence on its performance. Consider an engine operating at full throttle and constant speed with varying A/F ratio. Under these conditions, the A/F ratio will affect both the power output and the brake specific fuel consumption, as indicated by the typical curves shown in Fig.7.2. The mixture corresponding to the maximum output on the curve is called the *best power mixture* with an A/F ratio of approximately 12:1. The mixture corresponding to the minimum point on the *bsfc* curve is called the *best economy mixture*. The A/F ratio is approximately 16:1. It may be noted that the best power mixture is much richer than the chemically correct mixture and the best economy mixture is slightly leaner than the chemically correct.

Figure 7.2 is based on full throttle operation. The A/F ratios for the best power and best economy at part throttle are not strictly the same as at full throttle. If the A/F ratios for best power and best economy are constant over the full range of throttle operation

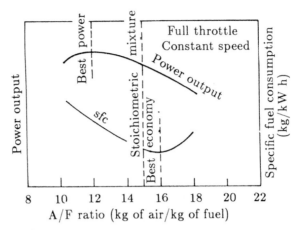

Fig.7.2 Variation of Power Output and Specific Fuel Consumption
with *A/F* Ratio for an SI Engine

and if the influence of other factors is disregarded, the ideal fuel metering device would be merely a two position carburettor. Such a
carburettor could be set for the best power mixture when maximum
performance is desired and for the best economy mixture when the
primary consideration is the fuel economy. These two settings are
indicated in Fig.7.3 by the solid horizontal lines X–X' and Z–Z', respectively. Actual engine requirements, however, again preclude the
use of such a simple and convenient arrangement. These requirements
are discussed in the succeeding section.

Under normal conditions it is desirable to run the engine on the
maximum economy mixture, viz., around 16:1 air-fuel ratio. For quick
acceleration and for maximum power, rich mixture, viz., 12:1 air-fuel
ratio is required.

7.6 ENGINE AIR–FUEL MIXTURE REQUIREMENTS

Actual air-fuel mixture requirements in an operating engine vary considerably from the ideal conditions discussed in the previous section.
For successful operation of the engine, the carburettor has to provide
mixtures which follow the general shape of the curve ABCD in Fig.7.3
which represents a typical engine requirement. The carburettor must
be suitably designed to meet the various engine requirements.

As indicated in Fig.7.3 there are three general ranges of throttle
operation. In each of these, the engine requirements differ. As a
result, the carburettor must be able to supply the required air-fuel
ratio to satisfy these demands. These ranges are:

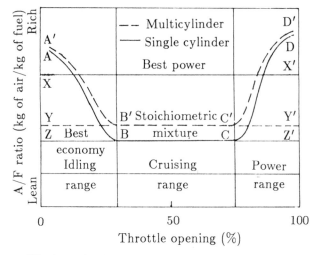

Fig.7.3 Carburettor Performance necessary
to fulfill Engine Requirements

(i) Idling   (mixture must be enriched)
(ii) Cruising   (mixture must be leaned)
(iii) High Power (mixture must be enriched)

7.6.1 Idling Range

An idling engine is one which operates at no load and with nearly
closed throttle. Under idling conditions, the engine requires a rich
mixture, as indicated by point A in Fig.7.3. This is due to the existing pressure conditions within the combustion chamber and the intake
manifold which cause exhaust gas dilution of the fresh charge. The
pressures indicated in Fig.7.4 are representative values which exist
during idling. The exhaust gas pressure at the end of the exhaust
stroke does not vary greatly from the value indicated in Fig.7.4, regardless of the throttle position. Since the clearance volume is constant, the mass of exhaust gas in the cylinder at the end of the exhaust stroke tends to remain fairly constant throughout the throttle
range. The amount of fresh charge brought in during idling, however, is much less than that during full throttle operation, due to the
restricted opening of the throttle. This results in a much larger proportion of exhaust gas being mixed with the fresh charge under idling
conditions. Furthermore, with the nearly closed throttle the pressure
in the intake manifold is considerably below atmospheric due to restriction to the air flow. When the intake valve opens, the pressure

differential between the combustion chamber and the intake manifold results in initial *backward* flow of exhaust gases into the intake manifold. As the piston proceeds down on the intake stroke, these exhaust gases are drawn back into the cylinder, along with the fresh charge. As a result, the final mixture of fuel and air in the combustion chamber is diluted more by exhaust gas. The presence of this exhaust gas tends to obstruct the contact of fuel and air particles — a requirement necessary for combustion. This results in poor combustion and, as a result, in loss of power. It is, therefore, necessary to provide more fuel particles by richening the air-fuel mixture. This richening increases the probability of contact between fuel and air particles and thus improves combustion.

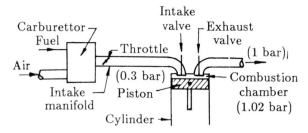

Fig.7.4 Schematic Diagram of Combustion Chamber and Induction System at the Start of Intake Stroke

As the throttle is gradually opened from A to B, (Fig.7.3), the pressure differential between the inlet manifold and the cylinder becomes smaller and the exhaust gas dilution of the fresh charge diminishes. Mixture requirements then proceed along line AB (Fig.7.3) to a leaner A/F ratio required for the cruising operation.

7.6.2 Cruising Range

In the cruising range from B to C (Fig.7.3), the exhaust gas dilution problem is relatively insignificant. The primary interest lies in obtaining the maximum fuel economy. Consequently, in this range, it is desirable that the carburettor provide the engine with the best economy mixture.

7.6.3 Power Range

During peak power operation the engine requires a richer mixture, as indicated by the line CD (Fig.7.3), for the following reasons

(i) *To provide best power*: Since high power is desired, it is logical to transfer the economy settings of the cruising range to that mixture which will produce the maximum power, or a setting in the vicinity of the best power mixture, usually in the range of 12:1.

(ii) *To prevent overheating of exhaust valve and the area near it*:
At high power, the increased mass of gas at higher temperatures passing through the cylinder results in the necessity of transferring greater quantities of heat away from critical areas such as those around the exhaust valve. Enriching the mixture reduces the flame temperature and the cylinder temperature, thereby reducing the cooling problem and lessening the tendency to damage exhaust valves at high power. In the cruising range, the mass of charge is smaller and the tendency to burn the exhaust valve is not as high.

In an automobile engine, indication of knocking is available in the form of an audible sound and the operator can make the engine operating conditions less stringent by releasing the throttle or by shifting to a lower gear. Furthermore, automobile engines generally operate well below full power and a complicated and expensive system for enrichment for this purpose is not economically feasible, although some means of richening at high power is usually used. For aircraft engine installations, however, the complication and expense is justified because of the necessity to increase power during take off.

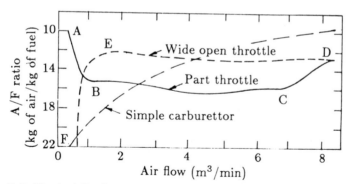

Fig.7.5 Typical Performance Curve of an Automobile Carburettor

Figure 7.3, then, is better representative of a typical engine requirements for the carburettor. Automobile engine requirements are similar in the idling and cruising ranges but tend to be relatively lower or less rich, in the power range (C to D in Fig.7.5). A more representative engine requirement curve for automobiles is shown in Fig.7.5.

The portion of the curve from D to E indicates the requirements after the throttle is wide open and the load is further increased.

7.7 THE WORKING PRINCIPLE OF CARBURETTOR

Both air and gasoline are drawn through the carburettor and into the engine cylinders by the suction created by the downward movement of the piston. This suction is due to an increase in the volume of the cylinder and a consequent decrease in the gas pressure in this chamber. It is the difference in pressure between the atmosphere and cylinder that causes the air to flow into the chamber. In the carburettor, air passing into the combustion chamber picks up fuel discharged from a tube. This tube has a fine orifice called carburettor jet which is exposed to the air path. The rate at which fuel is discharged into the air depends on the pressure difference or pressure head between the float chamber and the throat of the venturi and on the area of the outlet of the tube. In order that the fuel drawn from the nozzle may be thoroughly atomized, the suction effect must be strong and the nozzle outlet comparatively small. In order to produce a strong suction, the pipe in the carburettor carrying air to the engine is made to have a restriction. At this restriction called throat due to increase in velocity of flow, a suction effect is created. The restriction is made in the form of a venturi as shown in Fig.7.6 to minimise throttling losses. The end of the fuel jet is located at the venturi or throat of the carburettor.

A venturi tube is a passage as shown in Fig.7.6. It has a narrower path at the centre so that the flow area through which the air must pass is considerably reduced. As the same amount of air must pass through every point in the tube, its velocity will be greatest at the narrowest point. The smaller the area, the greater will be the velocity of the air, and thereby the suction is proportionately increased.

As mentioned earlier, the opening of the fuel discharge jet is usually located where the suction is maximum. Normally, this is just below the narrowest section of the venturi tube. The spray of gasoline from the nozzle and the air entering through the venturi tube are mixed together in this region and a combustible mixture is formed which passes through the intake manifold into the cylinders. Most of the fuel gets atomized and simultaneously a small part getting will be vapourized. Increased air velocity at the throat of the venturi helps the rate of evaporation of fuel. The difficulty of obtaining a mixture of sufficiently high fuel vapour-air ratio for efficient starting of the engine and for uniform fuel-air ratio in different cylinders (in case of multicylinder engine) cannot be fully overcome by the increased air velocity at the venturi throat.

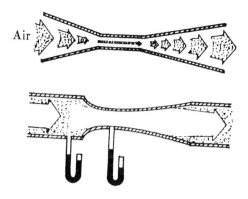

Fig.7.6 Operation of the Venturi Tube

7.8 THE SIMPLE CARBURETTOR

Carburettors are highly complicated. First, it is helpful to study the working principle of a simple or elementary carburettor which provides an air-fuel mixture for cruising or normal range at a single speed. Later, other mechanisms to provide for the various special requirements like starting, idling, variable load and speed operation and acceleration will be included.

Figure 7.7 shows a simple carburettor. It consists of a float chamber nozzle and a metering orifice, a venturi and a throttle valve. The float and a needle valve system maintains a constant level of gasoline in the float chamber. If the amount of fuel in the float chamber falls below the designed level, the float goes down, thereby opening the fuel supply valve and admitting fuel. When the designed level has been reached, the float closes the fuel supply valve thus stopping additional fuel flow from the supply system. Float chamber is vented either to the atmosphere or to the upstream side of the venturi.

During suction stroke air is drawn through the venturi. As already described, venturi is a tube of decreasing cross-section with a minimum area at the throat. Venturi tube is also known as the choke tube and is so shaped that it offers minimum resistance to the air flow. As the air passes through the venturi the velocity increases reaching a maximum at the venturi throat. Correspondingly, the pressure decreases reaching a minimum. From the float chamber, the fuel is fed to a discharge jet, the tip of which is located in the throat of the venturi. Because of the differential pressure between the float chamber

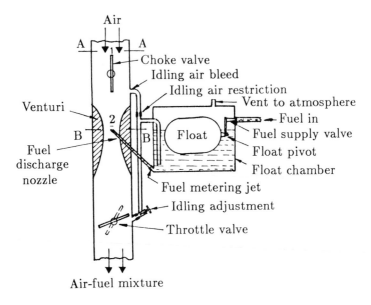

Fig.7.7 Simple Carburettor

and the throat of the venturi, known as *carburettor depression*, fuel is discharged into the air stream. The fuel discharge is affected by the size of the discharge jet and it is chosen to give the required air-fuel ratio. The pressure at the throat at the fully open throttle condition lies between 4 to 5 cm of Hg, below atmospheric and seldom exceeds 8 cm Hg below atmospheric. To avoid overflow of fuel through the jet, the level of the liquid in the float chamber is maintained at a level slightly below the tip of the discharge jet.

The gasoline engine is *quantity governed*, which means that when power output is to be varied at a particular speed, the amount of charge delivered to the cylinder is varied. This is achieved by means of a throttle valve usually of the butterfly type which is situated after the venturi tube. As the throttle is closed less air flows through the venturi tube and less is the quantity of air-fuel mixture delivered to the cylinder and hence power output is reduced. As the throttle is opened, more air flows through the choke tube resulting in increased quantity of mixture being delivered to the engine. This increases the engine power output.

A simple carburettor of the type described above suffers from a fundamental drawback in that it provides the required A/F ratio only at one throttle position. At the other throttle positions the mixture is either leaner or richer depending on whether the throttle is opened less or more. As the throttle opening is varied, the air flow varies and creates a certain pressure differential between the float chamber and the venturi throat. The same pressure differential regulates the flow of fuel through the nozzle. Therefore, the velocity of flow of air and fuel vary in a similar manner. At the same time, the density of air decreases as the pressure at the venturi throat decreases with increasing air flow whereas that of the fuel remains unchanged. This results in a simple carburettor producing a progressively rich mixture with increasing throttle opening. The mathematical analysis of the performance of a simple carburettor is given in the next section.

7.9 CALCULATION OF THE AIR–FUEL RATIO

A simple carburettor with the tip of the fuel nozzle h metres above the fuel level in the float chamber is shown in Fig.7.7. It may be noted that the density of air is not the same at the inlet to the carburettor (section A–A, point 1) and the venturi throat (section B–B, point 2). The calculation of exact air mass flow involves taking this change in density or compressibility of air into account.

Applying the steady flow energy equation to sections A–A and B–B and assuming unit mass flow of air, we have,

$$q - w = (h_2 - h_1) + \frac{1}{2}\left(C_2^2 - C_1^2\right) \tag{7.1}$$

Here q, w are the heat and work transfers from entrance to throat and h and C stand for enthalpy and velocity respectively.

Assuming an adiabatic flow, we get $q = 0$, $w = 0$ and $C_1 \approx 0$,

$$C_2 = \sqrt{2(h_1 - h_2)} \tag{7.2}$$

Assuming air to behave like ideal gas, we get $h = C_p T$. Hence, Eqn.7.2 can be written as,

$$C_2 = \sqrt{2C_p(T_1 - T_2)} \tag{7.3}$$

As the flow process from inlet to the venturi throat can be considered to be isentropic, we have

$$\frac{T_2}{T_1} = \left(\frac{p_2}{p_1}\right)^{\left(\frac{\gamma-1}{\gamma}\right)} \tag{7.4}$$

$$T_1 - T_2 = T_1\left[1 - \left(\frac{p_2}{p_1}\right)^{\left(\frac{\gamma-1}{\gamma}\right)}\right] \tag{7.5}$$

Substituting (7.5) in (7.3), we get

$$C_2 = \sqrt{2C_pT_1\left[1 - \left(\frac{p_2}{p_1}\right)^{\left(\frac{\gamma-1}{\gamma}\right)}\right]} \qquad (7.6)$$

Now, mass flow of air,

$$\dot{m}_a = \rho_1 A_1 C_1 = \rho_2 A_2 C_2 \qquad (7.7)$$

where A_1 and A_2 are the cross-sectional area at the air inlet (point 1) and venturi throat (point 2).

To calculate the mass flow rate of air at venturi throat, we have

$$p_1/\rho_1^\gamma = p_2/\rho_2^\gamma \qquad (7.8)$$

$$\rho_2 = (p_2/p_1)^{1/\gamma}\rho_1$$

$$\dot{m}_a = \left(\frac{p_2}{p_1}\right)^{1/\gamma}\rho_1 A_2\sqrt{2C_pT_1\left[\left(1 - \frac{p_2}{p_1}\right)^{\frac{\gamma-1}{\gamma}}\right]} \qquad (7.9)$$

$$= \left(\frac{p_2}{p_1}\right)^{1/\gamma}\frac{p_1}{RT_1}A_2\sqrt{2C_pT_1\left[1 - \left(\frac{p_2}{p_1}\right)^{\frac{\gamma-1}{\gamma}}\right]}$$

$$= \frac{A_2p_1}{R\sqrt{T_1}}\sqrt{2C_p\left[\left(\frac{p_2}{p_1}\right)^{\frac{2}{\gamma}} - \left(\frac{p_2}{p_1}\right)^{\frac{\gamma+1}{\gamma}}\right]} \qquad (7.10)$$

Substituting $C_p = 1.005$ kJ/kg K, $\gamma = 1.4$ and $R = 0.287$ kJ/kg K for air,

$$\dot{m}_a = 4.94\frac{A_2p_1}{\sqrt{T_1}}\sqrt{\left(\frac{p_2}{p_1}\right)^{1.43} - \left(\frac{p_2}{p_1}\right)^{1.71}}$$

$$= 4.94\frac{A_2p_1}{\sqrt{T_1}}\phi \quad \text{kg/s} \qquad (7.11)$$

where

$$\phi = \sqrt{\left(\frac{p_2}{p_1}\right)^{1.43} - \left(\frac{p_2}{p_1}\right)^{1.71}} \qquad (7.12)$$

Here, p is in N/m^2, A is in m^2 and T is in K.

Equation 7.11 gives the theoretical mass flow rate. To get the actual mass flow rate, the above equation should be multiplied by the co-efficient of discharge for the venturi, C_{da}.

$$\dot{m}_{a_{actual}} = 4.94 C_{da} \frac{A_2 p_1}{\sqrt{T_1}} \phi \tag{7.13}$$

Since C_{da} and A_2 are constants for a given venturi,

$$\dot{m}_{a_{actual}} \propto \frac{p_1}{\sqrt{T_1}} \phi \tag{7.14}$$

In order to calculate the air-fuel ratio, fuel flow rate is to be calculated. As the fuel is incompressible, applying Bernoulli's Theorem we get

$$\frac{p_1}{\rho_f} - \frac{p_2}{\rho_f} = \frac{C_f^2}{2} + gz \tag{7.15}$$

where ρ_f is the density of fuel, C_f is the fuel velocity at the nozzle exit and z is the height of the nozzle exit above the level of fuel in the float bowl

$$C_f = \sqrt{2\left[\frac{p_1 - p_2}{\rho_f} - gz\right]}$$

Mass flow rate of fuel,

$$\dot{m}_f = A_f C_f \rho_f \tag{7.16}$$

$$= A_f \sqrt{2\rho_f (p_1 - p_2 - gz\rho_f)} \tag{7.17}$$

where A_f is the area of cross-section of the nozzle and ρ_f is the density of the fuel

$$\dot{m}_{f_{actual}} = C_{df} A_f \sqrt{2\rho_f (p_1 - p_2 - gz\rho_f)} \tag{7.18}$$

where C_{df} is the coefficient of discharge for fuel nozzle

$$A/F \text{ ratio} = \frac{\dot{m}_{a_{actual}}}{\dot{m}_{f_{actual}}}$$

$$\frac{A}{F} = 4.94 \frac{C_{da}}{C_{df}} \frac{A_2}{A_f} \frac{p_1 \phi}{\sqrt{2T_1 \rho_f (p_1 - p_2 - gz\rho_f)}} \tag{7.19}$$

7.9.1 Air–Fuel Ratio Neglecting Compressibility of Air

When air is considered as incompressible, Bernoulli's theorem is applicable to air flow also. Hence, assuming $U_1 \approx 0$

$$\frac{p_1}{\rho_a} - \frac{p_2}{\rho_a} = \frac{C_2^2}{2} \tag{7.20}$$

$$C_2 = \sqrt{2\left[\frac{p_1 - p_2}{\rho_a}\right]} \tag{7.21}$$

$$\dot{m}_a = A_2 C_2 \rho_a$$

$$= A_2 \sqrt{2\rho_a (p_1 - p_2)} \tag{7.22}$$

$$\dot{m}_{a_{actual}} = C_{da} A_2 \sqrt{2\rho_a (p_1 - p_2)} \tag{7.23}$$

$$A/F \text{ ratio} = \frac{\dot{m}_a}{\dot{m}_f}$$

$$= \frac{C_{da} \, A_2}{C_{df} \, A_f} \sqrt{\frac{\rho_a (p_1 - p_2)}{\rho_f (p_1 - p_2 - gz\rho_f)}} \tag{7.24}$$

If $z = 0$

$$\frac{\dot{m}_a}{\dot{m}_f} = \frac{C_{da} \, A_2}{C_{df} \, A_f} \sqrt{\frac{\rho_a}{\rho_f}} \tag{7.25}$$

7.9.2 Air–Fuel Ratio Provided by a Simple Carburettor

1. It is clear from expression for $\dot{m}_f$ (7.17) that if $(p_1 - p_2)$ is less than $gz\rho_f$ there is no fuel flow and this can happen at very low air flow. As the air flow increases, $(p_1 - p_2)$ increases and when $(p_1 - p_2) > gz\rho_f$ the fuel flow begins and increases with increase in the differential pressure.

2. At high air flows where $(p_1 - p_2)$ is large compared to $gz\rho_f$ the fraction $gz\rho_f/(p_1 - p_2)$ becomes negligible and the air-fuel ratio approaches

$$\frac{C_{d_a} \, A_2}{C_{d_f} \, A_f} \sqrt{\frac{\rho_a}{\rho_f}}$$

3. A decrease in the density of air reduces the value of air-fuel (i.e., mixture becomes richer). It happens at

(i) high air flow rates when $(p_1 - p_2)$ becomes large and ρ_2 decreases.

(ii) high altitudes when the density of air is low.

7.9.3 Size of the Carburettor

The size of a carburettor is generally given in terms of the diameter of the venturi tube in mm and the jet size in hundredths of a millimeter. The calibrated jets have a stamped number which gives the flow in ml/min under a head of 500 mm of pure benzol.

For a venturi of 30 to 35 mm size (having a jet size which is one sixteenth of venturi size) the pressure difference $(p_1 - p_2)$ is about 50 mm of Hg. The velocity at throat is about 90 - 100 m/s and the coefficient of discharge for venturi C_{da} is usually 0.85.

7.10 ESSENTIAL PARTS OF A MODERN CARBURETTOR

The modern carburettor consists essentially of a main air passage through which the engine draws its supply of air, mechanisms to control the quantity of fuel discharged in relation to the flow of air, and a means of regulating the quantity of fuel-air mixture delivered to the engine cylinders.

The essential parts of a carburettor are

(i) float chamber
(ii) strainer
(iii) throttle
(iv) choke
(v) metering system
(vi) idling system
(vii) accelerating system
(viii) economizer system

7.10.1 The Float Chamber

The function of a float chamber in a carburettor is to supply the fuel to the nozzle at a constant pressure head. This is possible by maintaining a constant level of the fuel in the float bowl. The float in a carburettor is designed to control the level of fuel in the float chamber. This fuel level must be maintained slightly below the discharge nozzle outlet holes in order to provide the correct amount of fuel flow and to prevent leakage of fuel from the nozzle when the engine is not operating. The arrangement of a float mechanism in relation to the discharge nozzle is shown in Fig.7.7. When the float rises with the fuel coming in, the fuel supply valve closes and stops the flow of fuel into the chamber. At this point, the level of the fuel is correct for proper operation of the carburettor.

As shown in Fig.7.7, the float valve mechanism includes a fuel supply valve and a pivot. During the operation of the carburettor, the float assumes a position slightly below its highest level to allow a valve opening sufficient for replacement of the fuel as it is drawn out through the discharge nozzle.

7.10.2 The Fuel Strainer

The gasoline has to pass through a narrow nozzle exit. To prevent possible blockage of the nozzle by dust particles, the gasoline is filtered by installing a fuel strainer at the inlet to the float chamber. The strainer consists of a fine wire mesh or other type of filtering device, cone shaped or cylindrical shaped. The strainer is usually removable so that it can be taken out and thoroughly cleaned. It is retained in its seat by a strainer plug or a compression spring.

7.10.3 Throttle

The speed and the output of an engine is controlled by the use of the throttle valve, which is located on the downstream side of the venturi. The more the throttle is closed the greater is the obstruction to the flow of the mixture placed in the passage and the less is the quantity of mixture delivered to the cylinders. The decreased quantity of mixture gives a less powerful impulse to the pistons and the output of the engine is reduced accordingly. As the throttle is opened, the output of the engine increases. Opening the throttle usually increases the speed of the engine. But this is not always the case as the load on the engine is also a factor. For example, opening the throttle when the motor vehicle is starting to climb a hill may or may not increase the vehicle speed, depending upon the steepness of the hill and the extent of throttle opening. In short, the throttle is simply a means to regulate the output of the engine by varying the quantity of charge going into the cylinder.

7.10.4 Choke

A rich mixture is required to start the engine, specially when the engine is cold. To provide this rich mixture a choke valve (Fig.7.7) is inserted in the air intake passage of the carburettor. During starting, this valve is operated to shut off partially the supply of air to the carburettor, thus enriching the mixture supplied to the cylinders by the carburettor. These choke valves are operated automatically by thermostats in most passenger car carburettors. The valve is held in a partially closed position by the thermostat when the engine is being started and is opened automatically as the engine heats up, gradually leaning the mixture.

7.10.5 Metering System

The metering system of the carburettor controls the fuel feed for cruising and full throttle operations. It consists of three principal units:

(i) the main metering jet through which fuel is drawn from the float chamber

(ii) the main discharge nozzle

(iii) the passage leading to the idling system

The three functions of the main metering system are

(i) to proportion the fuel-air mixture

(ii) to decrease the pressure at the discharge nozzle

(iii) to limit the air flow at full throttle

In modern carburettors automatic compensating devices are provided to maintain the desired mixture proportions at the higher speeds. Metering rods, air-bleed jets, economizers, compound jets and auxiliary air valves are devices used to effect this compensation in commercial carburettors. The type of compensation mechanism used determines the metering system of the carburettor.

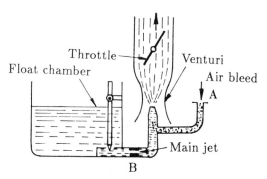

Fig.7.8 Air-Bleed Principle in a Typical Carburettor

Figure 7.8 shows a simplified picture of the air-bleed principle in a typical carburettor. As already discussed, it may be noted that with increase in throttle opening the mixture becomes richer in a simple carburettor. This tendency can be arrested by means of a restricted air-bleed system illustrated in Fig.7.8. The figure illustrates a schematic diagram of a carburettor which contains an air-bleed into the main nozzle and therefore is called a restricted air bleed jet carburettor (the most popular type of all). Air enters the nozzle through the small (restricted) hole at A, and enters the liquid stream through a larger hole at *B*. Thus the fluid stream becomes an emulsion of air and liquid with negligible viscosity and surface tension.

As a consequence the mass flow rate of fuel is increased considerably at low suctions. If the hole size is too large at *B* then the mixture will become richer and if the hole size is too small then there will not be any effect. However, by proper design of the hole size at *B* it is possible to maintain an approximately uniform mixture strength throughout the power range of engine operation.

Because of this, through the air-bleed nozzle, more air will pass through restricting the fuel flow.

The air flow through an opening of fixed size and the fuel flow through an air-bleed system respond to variations of pressure in approximately equal proportions. If the fuel discharge nozzle of the air-bleed system is located in the center of the venturi then both the air-bleed nozzle and the venturi are exposed to the suction of the engine to the same degree. This will enable to maintain an approximately uniform mixture of fuel and air throughout the power range of engine operations.

7.10.6 Idling System

Motor vehicle engines require a rich mixture for idling and low speed operation, usually about 12 parts of air by weight to one part of fuel (12:1 air-fuel ratio) as explained in section 7.6.1. To supply this mixture during idling, most modern carburettors incorporate in their construction a special idling system consisting of an idling fuel passage and idling ports. This system comes into action during starting, idling and low-speed operation and goes out of action when the throttle is opened beyond about 20%.

Figure 7.9 shows how an idling system, applied to a simple carburettor, operates when the throttle is closed or partially closed. With the throttle partially closed the limited air flow causes very little depression at the nozzle exit which will not be sufficient to draw any fuel through the main nozzle. At the same time the very low pressure on the downstream side of the throttle causes the fuel to rise in the idle tube and to be discharged through the idling discharge port directly into the engine intake manifold. This suction also draws air through the idling air-bleed which combines with the gasoline to help vaporize and atomize it as it passes through the idle passage. The fuel and air leaving the idle discharge port combine with the air stream that is going past the throttle in the manifold to produce the correct mixture. The idle air-bleed also helps in preventing the draining of the fuel from the float chamber through the idling passage due to siphon action when the engine is not operating.

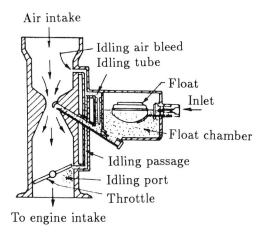

Air intake

Idling air bleed
Idling tube
Float
Inlet
Float chamber
Idling passage
Idling port
Throttle

To engine intake

Fig.7.9 Idling System of a Simple Carburettor

As the throttle is opened and the engine passes through the idling range of operation, the pressure at the exit of the idle discharge increases and the pressure difference between the manifold and the float chamber will not be sufficient to draw the fuel through the idling passage putting it out of action. At the same time the main air flow would have increased to a level that establishes a pressure differential across the main jet bringing it into action.

7.10.7 Accelerating System

In automobile engine situations arise when it is necessary to accelerate the vehicle. This requires a suddenly increased output from the engine. If the throttle is suddenly opened there is a corresponding increase in the air flow. However, because of the inertia of the liquid fuel, the fuel flow does not increase in proportion to the increase in air flow. This results in a *temporary lean mixture* causing the engine to misfire and a temporary reduction in power output. To prevent this condition, all modern carburettors are equipped with an accelerating system. This is either an accelerating pump or an accelerating well. The function of the accelerating system is to discharge an additional quantity of fuel into the carburettor air stream when the throttle is suddenly opened thus causing a temporary enrichment of the mixture and producing a smooth and positive acceleration of the vehicle.

The accelerating well is a space around the discharge nozzle connected by holes to the fuel passage leading to the discharge nozzle. The upper holes are located near the fuel level and are uncovered at the lowest pressure level that will draw fuel from the main discharge nozzle. Hence, they receive air during the time of the operation of main discharge nozzle.

Very little throttle opening is required at idling speeds. When the throttle is suddenly opened, air is drawn in to fill the intake manifold and which ever cylinder is on the intake stroke. This sudden rush of air temporarily creates a high suction at the main discharge nozzle, brings into operation the main metering system, and draws additional fuel from the accelerating well. Because of the throttle opening, the engine speed increases and the main metering system continues to function.

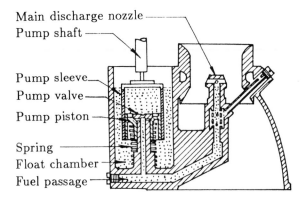

Fig.7.10 A Movable-Piston Type Accelerating Pump

The accelerating pump illustrated in Fig.7.10 is a sleeve type piston pump operated by the throttle. The piston is mounted on a stationary hollow stem screwed into the body of the carburettor. The hollow stem opens into the main fuel passage leading to the discharge nozzle. A movable cylinder or sleeve is mounted over the stem and piston is connected by the pump shaft to the throttle linkage. When the throttle is closed, the cylinder is raised and the space within the cylinder fills with the fuel through the clearance between the piston and the cylinder.

If the throttle is quickly moved to the open position, the cylinder is forced down as shown in Fig.7.11 and the increased fuel pressure also forces the piston part way down along the stem. As the piston moves down it opens the pump valve and permits the fuel to flow through the hollow stem into the main fuel passage. With the throttle fully open and the accelerating pump cylinder down, the spring pushes the piston up and forces most of the fuel out of the cylinder. When the piston reaches its highest position, it closes the valve and no more fuel

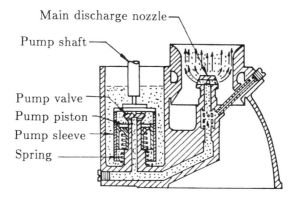

Main discharge nozzle

Pump shaft

Pump valve
Pump piston
Pump sleeve
Spring

Fig.7.11 Accelerating Pump in Operation

flows towards the main passage. If the throttle is slowly opened the pressure build-up is not high enough to force the piston downwards. Instead the fuel goes back to the fuel chamber through the clearance between the piston and the cylinder.

7.10.8 Economizer System

An economizer is a valve which remains closed at normal operation but is opened to provide an enriched mixture at full throttle operation. It supplies and regulates the additional fuel required for above the cruising range. In other words, an economizer is a device for enriching the mixture at increased throttle openings.

Three types of economizers are used for float-type carburettors namely,

(i) needle-valve type,

(ii) the piston type and

(iii) the manifold pressure – operated type.

Figure 7.12 shows the needle-valve type of economizer. This mechanism has a needle valve which is opened by the throttle linkage at a predetermined throttle position. This permits a quantity of fuel, in addition to the fuel from the main metering jet, to enter the discharge nozzle passage. As shown in the Fig.7.12, the economizer needle valve permits fuel to bypass the cruise valve metering jet.

The piston type economizer, shown in Fig.7.13, is also operated by the throttle. The lower piston serves as a fuel valve preventing any flow of fuel through the system at cruising speeds. The upper piston functions as an air valve allowing air to flow through the separate economizer discharge nozzle at part throttle. As the throttle

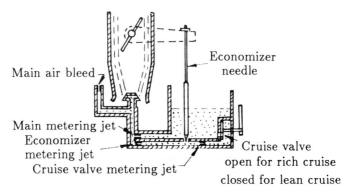

Main air bleed

Economizer
needle

Main metering jet
Economizer
metering jet
Cruise valve metering jet

Cruise valve
open for rich cruise
closed for lean cruise

Fig.7.12 Needle-Valve Type Economizer

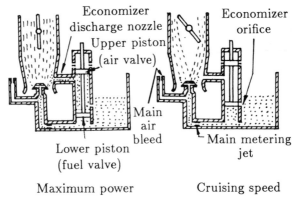

Economizer
discharge nozzle
Upper piston
(air valve)

Economizer
orifice

Main
air
bleed

Main metering
jet

Lower piston
(fuel valve)

Maximum power

Cruising speed

Fig.7.13 Piston Type Economizer

is opened to higher output positions the lower piston uncovers the fuel port leading the fuel into the economizer metering valve and the upper piston closes the air ports. Fuel fills the economizer well and is discharged into the carburettor venturi where it adds to the fuel from the main discharge nozzle. The upper piston of the economizer allows a small amount of air to bleed into the fuel thus assisting the atomization of the fuel from the economizer system. The space below the lower piston of the economizer acts as an accelerating well when the throttle is opened.

The manifold-pressure operated economizer is shown in Fig.7.14. It has a bellows which is compressed when the pressure from the engine blower rim produces a force greater than the compression spring in the bellows chamber. As engine speed increases, the blower pressure also increases. This pressure forces the bellows and causes the

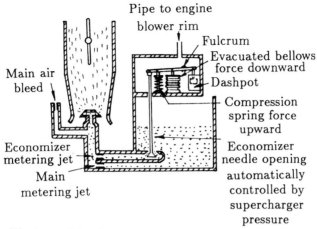

Fig.7.14 Manifold-Pressure Operated Economizer

economizer valve to open. Fuel then flows through the economizer metering jet to the main discharge system. The operation of the bellows and spring is stabilized by means of a dash pot as shown in the Fig.7.14.

7.11 TYPES OF CARBURETTORS

There are three general types of carburettors depending on the direction of flow of air. The first is the *updraught* type shown in Fig.7.15 in which the air enters at the bottom and leaves at the top so that the direction of its flow is upwards. The disadvantage of the updraught carburettor is that it must lift the sprayed fuel droplet by air friction. Hence, it must be designed for relatively small mixing tube and throat so that even at low engine speeds the air velocity is sufficient to lift and carry the fuel particles along. Otherwise, the fuel droplets tend to separate out providing only a lean mixture to the engine. On the other hand, the mixing tube is finite and small then it cannot supply mixture to the engine at a sufficiently rapid rate at high speeds.

In order to overcome this drawback the *downdraught* carburettor (Fig.7.16) is adopted. It is placed at a level higher than the inlet manifold and in which the air and mixture generally follow a downward course. Here the fuel does not have to be lifted by air friction as in the updraught carburettors but move into the cylinders by gravity even if the air velocity is low. Hence, the mixing tube and throat can be made large which makes high engine speeds and high specific outputs possible.

A *crossdraught* carburettor consists of a horizontal mixing tube with a float chamber on one side of it. By using a crossdraught

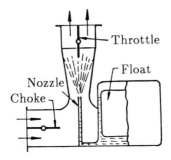

Fig.7.15 Updraught Type Carburettor

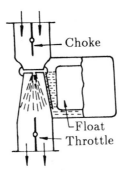

Fig.7.16 Downdraught Type Carburettor

carburettor in engines, one right-angled turn in the inlet passage is eliminated and the resistance to flow is reduced.

7.12 AUTOMOBILE CARBURETTORS

Some of the important carburettors used in modern automobile are

  (i) Solex carburettor
 (ii) Carter carburettor
(iii) S.U. carburettor
     The details of these carburettors are given in this section.

7.12.1 Solex Carburettor

The solex carburettor is famous for its ease of starting, good performance and reliability. It is made in various models and is used in many automobile engines.

The solex carburettor as shown in Fig.7.17(a) is a downdraught type of carburettor which has the provision for the supply of richer mixture required for starting and weaker mixture for high speed running of the engine. It consists of various fuel circuits such as starting, idling or low speed operation, normal running, acceleration etc.

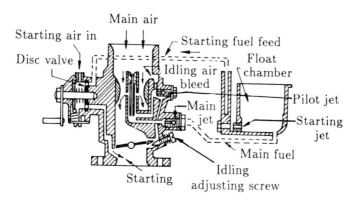

Fig.7.17(a) Solex Carburettor

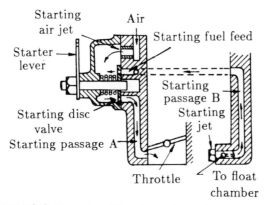

Fig.7.17(b) Starting Circuit of the Solex Carburettor

The starting circuit is shown in Fig.7.17(b). The fuel from float chamber enters the starting passage *B* through the starting jet. Then the fuel enters the starting passage *A* through starting valve. The starting valve is a disc having small hole which can be brought opposite the fuel supply hole by means of a starter lever. When the throttle valve is in closed position the engine suction is applied to starting passage A. The air enters from the starting air jet and fuel from the small hole provided in the disc valve. This mixture is sufficiently rich to start the engine. With the throttle valve opening the engine suction is applied to the main jet and starting lever is turned to the off position there by stopping the supply of the fuel through the disc valve.

The idle circuit of Solex Carburettor is shown in Fig.7.17(c). The fuel from main jet circuit enters the idle passage through the pilot jet and air enters this passage through the pilot air jet. At idling the throttle is almost closed and hence suction is applied at the outlet of the idle jets thereby supplying the mixture to the engine through the idle port provided immediately after the throttle on the downstream side. The opening of the idle port is controlled by the idle adjusting screw thereby obtaining the smooth running of the engine when the engine is on no load. When the throttle is opened for further running of the engine, the supply of the fuel through the idle circuit is restricted due to decrease in the suction at the idle passage.

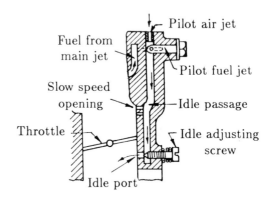

Fig.7.17(c) Idle Circuit of the Solex Carburettor

To provide an extra quantity of fuel during acceleration, this carburettor is provided with a diaphragm pump system as shown in Fig.7.17(d). When accelerator pedal is pressed for acceleration, the pump lever connected to it is also pressed. Due to this movement of the pump lever, the diaphragm which is connected to it by means of a push rod also moves towards left thereby compressing the fuel and forcing it into the main jet. When force from the pump lever is removed then diaphragm is released to its original position by means of a spring provided on the left side of the diaphragm. Due to this movement of the diaphragm a suction is created on the left side of

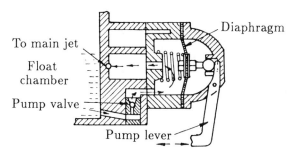

Fig.7.17(d) Diaphragm Pump System of the Solex Carburettor

the diaphragm thus opening the pump valve and admitting the fresh fuel into the pump circuit.

7.12.2 Carter Carburettor

A diagrammatic view of the Carter downdraught type carburettor is shown in Fig.7.18. The gasoline enters the float chamber which is of the conventional type. The air enters the carburettor from the top, a choke valve in the passage remains open during running. The Carter carburettor has a triple venturi diffusing type of choke it has three venturies, the smallest lies above the level in the float chamber, other two below the fuel level, one below other. At very low speeds, suction in the primary venturi is adequate to draw the fuel. The nozzle enters the primary venturi at an angle delivering the fuel upwards against the air stream securing an even flow of finely divided atomized fuel. The mixture from the primary venturi passes centrally through the second venturi where it is surrounded by a blanket of air stream and finally this leads to the third (main) venturi, where again the fresh air supply insulates the stream from the second venturi. The mixture reaches the engine cylinders in atomized form. Multiple venturies result in better formation of the mixture at very low speeds causing steady and smooth operation at very low as well as at very high engine speeds.

In this carburettor a mechanical metering method is used. In the fuel circuit there is a metering rod actuated by a mechanism connected with the main throttle. The metering rod has two or more steps of diameter. The area of opening between the metering rod jet and metering rod governs the amount of gasoline drawn into the engine. At high speed the metering rod is lifted, such that the smallest section of the rod is in the jet and the maximum quantity of gasoline flows out to mix with the maximum amount of air corresponding to full throttle opening.

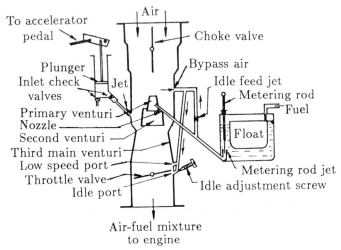

Fig.7.18 Carter Downdraught Type Carburettor

For starting, a choke valve is provided in the air circuit. The choke valve is a butterfly valve, one half of which is spring controlled. The valve is hinged at the centre. When the engine is fully choked, the whole of the engine suction is applied at the main nozzle, which then delivers fuel. The air flow rate is minimum and hence the mixture supplied is very rich. Once the engine is started, the spring-controlled half of the choke valve is sucked open to provide correct amount of air during warming up period.

The idling circuit is as shown in Fig.7.18. During idling a rich mixture is required in small quantity. In idling condition throttle valve is nearly closed. The whole of the engine suction is applied at the idle port. Consequently, the fuel is drawn through the idle feed jet and air through the first bypass and a rich idle mixture is supplied. During low speed operation the throttle valve is opened further. The main nozzle also starts supplying the fuel. At this stage the fuel is delivered both by the main venturi and low speed port through idle passage.

The acceleration pump is a pump and a plunger mechanism. The pump consists of a plunger working inside a cylinder consisting of inlet check valve and outlet check valve. The pump plunger is connected to accelerator pedal by throttle control rod. When the throttle is suddenly opened by pressing the accelerator pedal, the pump is actuated and a small quantity of petrol is delivered into the choke tube by a jet. When the accelerator pedal is released the pump piston moves up thereby sucking fuel from the float chamber for the next operation.

7.12.3 S. U. Carburettor

Carburettors in general are constant choke type. Zenith, Solex and Carter carburettors are examples of this type. S.U. carburettors differs completely from them being *constant vacuum or constant depression* type with automatic variable choke.

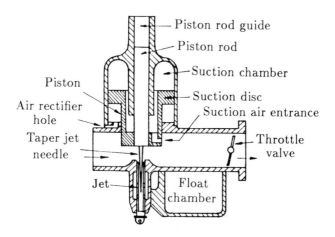

Fig.7.19 S.U. Carburettor

    A simplified sketch of this carburettor is shown in Fig.7.19. It consists of a sliding piston. The lower end of the piston is provided with a taper needle which is inserted into the main jet. When the piston is moved up and down the needle also moves up and down with the main jet. The upper end of the piston is given a flat form which is known as suction disc. The up and down movement of the piston and the suction disc is guided by means of piston rod and piston rod guide as shown in Fig.7.19.

    The piston always remains loaded by a helical spring. The movement of the piston controls the air passage. The portion above the disc is called suction chamber which connects the air passage by means of a slot provided in the piston. The main jet of the carburettor can be moved up and down along the taper needle by operating a lever from the dash board. This movement is required to adjust the mixture strength throughout the operating range of the carburettor. The carburettor consists of an ordinary butterfly type throttle valve. The lower portion of the suction disc is connected to the atmosphere by means of an air rectifier hole and the upper portion to the throttle air passage. The system does not have any separate idling slow running and accelerating system.

The weight of the piston is constant and always acts down. The vacuum in the suction chamber always tends to move the piston upwards. Therefore, at a particular instant the position of the piston is balanced by the weight and constant vacuum produced above the piston. In starting, a rich mixture is needed by the engine which can be obtained by pulling the jet downwards with the help of the lever attached to it. Therefore, as the throttle is opened more air is allowed to pass through the inlet due to upward movement of the piston. The upward movement of the taper needle also ensures more flow of fuel from the main jet. Thus the air and fuel passage are varied with different engine speeds and velocities of the fuel and air remains constant in this system.

7.13 ALTITUDE COMPENSATION

An inherent characteristic of the conventional float-type carburettor is to meter air and fuel by volume and not by weight as the basis of calculating combustible air-fuel ratio. The weight of one cubic meter of air decreases as altitude increases. Most automobile carburettors are calibrated at altitudes near sea level and similarly production carburettors are flow tested and adjusted in a controlled environment. When the atmospheric conditions are different from those at which the carburettor was calibrated, the air-fuel ratio changes. If the vehicle is operated at an altitude lower than the calibration altitude a lean mixture is obtained which results in poor drivability. At altitudes higher than the calibration altitude a rich mixture is supplied which causes incomplete combustion and emit hydrocarbon and carbon monoxide. The enrichment, E, due to variation of air density closely follows the relationship

$$E + 1 = \sqrt{\frac{\rho_0}{\rho}} = \sqrt{\frac{p_0 T}{p T_0}} \qquad (7.26)$$

where subscript '0' refers to the calibration conditions. Thus if the density ratio is (say) 0.84, then the enrichment E is given by

$$E + 1 = \sqrt{\frac{1}{0.84}} = 1.091$$

or

$$E = 0.091$$

$$= 9.1\%$$

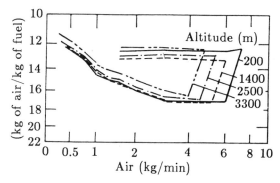

Fig.7.20 Flow Characteristics of a Typical Carburettor

This is the enrichment of the charge over the basic calibrated air-fuel ratio in an uncompensated carburettor. Figure 7.20 shows the flow characteristic of a typical carburettor at various altitude levels. It also illustrates the early enrichments caused by the effect of low manifold vacuum on the power enrichment system.

7.13.1 Altitude Compensation Devices

The problem of mixture enrichment is quite acute in aircraft carburettors. At higher altitudes, density of air is less and therefore the mass of the air taken into engine decreases and the power is reduced in approximately the same proportion. Since, the quantity of oxygen taken into the engine decreases, the fuel-air mixture becomes too rich for normal operation. The mixture strength delivered by the carburettor becomes richer at a rate inversely proportional to the square root of the density ratio as given by the Eqn.7.26.

If the pressure remains constant, the density of the air will vary according to temperature, increasing as the temperature drops. This will cause a leaning of the fuel-air mixture in the carburettor because the denser air contains more oxygen. The change in air pressure due to altitude is considerably more of a problem than the change in density due to temperature changes. At 6,000 m altitude, the air pressure is approximately one-half the pressure at sea level. Hence, in order to provide a correct mixture, the fuel flow would have to be reduced to almost one-half what it would be at sea level. The adjustment of fuel flow to compensate for changes in air pressure and temperature is a principal function of the mixture control.

Briefly, the *mixture-control system* can be described as a mechanism or device by means of which the richness of the mixture entering the engine during flight can be controlled to a reasonable extent. This control should exist through all normal altitudes of operation to prevent the mixture from becoming too rich at high altitudes and to economize on fuel during engine operation in the low-power range where cylinder temperature will not become excessive with the use of the leaner mixture.

Mixture-control systems may be classified according to their principles of operation as

(i)   *back suction type*, which reduces the effective suction on the metering system

(ii) *needle type*, which restricts the flow of fuel through the metering system; and

(iii) the *air-port type*, which allows additional air to enter the carburettor between the main discharge nozzle and the throttle valve.

It may be noted that piston engines have become obsolete in aircraft. Hence, the details of the altitude compensating mechanisms are not discussed.

Worked out Examples

7.1. A simple jet carburettor is required to supply 5 kg of air and 0.5 kg of fuel per minute. The fuel specific gravity is 0.75. The air is initially at 1 bar and 300 K. Calculate the throat diameter of the choke for a flow velocity of 100 m/s. Velocity coefficient is 0.8. If the pressure drop across the fuel metering orifice is 0.80 of that of the choke, calculate orifice diameter assuming, $C_{df} = 0.60$ and $\gamma = 1.4$.

SOLUTION:

$$\text{Vel. at throat, } C_2 \;=\; V_c\sqrt{2C_pT_1\left[1-\left(\frac{p_2}{p_1}\right)^{\frac{\gamma-1}{\gamma}}\right]}$$

$$100 \;=\; 0.8\times\sqrt{2\times 1005\times 300\times\left[1-\left(\frac{p_2}{p_1}\right)^{0.286}\right]}$$

$$\left(\frac{p_2}{p_1}\right)^{0.286} \;=\; 1-\left(\frac{100}{0.8}\right)^2\times\frac{1}{2\times 1005\times 300}$$

$$\;=\; 0.974$$

$$\frac{p_2}{p_1} \;=\; (0.974)^{1/0.286} \;=\; 0.912$$

$$p_2 \;=\; 0.912 \text{ bar}$$

$$v_1 \;=\; \frac{RT_1}{p_1} \;=\; \frac{0.287 \times 300}{10^5} \times 1000$$

$$=\; 0.861 \text{ m}^3$$

$$p_1 v_1^\gamma \;=\; p_2 v_2^\gamma$$

$$v_2 \;=\; v_1 \left(\frac{p_1}{p_2}\right)^{\frac{1}{\gamma}}$$

$$=\; 0.861 \times \left(\frac{1}{0.912}\right)^{0.714}$$

$$=\; 0.919 \text{ m}^3/\text{kg}$$

$$\text{Throat area } A_2 \;=\; \frac{\dot{m}_a \times v_2}{C_2}$$

$$=\; \frac{5}{60} \times \frac{0.919}{100} \times 10^4 \;=\; 7.658 \text{ cm}^2$$

$$d_2 \;=\; \sqrt{7.658 \times \frac{4}{\pi}} \;=\; 3.12 \text{ cm}$$

$$\Delta p_a \;=\; 1 - 0.912 \;=\; 0.088$$

$$\Delta p_f \;=\; 0.80 \times 0.088 \;=\; 0.07 \text{ bar}$$

$$\dot{m}_f \;=\; A_f C_f \sqrt{2 \rho_f \Delta p_f}$$

$$\frac{0.5}{60} \;=\; A_f \times 0.6 \times \sqrt{2 \times 750 \times 0.07 \times 10^5}$$

$$A_f \;=\; \frac{0.5 \times 10^4}{60 \times 0.6 \times 3.24 \times 10^3} \;=\; 0.0428 \text{ cm}^2$$

$$d_f \;=\; \mathbf{2.34 \text{ mm}} \qquad \underset{\Longleftarrow}{\text{Ans}}$$

7.2. A four-cylinder, four-stroke square engine running at 40 rev/s has a carburettor venturi with a 3 cm throat. Assuming the bore to be 10 cm, volumetric efficiency of 75%, the density of air to be

1.15 and coefficient of air flow to be 0.75. Calculate the suction at the throat.

SOLUTION:

$$Swept\ volume, V_s\ =\ \frac{\pi}{4} \times 10^2 \times 10 \times 10^{-6} \times 4$$

$$=\ 0.00314\ \text{m}^3$$

$$Volume\ sucked/s\ =\ \eta_v \times V_s \times n$$

$$=\ 0.75 \times 0.00314 \times \frac{40}{2}\ =\ 0.047\ \text{m}^3/\text{s}$$

$$\dot{m}_a\ =\ 0.047 \times 1.15\ =\ 0.054\ \text{kg/s}$$

Since the initial temperature and pressure is not given, the problem is solved by neglecting compressibility of the air

$$\dot{m}_a\ =\ C_d A_2 \sqrt{2\rho_a \Delta p_a}$$

$$\Delta p_a\ =\ \left(\frac{\dot{m}_a}{C_d A_2}\right)^2 \frac{1}{2\rho_a}$$

$$=\ \left(\frac{0.54}{0.75 \times \frac{\pi}{4} \times (0.03)^2}\right)^2 \times \frac{1}{2 \times 1.15}$$

$$=\ 4511\ \text{Pa}$$

$$=\ \textbf{0.04511 bar} \qquad \underset{\Longleftarrow}{\text{Ans}}$$

7.3. An experimental four-stroke gasoline engine of 1.7 litre capacity is to develop maximum power at 5000 revolutions per minute. The volumetric efficiency is 75% and the air-fuel ratio is 14:1. Two carburettors are to be fitted and it is expected that at maximum power the air speed at the choke is 100 m/s. The coefficient of discharge for the venturi is assumed to be 0.80 and that of main jet is 0.65. An allowance should be made for emulsion tube, the diameter of which can be taken as 1/3 of choke diameter. The gasoline surface is 6 mm below the choke at this engine condition. Calculate the sizes of a suitable choke and main jet. The specific gravity of the gasoline is 0.75. p_a and T_a are 1 bar and 300 K respectively.

SOLUTION:

Actual volume of air sucked per second

$$= \eta_v \times V_{cyl} \times \frac{rpm}{2} \times \frac{1}{60}$$

$$= 0.75 \times 1.7 \times 10^{-3} \times \frac{5000}{2} \times \frac{1}{60}$$

$$= 0.053125 \ \text{m}^3/\text{s}$$

Air flow through each carburettor at atmospheric conditions, V_1,

$$V_1 = \frac{0.053125}{2} = 0.0265 \ \text{m}^3/\text{s}$$

$$\rho_a = \frac{p_a}{RT_a} = \frac{1 \times 10^5}{0.287 \times 10^3 \times 300}$$

$$= 1.16 \ \text{kg/m}^3$$

$$\dot{m}_a = \rho_a V_1 = 1.16 \times 0.0265$$

$$= 0.0308 \ \text{kg/s}$$

The velocity of air at throat, C_2

$$C_2 = \sqrt{2C_p T_1 \left[1 - \left(\frac{p_2}{p_1}\right)^{\frac{\gamma-1}{\gamma}}\right]}$$

$$100^2 = 2 \times 1005 \times 300 \times \left[1 - \left(\frac{p_2}{p_1}\right)^{0.286}\right]$$

$$1 - \left(\frac{p_2}{p_1}\right)^{0.286} = 0.0166$$

$$\frac{p_2}{p_1} = (1 - 0.0166)^{1/0.286} = 0.943$$

Pressure at throat $= 0.943 \ \text{bar}$

Volume flow at choke, $V_2 = V_1 \left(\frac{p_1}{p_2}\right)^{1/\gamma}$

$$= \quad 0.0265 \times \left(\frac{1}{0.943}\right)^{1/1.4}$$

$$= \quad 0.0276 \text{ m}^3/\text{s}$$

If compressibility is neglected, $V_1 = V_2$

The nominal choke area, $A_2 \quad = \quad \dfrac{V_2}{C_2 \times C_{da}}$

$$= \quad \frac{0.0276}{100 \times 0.80} \times 10^4 \quad = \quad 3.45 \text{ cm}^2$$

If D is the diameter of choke tube and d is the diameter of the emulsion tube

$$A_2 \quad = \quad \frac{\pi}{4}\left(D^2 - d^2\right)$$

since $d = \frac{D}{3}$, we have

$$\frac{\pi}{4}\left[D^2 - \left(\frac{D}{3}\right)^2\right] \quad = \quad \frac{\pi}{4} \times \frac{8D^2}{9} \quad = \quad 3.45$$

$$D \quad = \quad \textbf{2.22 cm} \qquad\qquad \overset{\textbf{Ans}}{\Longleftarrow}$$

$$\dot{m}_f \quad = \quad \frac{\dot{m}_a}{14} \quad = \quad \frac{0.0308}{14}$$

$$= \quad 0.0022 \text{ kg/s}$$

$$\dot{m}_f \quad = \quad C_{df} A_f \rho_f \sqrt{\frac{2\Delta p_f}{\rho_f}}$$

$$= \quad C_{df} A_f \sqrt{2\rho_f \Delta p_f}$$

For gasoline the pressure difference across the main jet is given by

$$\Delta p_f \quad = \quad p_a - p_f - gh\rho_f$$

$$\dot{m}_f \quad = \quad C_{df} A_f \sqrt{2\rho_f (p_a - p_2 - gh\rho_f)}$$

$$0.0022 = 0.65 \times A_f \sqrt{2 \times 750\left[10^5(1 - 0.943) - 9.81 \times \frac{6}{10^3} \times 750\right]}$$

$$A_f \quad = \quad 1.162 \times 10^{-6} \text{ m}^2 \quad = \quad 0.01162 \text{ cm}^2$$

$$d \quad = \quad 0.122 \text{ cm}$$

$$= \quad \mathbf{1.22} \text{ mm} \qquad \qquad \underset{\Longleftarrow}{\underline{\text{Ans}}}$$

7.4. Determine the change of air-fuel ratio in an airplane-engine carburettor when it takes off from sea level to a height of 5000 m.
Carburettor is adjusted for 15:1 ratio at sea level where air temperature is 27°C and pressure 1 bar.

Assume the variation of temperature of air with altitude at
$t = t_s - 0.0065\,h$ where h is in m and t is in °C. The air pressure
decreases with altitude as per the relation $h = 19200 \log_{10}(1/p)$,
where p is in bar. Evaluate the variation of air-fuel ratio with
respect to altitude in steps of 1000 m on the trend. Show the
variation on a graph and discuss.

SOLUTION:

We shall illustrate the calculation for 5000 m

Temperature at 5000 m

$$t \quad = \quad 300 - 0.0065 \text{ h}$$

$$= \quad 300 - 0.0065 \times 5000 \quad = \quad 267.5 \text{ K}$$

$$h \quad = \quad 19200 \log\left(\frac{1}{p}\right)$$

$$10^{h/19200} \quad = \quad \frac{1}{p}$$

$$p \quad = \quad \frac{1}{10^{h/19200}} \quad = \quad 10^{-h/19200}$$

$$= \quad 0.549 \text{ bar}$$

$$\rho_{sl} \quad = \quad \frac{p_{sl}}{RT_{sl}}$$

$$= \quad \frac{1 \times 10^5}{287 \times 1000 \times 300} \quad = \quad 1.161$$

$$\rho_{al} \quad = \quad \frac{p_{al}}{RT_{al}}$$

$$= \frac{0.5490 \times 10^5}{0.287 \times 1000 \times 267.5} = 0.715$$

Therefore,

$$A/F_{al} = A/F_{sl} \times \sqrt{\frac{\rho_{al}}{\rho_{sl}}}$$

$$= 15 \times \sqrt{\frac{0.715}{1.16}}$$

$$= \mathbf{11.77} \qquad \text{Ans}$$

Similarly we can calculate A/F for various altitudes. The results are given in the table below.

h	0	1000	2000	3000	4000	5000
A/F	15	14.283	13.603	12.959	12.348	11.77

The variation is shown in the graph below.

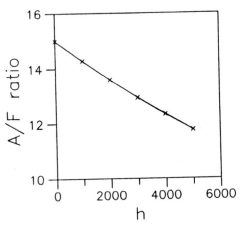

It is seen from the graph that with increase in altitude the mixture becomes richer and richer. Therefore some altitude compensating device should be incorporated. Otherwise proper combustion will not take place.

7.5. The venturi of a simple carburettor has a throat diameter of 20 mm and the coefficient of flow is 0.8. The fuel orifice has a diameter of 1.14 mm and the coefficient of fuel flow is 0.65. The gasoline surface is 5 mm below the throat, calculate

(i) the air-fuel ratio for a pressure drop of 0.08 bar when the nozzle tip is neglected

(ii) the air-fuel ratio when the nozzle tip is taken into account

(iii) the minimum velocity of air or critical air velocity required to start the fuel flow when the nozzle tip is provided.

Assume the density of air and fuel to be 1.20 kg/m³ and 750 kg/m³ respectively.

SOLUTION:

(i) When the nozzle tip is neglected

$$\dot{m}_a = C_{da} A_a \sqrt{2\rho_a \Delta p}$$

$$\dot{m}_f = C_{df} A_f \sqrt{2\rho_f \Delta p}$$

$$A/F = \frac{C_{da}}{C_{df}} \frac{A_a}{A_f} \sqrt{\frac{\rho_a}{\rho_f}}$$

$$= \frac{0.80}{0.65} \times \left(\frac{20}{1.14}\right)^2 \times \sqrt{\frac{1.20}{750}}$$

$$= \mathbf{15.15} \qquad \underset{\Longleftarrow}{\text{Ans}}$$

(ii) When the nozzle tip is taken into account

$\dot{m}_a$ will remain the same, But

$$\dot{m}_f = C_{df} A_F \sqrt{2\rho_f (\Delta p - \rho_f g h_f)}$$

$$\rho_f g h_f = \frac{5}{1000} \times 9.81 \times 750$$

$$= 36.79 \text{ N/m}^2$$

$$= 36.79 \times 10^{-5} = 0.00037 \text{ bar}$$

$$\Delta p - \rho_f g h_f = 0.08 - 0.00037 = 0.0796 \text{ bar}$$

$$A/F = \frac{C_{da}}{C_{df}} \frac{A_a}{A_f} \sqrt{\frac{\rho_a}{\rho_f}} \sqrt{\frac{\Delta p}{\Delta p - \rho_f g h_f}}$$

$$= \frac{0.8}{0.65} \times \left(\frac{20}{1.14}\right)^2 \times \sqrt{\frac{1.2}{750}} \times \sqrt{\frac{0.08}{0.0796}}$$

$$= \mathbf{15.19} \qquad \underset{\Longleftarrow}{\text{Ans}}$$

When there is a nozzle tip, the fuel flow will start, only when the minimum velocity of air required to create the requisite pressure difference for the flow of fuel to overcome nozzle tip effect.

The pressure difference Δp must be equal to $\rho_f g h_f$. Assuming velocity at the entrance of venturi $C_1 = 0$, we have

$$\frac{p_1}{\rho_a} = \frac{p_2}{\rho_a} + \frac{C_2^2}{2}$$

$$\frac{\Delta p}{\rho_a} = \frac{C_2^2}{2} = \frac{\rho_f g h_f}{\rho_a}$$

Therefore,

$$C_{min} = \sqrt{2 g h_f \frac{\rho_f}{\rho_a}}$$

$$= \sqrt{2 \times 9.81 \times \frac{5}{1000} \times \frac{750}{1.2}}$$

$$= \mathbf{7.83} \ \mathbf{m/s} \qquad \underset{\Longleftarrow}{\mathbf{Ans}}$$

The nozzle tip is the height of fuel nozzle in the throat above the gasoline surface in the carburettor. This is provided to avoid spilling of fuel due to vibration or slight inclination position of the carburettor. This would avoid wastage of fuel.

7.6. A carburettor, tested in the laboratory has its float chamber vented to atmosphere. The main metering system is adjusted to give an air-fuel ratio of 15:1 at sea level conditions. The pressure at the venturi throat is 0.8 bar. The atmospheric pressure is 1 bar. The same carburettor is tested again when an air cleaner is fitted at the inlet to the carburettor. The pressure drop to air cleaner is found to be 30 mm of Hg when air flow at sea level condition is 240 kg/h. Assuming zero tip and constant coefficient of flow, calculate (i) the throat pressure when the air cleaner is fitted and (ii) air-fuel ratio when the air cleaner is fitted.

SOLUTION:

$$\dot{m}_a = C_{da} A \sqrt{2 \rho_a \Delta p_a}$$

$$\Delta p_a = 1 - 0.8 = 0.2 \ \text{bar}$$

when air cleaner is fitted, let p_t' be the throat pressure, then

$$\Delta p'_a = (1 - 0.04 - p'_t)$$

$$= 0.96 - p'_t$$

For the same air flow and constant coefficients

$$\Delta p_a = \Delta p'_a$$

$$0.2 = 0.96 - p'_t$$

$$p'_t = 0.96 - 0.2$$

$$= \textbf{0.76 bar} \qquad \text{Ans}$$

Without air cleaner

$$\Delta p_a = 0.20 \text{ bar}$$

With air cleaner fitted with float chamber still vented to atmosphere

$$\Delta p_f = 1 - 0.76 = 0.24$$

Since Δp_f has increased more fuel will flow through, making the mixture richer

$$\text{New } A/F = 15 \times \sqrt{\frac{0.2}{0.24}}$$

$$= \textbf{13.7} \qquad \text{Ans}$$

7.7. A four-cylinder, four-stroke SI engine, having a bore of 10 cm and stroke 9 cm runs at 4000 rpm. The fuel used has a carbon content of 84.50 per cent and hydrogen content of 15.50 per cent by weight. The volumetric efficiency of the engine at 75% of full throttle and at 4000 rpm is 0.85 referred to 300 K and 1 bar. The engine is to be supplied with a mixture of air coefficient 0.95 when running at 75% of full throttle. Calculate the throat diameter of the venturi if the air velocity at throat is not to exceed 200 m/s under the above operating conditions. Also calculate the rate of fuel flow in kg/s at the pressure drop at venturi throat. Discharge coefficient for the venturi is 0.8 and the area ratio of the venturi is 0.8. Take R for air as 0.287 kJ/kg K and for fuel vapour is 0.09 kJ/kg K.

SOLUTION:

Volume flow rate of mixture/second

$$= \frac{\pi}{4} \times 10^2 \times 9 \times 10^{-6} \times \frac{4000}{2 \times 60} \times 4 \times 0.85$$

$$= 0.08 \ m^3/s$$

Air required for the stoichiometric mixture with 1 kg of fuel

$$= \left(\frac{0.845}{12} \times 32 + \frac{0.155}{2} \times 16 \right) \times \frac{100}{23.3}$$

$$= (2.25 + 1.24) \times \frac{100}{23.3}$$

$$= 14.98 \ kg/kg \ fuel$$

$$\dot{m}_a = 0.95 \times 14.98 = 14.23 \ kg/kg \ of \ fuel$$

$$F/A = \frac{1}{14.23} = 0.07$$

$$v_a = \frac{287 \times 300}{1 \times 10^5} = 0.861 \ m^3/kg$$

$$v_f = \frac{90 \times 300}{1 \times 10^5} \times 10^3 = 0.27 \ m^3/kg$$

Volume flow rate + Volume flow rate = Volume flow rate
     of air   of fuel   of mixture

Since A/F ratio is 14.23, we have

$$\dot{m}_a v_a + \dot{m}_f v_f = 0.08$$

$$\dot{m}_f = \frac{0.08}{14.23 \times 0.861 + 0.27}$$

$$= 0.00639 \ kg/s \qquad \underline{\underline{\textbf{Ans}}}$$

Density of air at inlet, ρ_1

$$= \frac{1 \times 10^5}{0.287 \times 300 \times 1000} = 1.16 \ kg/m^3$$

Pressure drop between inlet and venturi,

$$\Delta p = \frac{\rho_1}{2} C^2 \left[1 - \left(\frac{A_2}{A_1} \right)^2 \right]$$

$$= \frac{1.16}{2} \times 200^2 \times \left(1 - 0.8^2 \right) \times 10^{-5}$$

$$= 0.085 \text{ bar}$$

$$C_2 = \frac{\dot{m}_a}{A_2 C_{da} \rho_1}$$

$$200 = \frac{0.00639 \times 14.23}{A_2 \times 0.8 \times 1.16} = \frac{9.8}{A_2}$$

$$A_2 = \frac{0.098}{200} = 4.9 \times 10^4 \text{ m}^2$$

$$= 4.9 \text{ cm}^2$$

$$D_2 = \sqrt{\frac{4.9 \times 4}{\pi}}$$

$$= \textbf{2.50 cm} \qquad\qquad \underset{\Longleftarrow}{\text{Ans}}$$

Questions :-

7.1 Define carburetion.

7.2 Explain the various factors that affect the process of carburetion.

7.3 What are different air-fuel mixture on which an engine can be operated?

7.4 Explain the following:
     (i) rich mixture
     (ii) stoichiometric mixture
     (iii) lean mixture

7.5 How the power and efficiency of the SI engine vary with air-fuel ratio for different load and speed conditions?

7.6 By means of a suitable graph explain the necessary carburettor performance to fulfill engine requirements.

7.7 Briefly discuss the air-fuel ratio requirements of a petrol engine from no load to full load.

7.8 Explain why a rich mixture is required for the following:
   (i) idling
   (ii) maximum power
   (iii) sudden acceleration

7.9 Explain the principle of carburetion.

7.10 With a neat sketch explain the working principle of a simple carburettor.

7.11 Derive an expression for the air-fuel ratio of a simple carburettor.

7.12 Develop an expression for air-fuel ratio neglecting compressibility for a simple carburettor.

7.13 Explain why a simple carburettor cannot meet the engine requirement.

7.14 Describe the essential parts of a modern carburettor.

7.15 Describe with suitable sketches the following system of a modern carburettor:
   (i) main metering system
   (ii) idling system
   (iii) economizer system
   (iv) acceleration pump system
   (v) choke

7.16 What are the three basic types of carburettors? Explain.

7.17 With suitable sketches explain the various modern automobile carburettors.

7.18 With a suitable sketch explain the starting circuit of a solex carburettor.

7.19 Draw the sketch of a carter downdraught carburettor. How do the idle and low speed circuit work in this carburettor?

7.20 What are the special requirements of an air craft carburettor? What do you understand by altitude compensation? Explain.

Exercise:-

7.1. Determine the air-fuel ratio supplied at 4000 m altitude by a carburettor which is adjusted to give a stoichiometric air-fuel ratio at sea level where air temperature is 300 K and pressure is 1 bar.

Assume that the temperature of air decreases with altitude as given by

$$t \;=\; t_s - 0.00675h$$

where h is height in metres and t_s is sea level temperature in °C. The air pressure decreases with altitude as per the relation

$$h \;=\; 19000 \log_{10}\left(\frac{1}{p}\right)$$

where p is in bar. State any assumption made.

Ans: A/F at 4000 m = 12.35

7.2. Determine the sizes of fuel orifice to give a 13.5:1 air-fuel ratio, if the venturi throat has 3 cm diameter and the pressure drop in the venturi is 6.5 cm Hg. The air temperature and pressure at carburettor entrance are 1 bar and 27 °C respectively. The fuel orifice is at the same level as that of the float chamber. Take density of gasoline as 740 kg/m³ and discharge coefficient as unity. Assume atmospheric pressure to be 76 cm of Hg.

Ans: Orifice diameter = 1.62 mm

.3. A four-stroke gasoline engine with a swept volume of 5 litres has a volumetric efficiency of 75% when running at 3000 rpm. The engine is fitted with a carburettor which has a choke diameter of 35 mm. Assuming the conditions of a simple carburettor and neglecting the effects of compressibility, calculate the pressure and the air velocity at the choke. Take the coefficient of discharge at the throat as 0.85 and take the atmospheric conditions as 1 bar and 27 °C.   *Ans:* (i) Air velocity at the choke = 114.6 m/s
(ii) Pressure at the choke = 0.925 bar

7.4. A single jet carburettor is to supply 6 kg/min of air and 0.44 kg/min of petrol of specific gravity 0.74. The air is initially at 1 bar and 27 °C. Assuming an isentropic coefficient of 1.35 for air, determine (i) the diameter of the venturi if the air speed is 90 m/s and the velocity coefficient for venturi is 0.85 (ii) the diameter of the jet if the pressure drop at the jet is 0.8 times the

pressure drop at the venturi, and the coefficient of discharge for the jet is 0.66. *Ans*: (i) Diameter of the venturi = 3.58 cm
(ii) Diameter of the jet = 2.2 mm

7.5. A 8.255 cm × 8.9 cm four-cylinder, four-stroke cycle SI engine is to have a maximum speed of 3000 rpm and a volumetric efficiency of 80 per cent. If the maximum venturi depression is to be 0.1 bar (i) what must be the size of the venturi? (ii) Determine the size of fuel orifice if an air-fuel ratio of 12 to 1 is desired. Assume the density of air as 1.2 kg/m^3.

Ans: (i) Diameter of the venturi = 2.96 cm
(ii) Diameter of the jet = 1.45 mm

7.6. A six-cylinder, four-stroke cycle engine has a bore of 8 cm and a stroke of 11 cm and operates at a speed of 2000 rpm with volumetric efficiency of 80 per cent. If the diameter of the venturi section is 2.5 cm, what should be the diameter of the fuel orifice to obtain an A/F ratio of 12 to 1? Assume air density as 1.16 kg/m^3.

Ans: Diameter of the fuel orifice = 1.87 mm

7.7. An automobile carburettor having its float chamber vented to the atmosphere is tested in the factory without an air cleaner at sea-level conditions, the main metering system of this carburettor is found to yield a fuel-air ratio of 0.065. The venturi throat pressure is 0.84 bar. This carburettor is now installed in an automobile and an air cleaner is placed on the inlet to the carburettor. With the engine operating at 230 kg/h air consumption and at sea-level conditions, there is found to be a pressure drop through the air filter of 0.035 bar. Assuming $z = 0$ and orifice coefficients constant, calculate
(i) the venturi throat pressure with air cleaner
(ii) fuel-air ratio with air cleaner
Assume that the flow through the carburettor is incompressible.

Ans: (i) Venturi throat pressure with air cleaner = 0.805 bar
(ii) Fuel-air ratio with air cleaner = 0.0717

8

INJECTION

8.1 INTRODUCTION

The fuel-injection system is the most vital component in the working of CI engines. The engine performance viz., power output, economy etc. is greatly dependent on the effectiveness of the fuel-injection system. The injection system has to perform the important duty of initiating and controlling the combustion process.

Basically, the purpose of carburetion and fuel-injection is the same viz., preparation of the combustible charge. But in case of carburetion fuel is atomized by processes relying on the air speed greater than fuel speed at the fuel nozzle, whereas, in fuel-injection the fuel speed at the point of delivery is greater than the air speed to atomize the fuel. In carburettors, as explained in Chapter 7, air flowing through a venturi picks up fuel from a nozzle located there. The amount of fuel drawn into the engine depends upon the air velocity in the venturi. In a fuel-injection system, the amount of fuel delivered into the air stream going to the engine is controlled by a pump which forces the fuel under pressure.

When the fuel is injected into the combustion chamber towards the end of compression stroke, it is atomized into very fine droplets. These droplets vaporize due to heat transfer from the compressed air and form a fuel-air mixture. Due to continued heat transfer from hot air to the fuel, the temperature reaches a value higher than its self-ignition temperature. This causes the fuel to ignite spontaneously initiating the combustion process.

8.2 FUNCTIONAL REQUIREMENTS OF AN INJECTION SYSTEM

For a proper running and good performance from the engine, the following requirements must be met by the injection system:

(i) Accurate metering of the fuel injected per cycle. This is very critical due to the fact that very small quantities of fuel being handled. Metering errors may cause drastic variation from the desired output. The quantity of the fuel metered should vary to meet changing speed and load requirements of the engine.

(ii) Timing the injection of the fuel correctly in the cycle so that maximum power is obtained ensuring fuel economy and clean burning.

(iii) Proper control of rate of injection so that the desired heat-release pattern is achieved during combustion.

(iv) Proper atomization of fuel into very fine droplets.

(v) Proper spray pattern to ensure rapid mixing of fuel and air.

(vi) Uniform distribution of fuel droplets throughout the combustion chamber.

(vii) To supply equal quantities of metered fuel to all cylinders in case of multi cylinder engines.

(viii) No lag during beginning and end of injection i.e., to eliminate dribbling of fuel droplets into the cylinder.

8.3 CLASSIFICATION OF INJECTION SYSTEMS

In a constant-pressure cycle or diesel engine, pure air is compressed in the cylinder and then fuel is injected into the cylinder by means of a fuel-injection system. For producing the required pressure for atomizing the fuel either air or a mechanical means is used. Accordingly the injection systems can be classified as:

(i) Air injection systems

(ii) Solid injection systems

8.3.1 Air Injection System

In this system, fuel is forced into the cylinder by means of compressed air. This system is little used nowadays, because it requires a bulky multi-stage air compressor. This causes an increase in engine weight and reduces the brake power output further. One advantage that is claimed for the air injection system is good mixing of fuel with the air with resultant higher mean effective pressure. Another is the ability to utilize fuels of high viscosity which are less expensive than those used by the engines with solid injection systems. These advantages are off-set by the requirement of a multistage compressor thereby making the air-injection system obsolete.

8.3.2 Solid Injection System

In this system the liquid fuel is injected directly into the combustion chamber without the aid of compressed air. Hence, it is also called *airless mechanical injection* or *solid injection system*. Solid injection systems can be classified into four types.

  (i) Individual pump and nozzle system

 (ii) Unit injector system

(iii) Common rail system

(iv) Distributor system

    All the above systems comprise mainly of the following components.

   (i) fuel tank,

  (ii) fuel feed pump to supply fuel from the main fuel tank to the injection system,

 (iii) injection pump to meter and pressurize the fuel for injection,

 (iv) governor to ensure that the amount of fuel injected is in accordance with variation in load,

  (v) injector to take the fuel from the pump and distribute it in the combustion chamber by atomizing it into fine droplets,

 (vi) fuel filters to prevent dust and abrasive particles from entering the pump and injectors thereby minimising the wear and tear of the components.

    A typical arrangement of various components for the solid injection system used in a CI engine is shown in Fig.8.1. Fuel from the fuel tank first enters the coarse filter from which is drawn into the plunger feed pump where the pressure is raised very slightly. Then the fuel enters the fine filter where all the dust and dirt particles are removed. From the fine filter the fuel enters the fuel pump where it is pressurized to about 200 bar and injected into the engine cylinder by means of the injector. Any spill over in the injector is returned to the fine filter. A pressure relief valve is also provided for the safety of the system. The above functions are achieved with the components listed above.

    The types of solid injection system described below differ only in the manner of operation and control of the above mentioned components.

8.3.3 Individual Pump and Nozzle System

The details of the individual pump and nozzle system are shown in Fig.8.2(a) and (b). In this system, each cylinder is provided with one pump and one injector. In this arrangement a separate metering

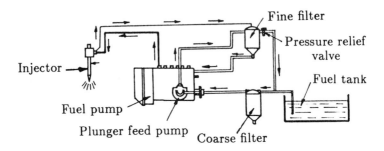

Fig.8.1 Typical Fuel Feed System for a CI Engine

and compression pump is provided for each cylinder. The pump may be placed close to the cylinder as shown in Fig.8.2(a) or they may be arranged in a cluster as shown in Fig.8.2(b). The high pressure pump plunger is actuated by a cam, and produces the fuel pressure necessary to open the injector valve at the correct time. The amount of fuel injected depends on the effective stroke of the plunger.

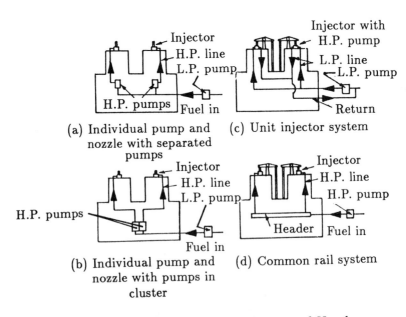

(a) Individual pump and
    nozzle with separated
    pumps

(c) Unit injector system

(b) Individual pump and
    nozzle with pumps in
    cluster

(d) Common rail system

Fig.8.2 Injection Systems with Pump and Nozzle
Arrangements used in CI Engines

8.3.4 Unit Injector System

The unit injector system, Fig.8.2(c), is one in which the pump and the injector nozzle are combined in one housing. Each cylinder is provided with one of these unit injectors. Fuel is brought up to the injector by a low pressure pump, where at the proper time, a rocker arm actuates the plunger and thus injects the fuel into the cylinder. The amount of fuel injected is regulated by the effective stroke of the plunger. The pump and the injector can be integrated in one unit as shown in Fig.8.2(c).

8.3.5 Common Rail System

In the common rail system, Fig.8.2.(d), a HP pump supplies fuel, under high pressure, to a fuel header. High pressure in the header forces the fuel to each of the nozzles located in the cylinders. At the proper time, a mechanically operated (by means of a push rod and rocker arm) valve allows the fuel to enter the proper cylinder through the nozzle. The pressure in the fuel header must be that, for which the injector system was designed, i.e., it must enable to penetrate and disperse the fuel in the combustion chamber. The amount of fuel entering the cylinder is regulated by varying the length of the push rod stroke.

A high pressure pump is used for supplying fuel to a header, from where the fuel is metered by injectors (assigned one per cylinder). The details of the system are illustrated in Fig.8.2(d).

8.3.6 Distributor System

Figure 8.3 shows a schematic diagram of a distributor system. In this system the pump which pressurizes the fuel also meters and times it. The fuel pump after metering the required amount of fuel supplies it to a rotating distributor at the correct time for supply to each cylinder. The number of injection strokes per cycle for the pump is equal to the number of cylinders. The details of the system are given in Fig.8.3.

Since there is one metering element in each pump, a uniform distribution is automatically ensured. Not only that, the cost of the fuel-injection system also reduces to a value less than two-thirds of that for individual pump system. A comparison of various fuel-injection systems is given in Table 8.1.

8.4 FUEL FEED PUMP

A schematic sketch of fuel feed pump is shown in Fig.8.4. It is of spring loaded plunger type. The plunger is actuated through a push rod from the cam shaft.

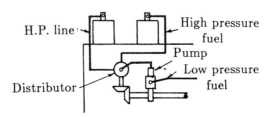

Fig.8.3 Schematic Diagram of Distributor System

Table 8.1 *Comparison of Various Fuel-Injection Systems*

Job	Air injection system	Solid injection system		
		Individual pump	Common rail	Distributor
Metering	Pump	Pump	Injection valve	Pump
Timing	Fuel cam	Pump cam	Fuel cam	Fuel cam
Injection rate	Spray valve	Pump cam	Spray valve	Fuel cam
Atomization	Spray valve	Spray tip	Spray tip	Spray tip
Distribution	Spray valve	Spray tip	Spray tip	Spray tip

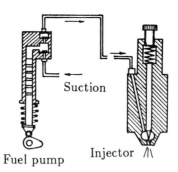

Fig.8.4 Schematic Diagram of Fuel Feed Pump

At the minimum lift position of the cam the spring force on the plunger creates a suction which causes fuel flow from the main tank into the pump. When the cam is turned to its maximum lift position, the plunger is lifted upwards. At the same time the inlet valve is closed and the fuel is forced through the outlet valve.

When the operating pressure gets released, the plunger return spring ceases to function resulting in varying of the pumping stroke under varying engine loads according to the quantity of fuel required by the injection pump.

8.5 INJECTION PUMP

The main objectives of fuel-injection pump is to deliver accurately metered quantity of fuel under high pressure (in the range from 120 to 200 bar) at the correct instant to the injector fitted on each cylinder.

Injection pumps are generally of two types, viz.,

(i) Jerk type pumps (ii) Distributor type pumps

8.5.1 Jerk Type Pump

It consists of a reciprocating plunger inside a barrel. The plunger is driven by a camshaft.

The working principle of jerk pump is illustrated in Fig.8.5.

(a) A sketch of a typical plunger is shown.

(b) A schematic diagram of the plunger within the barrel is shown. Near the port A, fuel is always available under relatively low pressure. While the axial movement of the plunger is through cam shaft, its rotational movement about its axis by means of rack D. Port B is the orifice through which fuel is delivered to the injector. At this stage it is closed by means of a spring loaded check valve.

When the plunger is below port A, the fuel gets filled in the barrel above it. As the plunger rises and closes the port A the fuel will flow out through port C. This is because it has to overcome the spring force of the check valve in order to flow through port B. Hence it takes the easier way out via port C.

(c) At this stage rack rotates the plunger and as a result port C also closes. The only escape route for the fuel is past the check valve through orifice B to the injector. This is the beginning of injection and also the effective stroke of the plunger.

(d) The injection continues till the helical indentation on the plunger uncovers port C. Now the fuel will take the easy way out through C and the check valve will close the orifice B. The fuel-injection stops and the effective stroke ends. Hence the effective stroke of the plunger is the axial distance traversed between the time port A is closed off and the time port A is uncovered.

(e) & (f) The plunger is rotated to the position shown. The same sequence of events occur. But in this case port C is uncovered sooner. Hence the effective stroke is shortened.

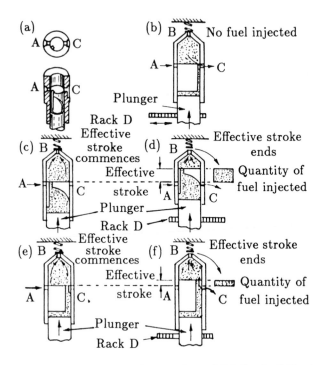

Fig.8.5 Diagrams illustrating an Actual Method of Controlling Quantity of Fuel Injected in a CI Engine

It is important to remember here that though the axial distance traversed by the plunger is same for every stroke, the rotation of the plunger by the rack determining the length of the effective stroke and thus the quantity of fuel injected.

A typical example of this type of pump is the Bosch fuel-injection pump shown in Fig.8.6.

8.5.2 Distributor Type Pump

This pump has only a single pumping element and the fuel is distributed to each cylinder by means of a rotor (Fig.8.7)

There is a central longitudinal passage in the rotor and also two sets of radial holes (each equal to the number of engine cylinders) located at different pressures. One set is connected to pump inlet via central passage whereas the second set is connected to delivery lines leading to injectors of the various cylinders.

The fuel is drawn into the central rotor passage from the inlet port when the pump plunger move away from each other. Wherever,

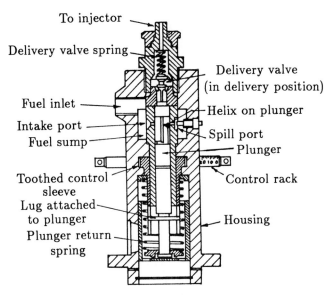

Fig.8.6 Single Cylinder Jerk Pump Type Fuel-Injection System

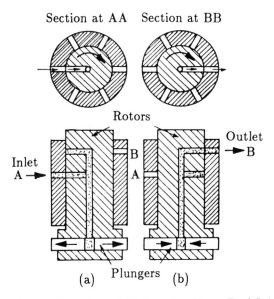

Fig.8.7 Principle of Working of Distributor Type Fuel-Injection Pump

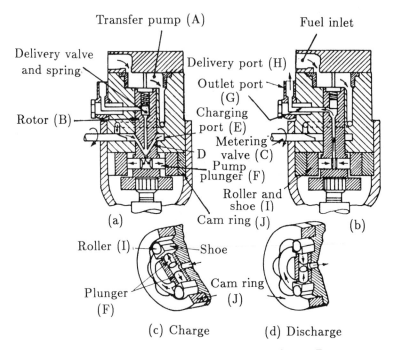

Transfer pump (A)   Fuel inlet

Delivery valve
and spring   Delivery port (H)

Outlet port
(G)

Charging
Rotor (B)   port (E)

Metering
D   valve (C)

Pump
plunger (F)

Roller and
shoe (I)

(a)   Cam ring (J)   (b)

Roller (I)   Shoe

Cam ring
Plunger   (J)
(F)

(c) Charge   (d) Discharge

Fig.8.8 Schematic of Roosa Master Distributor Pump

the radial delivery passage in the rotor coincides with the delivery port for any cylinder the fuel is delivered to each cylinder in turn.

Main advantages of this type of pump lies in its small size and its light weight.

A schematic diagram of Roosa Master distributor pump is shown in Fig.8.8.

8.6 INJECTION PUMP GOVERNOR

In a CI Engine the fuel delivered is independent of the injection pump characteristic and the air intake.

Fuel delivered by a pump increases with speed whereas the opposite is true about the air intake. This results in over fueling at higher speeds. And at idling speeds (low speeds) the engine tends to stall due to insufficiency of fuel.

Quantity of fuel delivered increases with load causing excessive carbon deposits and high exhaust temperature. Drastic reduction in load will cause over speeding to dangerous values.

It is the duty of an injection pump governor to take care of the above limitations.

Governors are generally of two types, viz.,

(i) Mechanical governor

(ii) Pneumatic governor

8.7 MECHANICAL GOVERNOR

The working principle of mechanical governor is illustrated in Fig.8.9. When the engine speed tends to exceed the limit the weights fly apart. This causes the bell crank levers to raise the sleeve and operate the control lever in downward direction. This actuates the control rack on the fuel-injection pump in a direction which reduces the amount of fuel delivered. Lesser fuel causes the engine speed to decrease. The reverse happens when engine speed tends to decrease.

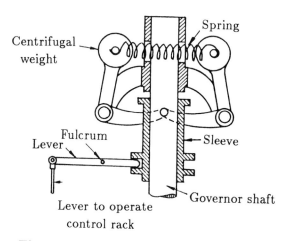

Fig.8.9 Principle of Mechanical Governor

8.8 PNEUMATIC GOVERNOR

The details of a pneumatic governor is shown in Fig.8.10. The amount of vacuum applied to the diaphragm is controlled by the accelerator pedal through the position of the butterfly valve in the venturi unit. A diaphragm is connected to the fuel pump control rack. Therefore, position of the accelerator pedal also determines the position of the pump control rack and hence the amount of fuel injected.

8.9 FUEL INJECTOR

Quick and complete combustion is ensured by a well designed fuel injector. By atomizing the fuel into very fine droplets, it increases the

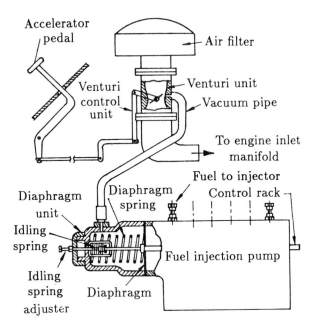

Fig.8.10 Principle of Pneumatic Governor

surface area of the fuel droplets resulting in better mixing and subsequent combustion. Atomization is done by forcing the fuel through a small orifice under high pressure.

The injector assembly consists of

(i) a needle valve

(ii) a compression spring

(iii) a nozzle

(iv) an injector body

A cross sectional view of a typical Bosch fuel injector is shown in Fig.8.11.

When the fuel is supplied by the injection pump it exerts sufficient force against the spring to lift the nozzle valve, fuel is sprayed into the combustion chamber in a finely atomized particles. After, fuel from the delivery pump gets exhausted, the spring pressure pushes the nozzle valve back on its seat. For proper lubrication between nozzle valve and its guide a small quantity of fuel is allowed to leak through the clearance between them and then drained back to fuel tank through leak off connection. The spring tension and hence the valve opening pressure is controlled by adjusting the screw provided at the top.

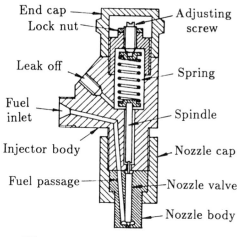

Fig.8.11 Fuel Injector (Bosch)

8.10 NOZZLE

Nozzle is that part of an injector through which the liquid fuel is sprayed into the combustion chamber.

The nozzle should fulfill the following functions:

(i) *Atomization* : This is a very important function since it is the first phase in obtaining proper mixing of the fuel and air in the combustion chamber.

(ii) *Distribution of Fuel* : Distribution of fuel to the required areas within the combustion chamber. Factors affecting this are:

(a) *Injection Pressure* : Higher the injection pressure better the dispersion and penetration of the fuel into all the desired locations in combustion chamber.

(b) *Density of Air in the cylinder* : If the density of compressed air in the combustion chamber is high then the resistance to the movement of the droplets is higher and dispersion of the fuel is better.

(c) *Physical Properties of Fuel* : The properties like self-ignition temperature, vapour pressure, viscosity, etc. play an important role in the distribution of fuel.

(iii) *Prevention of impingement on walls* : Prevention of the fuel from impinging directly on the walls of combustion chamber or piston. This is necessary because fuel striking the walls decomposes and produces carbon deposits. This causes smoky exhaust as well as increase in fuel consumption.

(iv) *Mixing* : Mixing the fuel and air in case of non-turbulent type of combustion chamber should be taken care of by the nozzle.

8.10.1 Types of Nozzle

The design of the nozzle must be such that the liquid fuel forced through the nozzle will be broken up into fine droplets, or atomized, as it passes into the combustion chamber. This is the first phase in obtaining proper mixing of the fuel and air in the combustion chamber.

The fuel must then be properly distributed, or dispersed, in the desired areas of the chamber. In this phase, the injection pressure, the density of the air in the cylinder and the physical qualities of the fuel in use, as well as the nozzle design, become important factors. Higher injection pressure results in better dispersion as well as greater penetration of the fuel into all locations in the chamber where is presence is desired. It also produces finer droplets which tend to mix more readily with the air. The greater the density of the compressed air in the combustion chamber, the greater the resistance offered to the travel of the fuel droplets across the chamber, with resultant better dispersion of the fuel. The physical qualities of the fuel itself, such as viscosity, surface tension, etc. also enter into the dispersion of the fuel.

The nozzle must spray the fuel into the chamber in such a manner as to minimize the quantity of fuel reaching the surrounding walls. Any fuel striking the walls tends to decompose, producing carbon deposits, unpleasant odour and a smoky exhaust, as well as an increase in fuel consumption.

The design of the nozzle is closely interrelated to the type of combustion chamber used. It is sufficient to state here that the *turbulent* type of combustion chamber depends upon chamber turbulence to produce the required mixing of the fuel and air. The *non-turbulent* type of combustion chamber, on the other hand, depends almost entirely on both the nozzle design and injection pressure to secure the desired *mixing* in the combustion chamber; consequently, with this type of chamber, the nozzle must accomplish the additional function of *mixing* the fuel and air.

Various types of nozzles are used in CI engines. These types are shown in Fig.8.12. The most common types are:

(i)   the pintle nozzle,   (ii) the single hole nozzle

(iii) the multi-hole nozzle, (iv) pintaux nozzle

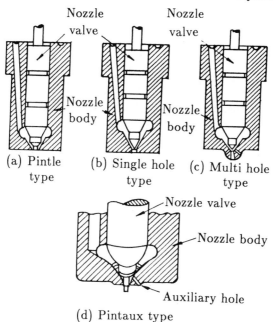

(a) Pintle type   (b) Single hole type   (c) Multi hole type

(d) Pintaux type

Fig.8.12 Types of Nozzles

(i) *Pintle Nozzle :* The stem of the nozzle valve is extended to form a pin or pintle which protrudes through the mouth of the nozzle [Fig.8.12(a)]. The size and shape of the pintle can be varied according to the requirement. It provides a spray operating at low injection pressures of 8-10 MPa. The spray cone angle is generally 60°. Advantage of this nozzle is that it avoids weak injection and dribbling. It prevents the carbon deposition on the nozzle hole.

(ii) *Single Hole Nozzle :* At the centre of the nozzle body there is a single hole which is closed by the nozzle valve [Fig.8.12(b)]. The size of the hole is usually of the order of 0.2 mm. Injection pressure is of order of 8-10 MPa and spray cone angle is about 15°. Major disadvantage with such nozzle is that they tend to dribble. Besides, their spray angle is too narrow to facilitate good mixing unless higher velocities are used.

(iii) *Multi-hole Nozzle :* It consists of a number of holes bored in the tip of the nozzle [Fig.8.12(c)]. The number of holes varies from 4 to 18 and the size from 35 to 200 μm. The hole angle may be from 20° upwards. These nozzles operate at high injection pressures of the order of 18 MPa. Their advantage lies in the ability to distribute the fuel properly even with lower air motion available in open combustion chambers.

(iv) *Pintaux Nozzle :*   It is a type of pintle nozzle which has an auxiliary hole drilled in the nozzle body [Fig.8.12(d)]. It injects a small amount of fuel through this additional hole (pilot injection) in the upstream direction slightly before the main injection. The needle valve does not lift fully at low speeds and most of the fuel is injected through the auxiliary hole. Main advantage of this nozzle is better cold starting performance. (20 to 25 °C lower than multi hole design). A major drawback of this nozzle is that its injection characteristics are poorer than the multihole nozzle.

8.10.2 Spray Formation

The various phases of spray formation as the fuel is injected through the nozzle are shown in Fig.8.13.

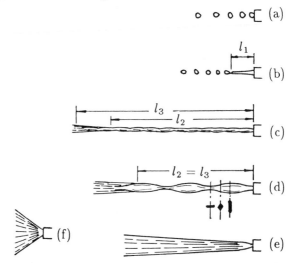

Fig.8.13 Successive Phases of Spray Formation

At the start of the fuel-injection the pressure difference across the orifice is low. Therefore single droplets are formed as in Fig.8.13(a). As the pressure difference increases the following process occur one after the other.

(i) A stream of fuel emerges from the nozzle, [Fig.8.13 (b)].

(ii) The stream encounters aerodynamic resistance from the dense air present in the combustion chamber (12 to 14 times the ambient pressure) and breaks into a spray, say at a distance of l_3, [Fig.8.13(c)]. The distance of this point where this event occurs from the orifice is called the *break-up distance.*

(iii) With further and further increase in the pressure difference, the break-up distance decreases and the cone angle increases until the apex of the cone practically coincides with the orifice [Fig.8.13(d), (e) and (f)].

At the exit of the orifice the fuel jet velocity, V_f, is of the order of 400 m/s. It is given by the following equation

$$V_f \;=\; C_d \sqrt{\frac{2\,(p_{inj} - p_{cyl})}{\rho_f}}$$

where
C_d = coefficient of discharge for the orifice
p_{inj} = fuel pressure at the inlet to injector, N/m^2
p_{cyl} = pressure of charge inside the cylinder, N/m^2
ρ_f = fuel density, kg/m^3

The spray from a circular orifice has a denser and compact core, surrounded by a cone of fuel droplets of various sizes and vaporized liquid. Larger droplets provide a higher penetration into the chamber but smaller droplets are required for quick mixing and evaporation of the fuel. The diameter of most of the droplets in a fuel spray is less than 5 microns. The droplet sizes depends on various factors which are listed below:

(i) Mean droplet size decreases with increase in injection pressure.

(ii) Mean droplet size decreases with increase in air density.

(iii) Mean droplet size increases with increase in fuel viscosity.

(iv) Size of droplets increases with increase in the size of the orifice.

8.10.3 Quantity of Fuel and the Size of Nozzle Orifice

The quantity of the fuel injected per cycle depends to a great extent upon the power output of the engine. As already mentioned the fuel is supplied into the combustion chamber through the nozzle holes and the velocity of the fuel for good atomization is of the order of 400 m/s. The velocity of the fuel through nozzle orifice in terms of h can be given by

$$V_f \;=\; C_d \sqrt{2gh}$$

where h is the pressure difference between injection and cylinder pressure, measured in m of fuel column.

The volume of the fuel injected per second, Q, is given by

Q = Area of all orifices × fuel jet velocity × time of one injection × number of injections per second for one orifice

$$Q \;=\; \left(\frac{\pi}{4}d^2 \times n\right) \times V_f \times \left(\frac{\theta}{360} \times \frac{60}{N}\right) \times \left(\frac{N_i}{60}\right)$$

where N_i for four-stroke engine is rpm/2 and for a two-stroke engine N_i is rpm itself and d is the diameter of one orifice in m, n is the number of orifices, θ is the duration of injection in crank angle degrees and N_i is the number of injections per minute.

Usually the rate of fuel-injection is expressed in mm^3/degree crank angle/litre cylinder displacement volume to normalize the effect of engine size.

The rate of fuel injected/degree of crankshaft rotation is a function of injector camshaft velocity, the diameter of the injector plunger, and flow area of the tip orifices. Increasing the rate of injection decreases the duration of injection for a given fuel input and subsequently introduces a change in injection timing. A higher rate of injection may permit injection timing to be retarded from optimum value. This helps in maintaining fuel economy without excessive smoke emission. However, an increase in injection rate requires an increased injection pressure and increases the load on the injector push rod and the cam. This may affect the durability of the engine.

8.11 INJECTION IN SI ENGINE

Fuel-injection systems are commonly used in CI engines. Presently gasoline injection system is coming into vogue in SI engines because of the following drawbacks of the carburetion.

   (i) Non uniform distribution of mixture in multicylinder engines.

  (ii) Loss of volumetric efficiency due to the restrictions imposed by the presence of carburettor and other components.

 (iii) Possibility of back firing.

A gasoline injection system eliminates all these drawbacks. The injection of fuel into an SI engine can be done by employing any of the following methods which are shown in Fig.8.14.

  (a) direct injection of fuel into the cylinder

  (b) injection of fuel close to the inlet valve

  (c) injection of fuel into the inlet manifold

      There are two types of gasoline injection systems, viz.,

   (i) *Continuous Injection* : Fuel is continuously injected. It is adopted when manifold injection is contemplated.

  (ii) *Timed Injection* : Fuel is injected only during induction stroke over a limited period. Injection timing is not a critical factor in SI engines.

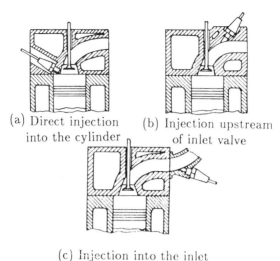

(a) Direct injection (b) Injection upstream
into the cylinder of inlet valve

(c) Injection into the inlet
manifold

Fig.8.14 Location of Injection Nozzle

Major advantages of fuel-injection in an SI engine are:

(i) Increased volumetric efficiency

(ii) Better thermal efficiency

(iii) Lower exhaust emissions

(iv) High quality fuel distribution

The use of petrol injection is limited by its high initial cost, complex design and increased maintenance requirements. It is believed that the petrol injection has a promising future compared to carburetion and may replace carburettor in the near future.

Worked out Examples

8.1. A six cylinder, four-stroke diesel engine develops 125 kW at 3000 rpm. Its brake specific fuel consumption is 200 gm/kW h. Calculate the quantity of fuel to be injected per cycle per cylinder. Specific gravity of the fuel may be taken as 0.85.

SOLUTION:

$$Fuel\ consumed/hour \quad = \quad bsfc \times Power\ output$$

$$= \quad 200 \times 10^{-3} \times 125 \quad = \quad 25\ \text{kg}$$

$$\text{Fuel consumption/cylinder} \quad = \quad \frac{25}{6} \quad = \quad 4.17 \text{ kg/h}$$

$$\text{Fuel consumption/cycle} \quad = \quad \frac{\text{Fuel consumption/minute}}{n}$$

where $n = N/2$ for four-stroke cycle engines

$$= \quad \frac{4.17/60}{3000/2} \quad = \quad 4.63 \times 10^{-5} \text{ kg}$$

$$= \quad 0.0463 \text{ gm}$$

$$\text{Volume of fuel injected} \quad = \quad \frac{0.0463}{0.85}$$

$$= \quad \textbf{0.0545 cc/cycle} \qquad \overset{\textbf{Ans}}{\Longleftarrow}$$

8.2. Calculate the diameter of the fuel orifice of a four-stroke engine which develops 25 kW per cylinder at 2500 rpm. The specific fuel consumption using 0.3 kg/kW h fuel of 30°API. The fuel is injected at a pressure of 150 bar over a crank travel of 25°. The pressure in the combustion chamber is 40 bar. Coefficient of velocity is 0.875 and specific gravity is given by

$$S.G. \quad = \quad \frac{141.5}{131.5 + {}^\circ API}$$

SOLUTION:

$$\text{Duration of injection} \quad = \quad \frac{25}{360 \times 2500/60}$$

$$= \quad 1.667 \times 10^{-3} \text{ s}$$

$$S.G. \quad = \quad \frac{141.5}{131.5 + 30^\circ} \quad = \quad 0.8762$$

$$\text{Velocity of injection, } V_{inj} \quad = \quad C_d \sqrt{\frac{2(p_{inj} - p_{cyl})}{\rho_f}}$$

$$= \quad 0.875 \times \sqrt{\frac{2 \times (150 - 40) \times 10^5}{876.2}}$$

$$= \quad 138.65 \text{ m/s}$$

$$\text{Vol. of fuel injected/cycle} = \frac{(bsfc/60) \times output}{(rpm/2) \times \rho_{fuel}}$$

$$= \frac{(0.3/60) \times 25}{(2500/2) \times 876.2}$$

$$= 0.114 \times 10^{-6} \text{ m}^3/\text{cycle}$$

$$\text{Nozzle orifice area, } A_f = \frac{Volume \ of \ fuel \ injected/cycle}{Inj. \ velocity \times \ Inj. \ time}$$

$$= \frac{0.114 \times 10^{-6}}{138.65 \times 1.667 \times 10^{-3}}$$

$$= 0.4932 \times 10^{-6} \text{ m}^2$$

$$\text{Area of the orifice} = \frac{\pi}{4} \times d^2 = 0.4932 \times 10^{-6}$$

$$d = \sqrt{\frac{4 \times 0.4932 \times 10^{-6}}{\pi}}$$

$$= 0.792 \times 10^{-3} \text{ m}$$

$$= \mathbf{0.792} \text{ mm} \qquad \underset{\Leftarrow}{\underline{\mathbf{Ans}}}$$

8.3. A four-cylinder, four-stroke diesel engine develops a power of 180 kW at 1500 rpm. The *bsfc* is 0.2 kg/kW h. At the beginning of injection pressure is 30 bar and the maximum cylinder pressure is 50 bar. The injection is expected to be at 200 bar and maximum pressure at the injector is set to be about 500 bar. Assuming the following:

C_d for injector	=	0.7
S.G. of fuel	=	0.875
Atmospheric pressure	=	1 bar
Effective pressure difference	=	Average pressure difference over the injection period

Determine the total orifice area required per injector if the injection takes place over 15° crank angles.

SOLUTION:

$$\text{Power output/cylinder} = \frac{180}{4} = 45 \text{ kW}$$

$$\text{Fuel consumption/cylinder} = 45 \times bsfc$$

$$= 45 \times 0.2 = 9 \text{ kg/h}$$

$$= 0.15 \text{ kg/min}$$

$$\text{Fuel injected/cycle} = \frac{0.15}{(rpm/2)} = \frac{0.15}{(1500/2)}$$

$$= 2 \times 10^{-4} \text{ kg}$$

$$\text{Time for injection} = \frac{\theta}{360 \times rpm/60}$$

$$= \frac{15 \times 60}{360 \times 1500}$$

$$= 1.667 \times 10^{-3} \text{ s}$$

$$\text{Pressure difference at beginning} = 200 - 30$$

$$= 170 \text{ bar}$$

$$\text{Pressure difference at end} = 500 - 50$$

$$= 450 \text{ bar}$$

$$\text{Average pressure difference} = \frac{450 + 170}{2}$$

$$= 310 \text{ bar}$$

$$\text{Velocity of injection, } V_{inj} = C_d \sqrt{\frac{2(p_{inj} - p_{cyl})}{\rho_f}}$$

$$= 0.7 \times \sqrt{\frac{2 \times 310 \times 10^5}{875}}$$

$$= 186.33 \text{ m/s}$$

$$\text{Volume of fuel injected/cycle} = \frac{2 \times 10^{-4}}{875}$$

$$= 0.2286 \times 10^{-6} \text{ m}^3\text{/cycle}$$

$$A_f = \frac{0.2286 \times 10^{-6}}{186.33 \times 1.667 \times 10^{-3}}$$

$$= \mathbf{0.736 \times 10^{-6}} \ \mathbf{m^2} \qquad \underset{=}{\mathbf{Ans}}$$

8.4. A closed type injector has a nozzle orifice diameter of 0.9 mm and the maximum cross sectional area of the passage between the needle cone and the seat is 1.75 mm². The discharge coefficient for the orifice is 0.85 and for the passage is 0.80. The injection pressure is 175 bar and the average pressure of charge during injection is 25 bar, when the needle cone is fully lifted up. Calculate the volume rate of flow per second of fuel through the injector and the velocity of jet at that instant. Density of fuel is 850 kg/m³.

SOLUTION:

$$p_{inj} - p = \frac{\dot{V}_f^2 \times \rho_f}{2 \times 10^5} \frac{1}{(C_d A)^2} \ \text{bar}$$

Now we have two equations

$$175 - p = \frac{\dot{V}_f^2 \times 850}{2 \times 10^5} \times \frac{1}{(0.8 \times 1.75 \times 10^{-6})^2} \ \text{bar}$$

$$= 2.1684 \times 10^9 \times \dot{V}_f^2 \qquad (1)$$

$$p - 25 = \frac{\dot{V}_f^2 \times 850}{2 \times 10^5} \times \frac{1}{(0.85 \times 0.7584 \times 0.9 \times 10^{-6})^2}$$

$$= 12.6261 \times 10^9 \times \dot{V}_f^2 \ \text{bar} \qquad (2)$$

Adding (1) and (2), we get

$$150 = \dot{V}_f^2 \times (2.1684 + 12.6261) \times 10^9$$

$$= 14.7945 \times 10^9 \times V_f^2$$

$$\dot{V}_f = \sqrt{\frac{150}{14.7945 \times 10^9}}$$

$$= 1.007 \times 10^{-4} \ \text{m}^3/\text{s}$$

$$= \mathbf{100.7 \ cc/s} \qquad \underset{=}{\mathbf{Ans}}$$

Now, from Eqn.1,

$$p = 175 - 2.1684 \times 10^9 \times (1.007 \times 10^{-4})^2$$

$$= 153.01 \text{ bar}$$

Velocity of fuel through orifice

$$= \sqrt{\frac{2 \times (153.68 - 25)}{850}} \times 10^5$$

$$= \textbf{173.55 m/s} \qquad \underset{\Longleftarrow}{\textbf{Ans}}$$

8.5. At injection pressure of 150 bar a spray penetration of 25 cm in 20 milliseconds is obtained. If an injectior pressure of 250 bar had been used, what would have been the time taken to penetrate the same distance. Assume the same orifice and combustion chamber density. The combustion chamber pressure is 25 bar.

Use the relation

$$S \propto t\sqrt{\Delta p}$$

Where   S is penetration in cm
t is time in millisecond
Δp is the pressure difference between injection pressure
and combustion chamber pressure

SOLUTION:

$$\frac{S_1}{S_2} = \frac{t_1\sqrt{\Delta p_1}}{t_2\sqrt{\Delta p_2}}$$

$$\Delta p_1 = p_{inj,1} - p_{cyl} = 150 - 25$$

$$\Delta p_2 = p_{inj,2} - p_{cyl} = 250 - 25$$

$$t_2 = \frac{S_2}{S_1} t_1 \frac{\sqrt{\Delta p_1}}{\sqrt{\Delta p_2}}$$

$$= 1 \times 20 \times \sqrt{\frac{125}{225}}$$

$$= \textbf{14.91 milliseconds} \qquad \underset{\Longleftarrow}{\textbf{Ans}}$$

8.6. A six cylinder diesel engine produces 100 kW at 1500 rpm. The specific fuel consumption of the engine is 0.3 kg/kW h. Each cylinder has a separate fuel pump, injector and pipe line. At the beginning of effective plunger stroke of one fuel pump, the fuel in the pump barrel is 4 cc, fuel inside the injector is 2 cc and fuel in the pipe line is 3 cc. If the average injector pressure is 300 bar and average pressure of charge during injection is 40 bar, calculate the displacement volume of one plunger per cycle and power lost in pumping fuel to the engine (for all cylinders). Specific gravity of fuel is 0.9 and the fuel enter the pump barrel at 1 bar. Coefficient of compressibility of fuel may be taken as 80×10^{-6} per bar.

SOLUTION:

$$\text{Fuel consumed} \quad = \quad \frac{100}{6} \times 0.3 \times \frac{1}{3600}$$

$$= \quad 1.3888 \times 10^{-3} \text{ kg/s}$$

$$\dot{V}_f \quad = \quad \frac{1.3888 \times 10^{-3}}{\left(\frac{1500}{2 \times 60}\right)} \times \frac{1}{0.9 \times 10^3}$$

$$= \quad 0.12345 \times 10^{-6} \text{ m}^3/\text{s}$$

$$= \quad 0.12345 \text{ cc/s}$$

Coefficient of compressibility is defined as

$$K_{comp} \quad = \quad \frac{\text{Change in volume/unit volume}}{\text{Difference in pressure causing the compression}}$$

$$= \quad \frac{(V_1 - V_2)/V_1}{(p_2 - p_1)}$$

Here, $\qquad K_{comp} \quad = \quad 80 \times 10^{-6}$ *per bar*

$$V_1 \quad = \quad \textit{Fuel in pump barrel} + \textit{Fuel inside the}$$

$$\textit{injector} + \textit{Fuel in pipe line}$$

$$= \quad 4 + 2 + 3 \quad = \quad 9 \text{ cc}$$

$$V_1 - V_2 \;=\; K_{comp} \times V_1 \times (p_2 - p_1)$$

$$=\; 80 \times 10^{-6} \times 9 \times (300 - 1) \;=\; 0.21528 \text{ cc}$$

Plunger displacement volume

$$=\; 0.12345 + 0.21528$$

$$=\; \mathbf{0.339 \text{ cc}} \qquad\qquad \underset{\Longleftarrow}{\text{Ans}}$$

$$\text{Pump Work, } W_P \;=\; \frac{1}{2}(p_{inj} - p_o)(V_1 - V_2) + (p_{inj} - p_{cyl}) \times V_f$$

$$=\; \frac{1}{2}(300 - 1) \times 10^5 \times 0.2153 \times 10^{-6} +$$

$$(300 - 40) \times 10^5 \times 0.12345 \times 10^{-6}$$

$$=\; 3.22 + 3.21$$

$$=\; 6.43 \text{ J}$$

Power lost for pumping the fuel

$$=\; \frac{6.43}{1000} \times \frac{1500}{2 \times 60}$$

$$=\; \mathbf{0.08 \text{ kW}} \qquad\qquad \underset{\Longleftarrow}{\text{Ans}}$$

8.7. Before commencement of the effective stroke the fuel in the pump barrel of a diesel fuel injection pump is 6 cc. The diameter of the fuel line from pump to injector is 2.5 mm and is 600 mm long. The fuel in the injection valve is 2 cc.

 (i) To deliver 0.10 cc of fuel at a pressure of 150 bar, how much displacement the plunger undergoes. Assume a pump inlet pressure of 1 bar.

 (ii) What is the effective stroke of the plunger if its diameter is 7 mm.

Assume coefficient of compressibility of oil as 75×10^{-6} per bar at atmospheric pressure.

SOLUTION:

$$K \;=\; \frac{(V_1 - V_2)}{V_1(p_2 - p_1)}$$

Change in volume due to compression,

$$V_1 \quad = \quad \text{Total initial fuel volume}$$

$$= \quad \text{Volume of fuel in barrel}$$

$$+ \text{ Volume of fuel in delivery line}$$

$$+ \text{ Volume of fuel in injection valve}$$

$$= \quad 6 + \frac{\pi}{4} \times (0.25)^2 \times \frac{600}{10} + 2$$

$$= \quad 10.95 \text{ cc}$$

$$V_1 - V_2 \quad = \quad K \times V_1 \times (p_2 - p_1)$$

$$= \quad 75 \times 10^{-6} \times 10.95 \times (150 - 1)$$

$$= \quad 0.122 \text{ cc}$$

Total displacement of plunger

$$= \quad (V_1 - V_2) + 0.10$$

$$= \quad 0.122 + 0.10$$

$$= \quad \mathbf{0.222} \text{ cc} \qquad \underset{\Longleftarrow}{\text{Ans}}$$

$$\frac{\pi}{4} d^2 \times l \quad = \quad 0.222$$

Effective stroke of plunger $\quad = \quad 0.222 \times \dfrac{4}{\pi} \times \dfrac{1}{(0.7)^2}$

$$= \quad 0.577 \text{ cm}$$

$$= \quad \mathbf{5.77} \text{ mm} \qquad \underset{\Longleftarrow}{\text{Ans}}$$

Questions :-

8.1 What are the functional requirements of an injection system?

8.2 How are the injection system classified? Describe them briefly. Why the air injection system is not used nowadays?

8.3 How is the solid injection further classified? With a neat sketch explain the various components of a fuel feed system of a CI engine.

8.4 Explain (i) individual pump and nozzle system (ii) unit injector system (iii) common rail system (iv) distributor system.

8.5 Draw a schematic diagram of fuel feed pump and explain its working principle.

8.6 What are the main functions of an injection pump? What are two types of injection pump that are commonly used?

8.7 With a neat sketch explain the jerk pump type injection system.

8.8 With a neat sketch explain the working principle of a distributor type fuel-injection pump.

8.9 What is the purpose of using a governor in CI engines? What are the two major type of governors?

8.10 Explain with a neat sketch the working principle of a mechanical governor.

8.11 With a neat diagram bring out clearly the working principle of a pneumatic governor.

8.12 What is the purpose of a fuel injector? Mention the various parts of a injector assembly.

8.13 What are the functions of a nozzle? With sketches explain the various types of nozzles.

8.14 Give a brief account of the injection in SI engine.

8.15 With sketches explain the possible locations of the injection nozzles in SI engines.

8.16 Draw a sketch of pintaux nozzle and discuss it merits and demerits.

8.17 Derive an expression for the velocity of injection.

8.18 Develop an equation for the amount of fuel injected per cycle in terms of brake horse power and speed of a four-stroke CI engine.

Exercise:-

8.1. A four-cylinder, four-stroke diesel engine develops 100 kW at 3500 rpm. Its brake specific fuel consumption is 180 gm per kW h. Calculate the quantity of fuel to be injected per cycle per cylinder. Specific gravity of the fuel may be taken as 0.88.

Ans: Volume of the fuel injected = 0.0487 cc

8.2. Calculate the diameter of the fuel orifice of a four-stroke engine which develops 20 kW per cylinder at 2000 rpm. The specific fuel consumption is 0.25 kg/kW h of fuel with 30°API. The fuel is injected at a pressure of 180 bar over a crank travel of 25°. The pressure in the combustion chamber is 38 bar. Coefficient of velocity is 0.85 and specific gravity is given by

$$S.G. \quad = \quad \frac{141.5}{131.5 + °API}$$

Ans: Diameter of fuel orifice = 0.6164 mm

8.3. A six-cylinder, four-stroke diesel engine develops a power of 200 kW at 2000 rpm. The *bsfc* is 0.2 kg/kW h. At the beginning of injection, pressure is 35 bar and the maximum cylinder pressure is 55 bar. The injection is expected to be at 180 bar and maximum pressure at the injector is set to be about 520 bar. Assuming the following:

C_d for injector	=	0.75
S.G. of fuel	=	0.85
Atmospheric pressure	=	1 bar
Effective pressure difference	=	Average pressure difference over the injection period

Determine the total orifice area required per injector if the injection takes place over 16° crank angles.

Ans: Orifice area = 0.4747 × 10⁻⁶ m²

8.4. A closed type injector has a nozzle orifice diameter of 1.0 mm and the maximum cross sectional area of the passage between the needle cone and the seat is 2.0 mm². The discharge coefficient for the orifice is 0.85 and for the passage is 0.80. The injection pressure is 200 bar and the average pressure of charge during injection is 25 bar, when the needle cone is fully lifted up. Calculate the volume rate of flow per second of fuel through the injector and the velocity of jet at that instant. Specific gravity of fuel is 0.85.

Ans: (i) Volume rate of flow of fuel = 121.3 cc/s
(ii) Velocity of jet = 188.23 m/s

8.5. At injection pressure of 180 bar a spray penetration of 30 cm in 22 milliseconds is obtained. If an injection pressure of 240 bar had been used, what would have been the time taken to penetrate the same distance. Assume the same orifice and combustion chamber density. The combustion chamber pressure is 30 bar. Use the relation:

$$S \quad \propto \quad t\sqrt{\Delta p}$$

where S is penetration in cm
t is time in millisecond
Δp is the pressure difference between injection pressure and combustion chamber pressure
Ans: (i) Time taken for penetration = 18.593 milliseconds

8.6. A four-cylinder diesel engine produces 100 kW at 1500 rpm. The specific fuel consumption of the engine is 0.28 kg/kW h. Each cylinder has a separate fuel pump, injector and pipe line. At the beginning of effective plunger stroke of one fuel pump, the fuel in the pump barrel is 4 cc, fuel inside the injector is 2 cc and fuel in the pipe line is 4 cc. If the average injector pressure is 200 bar and average pressure of charge during injection is 30 bar, calculate the displacement volume of one plunger per cycle and power lost in pumping fuel to the engine (for all cylinders). Specific gravity of fuel is 0.85 and the fuel enter the pump barrel at 1 bar. Coefficient of compressibility of fuel is 80×10^{-6} per bar.

Ans: (i) Displacement volume of each plunger/cycle = 0.3322 cc
(ii) Power lost in pumping = 0.056 kW

8.7. Before commencement of the effective stroke the fuel in the pump barrel of a diesel fuel injection pump is 7 cc. The diameter of the fuel line from pump to injector is 3 mm and is 600 mm long. The fuel in the injection valve is 3 cc.

(i) To deliver 0.10 cc of fuel at a pressure of 150 bar, how much displacement the plunger undergoes? Assume a pump inlet pressure of 1 bar.

(ii) What is the effective stroke of the plunger if its diameter is 6 mm? Assume coefficient of compressibility of oil as 78×10^{-6} at atmospheric pressure.

Ans: (i) Plunger displacement = 0.2655 cc
(ii) Effective stroke of plunger = 9.39 mm

9

IGNITION

9.1 INTRODUCTION

The electrical discharge produced between the two electrodes of a spark plug by the ignition system starts the combustion process in a spark-ignition engine. This takes place close to the end of the compression stroke. The high temperature plasma kernel created by the spark, develops into a self sustaining and propagating flame front. In this thin reaction sheet certain exothermic chemical reactions occur. The function of the ignition system is to initiate this flame propagation process. It must be noted that the spark is to be produced in a repeatable manner viz., cycle-by-cycle, over the full range of load and speed of the engine at the appropriate moment in the engine cycle. By implication, ignition is merely a prerequisite for combustion. Therefore, the study of ignition must begin with the phenomenon of combustion so that a criterion may be established to decide whether ignition has occurred. Although the ignition process is intimately connected with the initiation of combustion, it is not associated with the gross behaviour of combustion. Instead, it is a local small-scale phenomenon that takes place within a small zone in the combustion chamber.

In terms of its simplest definition, ignition has no degree, intensively or extensively. Either the combustion of the medium is initiated or it is not. Therefore, it is reasonable to consider ignition from the standpoint of the beginning of the combustion process that it initiates.

9.2 ENERGY REQUIREMENTS FOR IGNITION

The total enthalpy required to cause the flame to be self sustaining and promote ignition, is given by the product of the surface area of the spherical flame and the enthalpy per unit area. It is reasonable to assume that the basic requirement of the ignition system is that it should supply this energy within a small volume. Further, ignition should occur in a time interval sufficiently short to ensure that only a negligible amount of energy is lost other than to establish the flame. In view of this last mentioned condition, it is apparent that the rate of supply of energy is as important a factor as the total energy supplied.

A small electric spark of short duration would appear to meet most of the requirements for ignition. A spark can be caused by applying a sufficiently high voltage between two electrodes separated by a gap, and there is a critical voltage below which no sparking occurs. This critical voltage is a function of the dimension of the gap between the electrodes, the fuel-air ratio and the pressure of the gas. Additionally, the manner in which the voltage is raised to the critical value and the configuration and the condition of the electrodes are important in respect of the energy required.

An ignition process obeys the law of conservation of energy. Hence, it can be treated as a balance of energy between:

(i) that provided by an external source
(ii) that released by chemical reaction and
(iii) that dissipated to the surroundings by means of thermal conduction, convection and radiation

9.3 THE SPARK ENERGY AND DURATION

With a homogeneous mixture in the cylinder, spark energy of the order of 1 mJ and a duration of a few micro-seconds would suffice to initiate the combustion process. However, in practice, circumstances are less than the ideal. The pressure, temperature and density of the mixture between the spark plug electrodes have a considerable influence on the voltage required to produce a spark. Therefore, the spark energy and duration are to be of sufficient order to initiate combustion under the most unfavourable conditions expected in the vicinity of the spark plug over the complete range of engine operation. Usually, if the spark energy exceeds 40 mJ and the duration is longer than 0.5 ms, reliable ignition is obtained. If the resistance of the deposits on the spark plug electrodes is sufficiently high, the loss of electrical energy through these deposits may prevent the spark discharge.

9.4 IGNITION SYSTEM

In principle, a conventional ignition system should provide sufficiently large voltage across the spark plug electrodes to effect the spark discharge. Further, it should supply the required energy for the spark to ignite the combustible mixture adjacent to the plug electrodes under all operating conditions. It may be noted that for a given engine design, the optimum spark timing varies with engine speed, inlet manifold pressure and mixture composition. The design of a conventional ignition system should take these factors into account to provide the spark of proper energy and duration at the appropriate time.

As air is a poor conductor of electricity an air gap in an electric circuit acts as a high resistance. But when a high voltage is applied across the electrodes of a spark plug it produces a spark across the gap. When such a spark is produced to ignite a homogeneous air-fuel mixture in the combustion chamber of an engine it is called the spark-ignition system. The ignition systems are classified depending upon how the primary energy for operating the circuit is made available as:

(i) battery ignition systems

(ii) magneto ignition systems

9.5 REQUIREMENTS OF AN IGNITION SYSTEM

A smooth and reliable functioning of an ignition system is essential for reliable working of an engine. The requirements of such an ignition system are:

(i) It should provide a good spark between the electrodes of the plugs at the correct timing.

(ii) It should function efficiently over the entire range of engine speed.

(iii) It should be light, effective and reliable in service.

(iv) It should be compact and easy to maintain.

(v) It should be cheap and convenient to handle.

(vi) The interference from the high voltage source should not affect the functioning of the radio and television receivers inside an automobile.

9.6 BATTERY IGNITION SYSTEM

Most of the modern spark-ignition engines use battery ignition system. In this system, the energy required for producing spark is obtained from a 6 or 12 volt battery. The construction of a battery ignition system is extremely varied. It depends on the type of ignition energy storage as well as on the ignition performance which is required by the particular engine. The reason for this is that an ignition system is not an autonomous machine, that is, it does not operate completely by itself, but instead it is but one part of the internal combustion engine, the *heart of the engine*. It is therefore extremely important that the ignition system be matched sufficiently well to its engine.

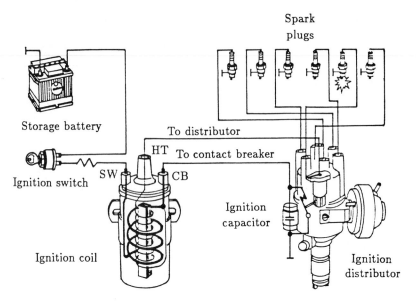

Fig.9.1 Battery Ignition System for a Six-Cylinder Engine

Passenger cars, light trucks, some motorcycles and large stationary engines are fitted with battery ignition systems. The details of the battery ignition system of a six-cylinder engine are shown in Fig.9.1.

The essential components of the system are:

  (i) battery
 (ii) ignition switch
(iii) ballast resistor
 (iv) ignition coil
  (v) contact breaker
 (vi) capacitor
(vii) distributor
(viii) spark plug

In the above list the first three components are housed in the primary side of the ignition coil whereas the last four are in the secondary side. The details of the various components are described briefly in the following sections.

9.6.1 Battery

To provide electrical energy for ignition, a storage battery is used. It is charged by a dynamo driven by the engine. Owing to the electro-chemical reactions, it is able to convert the chemical energy into electrical energy. The battery must be mechanically strong to withstand the strains to which it is constantly subjected to. Given reasonable care and attention two years or more trouble-free life may be obtained from a battery.

A lead acid battery consists of a number of cells connected together in series and each having a nominal potential of 2 volts when fully charged. A six volt battery has three such cells and a 12 volt battery has six. Figure 9.2 illustrates how six cells are coupled together to form a 12 volt battery and shows that for this coupling in series the positive of one cell is connected to the negative of the next.

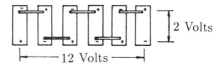

Fig.9.2 Cell Connections for 12 Volt Battery

Two types of batteries are used for spark-ignition engines, the lead acid battery and the alkaline battery. The former is used in light duty commercial vehicles and the later on heavy duty commercial vehicles.

9.6.2 Ignition Switch

Battery is connected to the primary winding of the ignition coil through an ignition switch and ballast resistor. With the help of the ignition switch the ignition system can be turned on or off.

9.6.3 Ballast Resistor

A ballast resistor is provided in series with the primary winding to regulate the primary current. The object of this is to prevent injury to the spark coil by overheating if the engine should be operated for a long time at low speed, or should be stalled with the breaker in the *closed* position. This coil is made of iron wire, and iron has the property that its electrical resistance increases very rapidly if a certain temperature is exceeded. The coil is therefore made of wire of such size that if the primary current flows nearly continuously, the ballast coil reaches a temperature above that where this rapid increase in resistance occurs. This additional resistance in the primary circuit holds the primary current down to a safe value. For starting from cold this resistor is by-passed to allow more current to flow in the primary circuit.

9.6.4 Ignition Coil

Ignition coil is the source of ignition energy in the conventional ignition system. This coil stores the energy in its magnetic field and delivers it at the appropriate time in the form of a ignition pulse through the high-tension ignition cables to the respective spark plug. The purpose of the ignition coil is to step up the 6 or 12 volts of the battery to a high voltage, sufficient to induce an electric spark across the electrodes of the spark plug. The ignition coil consists of a magnetic core of soft iron wire or sheet and two insulated conducting coils, called primary and the secondary windings.

The secondary coil consists of about 21,000 turns of 38-40 gauge enameled copper wire sufficiently insulated to withstand the high voltage. It is wound close to the core with one end connected to the secondary terminal and the other end grounded either to the metal case or the primary coil.

The primary winding, located outside the secondary coil is generally formed of 200-300 turns of 20 gauge wire to produce a resistance of about 1.15Ω. The ends are connected to exterior terminals. More heat is generated in the primary than in the secondary and with the primary coil wound over the secondary coil, it is easier to dissipate the heat.

The entire unit when assembled, is enclosed in a metal container and forms a neat and compact unit. On the top of the coil assembly is the heavily insulated terminal block, which supports three terminals. To the two smaller terminals (Fig.9.3), usually marked SW (switch wire) and CB (contact breaker) the two ends of the primary are connected.

As can be seen in Fig.9.3, one end of the secondary winding is connected to the central high-tension terminal in the moulded cover of the distributor. The other end is connected to the primary. An external high tension wire connects this central terminal to the central terminal of the distributor.

9.6.5 Contact Breaker

This is a mechanical device for making [Fig.9.4(a)] and breaking [Fig.9.4(b)] the primary circuit of the ignition coil. It consists essentially of a fixed metal point against which, another metal point bears which is being on a spring loaded pivoted arm. The metal used is invariably one of the hardest metals, usually tungsten and each point has a circular flat face of about 3 mm dia. The fixed contact point is earthed by mounting it on the base of the contact breaker assembly whereas the arm to which the movable contact point is attached, is electrically insulated. When the points are closed the current flows

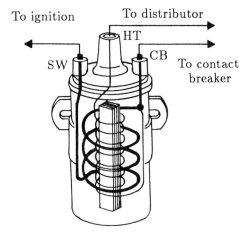

Fig.9.3 Ignition Coil

and when they are open, the circuit is broken and the flow of current stops. The pivoted arm has, generally, a heel or a rounded part of some hard plastic material attached in the middle and this heel bears on the cam which is driven by the engine. Consequently, every time the cam passes under the heel, the points are forced apart and the circuit is broken.

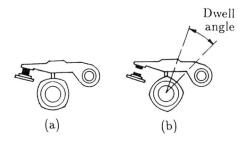

Fig.9.4 Contact Breaker

The pivoted arm is spring loaded, so that when the points are not separated by the action of the cam, they are held together by the pressure of the spring thereby closing the primary circuit.

The condition and adjustment of the contact breaker points are important. The points are subjected to a very severe hammering during their period of service. Uneven wear of the points may require a refacing or replacing depending upon the condition of the points.

An eight cylinder engine running at 3000 rpm requires 12000 sparks per minute, i.e., 200 sparks per second. If the breaker is to operate satisfactorily at this speed, the travel of the breaker arm must be held down to the minimum to ensure a positive spark and the breaker arm must be made very light.

9.6.6 Capacitor

The principle of construction of the ignition capacitor is the same as that of every electrical capacitor, which is very simple: two metal plates – separated by an insulating material – are placed face to face. The insulation is often only air (for example, in the case of air capacitors), but in most cases it consists of some high-quality insulating material suitable for the particular technical requirements, material which because of space limitation must be as thin as possible but nevertheless capable of withstanding electrostatic stresses without suffering damage. The metal plates themselves are usually replaced by metal foil or by metallic layers deposited by evaporation on the insulating material itself. In order to save space, these thin strips, for example consisting of two strips of aluminium foil and several layers of special capacitor paper, are rolled up in a solid roll. Contacts are attached to the two metal strips and the entire roll is first impregnated in an oily or waxy material to improve the insulating properties of the paper, and then the roll is inserted into a metal shell for protection against moisture, external physical contact and damage.

9.6.7 Distributor

The function of the distributor is to distribute the ignition surges to the individual spark plugs in the correct sequence and at the correct instants in time. Depending on whether a particular engine has 4, 6 or 8 cylinders, there are 4, 6 or 8 ignition pulses (surges) generated for every rotation of the distributor shaft. The use of a distributor represents a considerable simplification in a battery ignition system because in most cases we want to use only a single ignition circuit. The contact breaker and the spark advance mechanism are combined with the distributor in a single unit because of the absolute necessary that the distributor operate in synchronism with the crankshaft.

There are two types of distributors, the brush type and the gap type. In the former carbon brush carried by the rotor arm slides over metallic segments embedded in the distributor cap of molded insulating material, thus establishing electrical connection between the secondary winding of the coil and the spark plug, while in the latter the electrode of the rotor arm pass close to, but does not actually contact the segments in the distributor cap. With the latter type of distributor, there will not be any appreciable wear of the electrodes.

The distributor unit also consists of several other auxiliary units. In the lower part of the housing there is a speed sensitive device or governor, whose function is to advance the spark with increase in engine speed. Above this unit is the contact breaker assembly which can be rotated to adjust the timing of spark. In the upper part of the housing is located the high tension distributor. It also carries the vacuum ignition governor, which serves to retard the spark as the load on the engine increases.

Each of the segments of the distributor is connected to a sparking plug and as the rotor presses it, the contact breaker opens, the high tension current is passed through the rotor and brass segment through the high tension wiring to the appropriate spark plug. Obviously, the order in which the sparking plugs are connected to the distributor head will depend on the firing order of the engine.

9.6.8 Spark Plug

The spark plug provides the two electrodes with a proper gap across which the high potential discharges to generate a spark and ignite the combustible mixture within the combustion chamber.

A sectional view of a spark plug is shown in Fig.9.5. A spark plug consists essentially of a steel shell, an insulator and two electrodes. The central electrode to which the high tension supply from the ignition coil is connected, is well insulated with porcelain or other ceramic materials. The other electrode is welded to the steel shell of the plug and thereby is automatically grounded when the plug is installed on the cylinder head of the engine. The electrodes are usually made of high nickel alloy to withstand the severe erosion and corrosion to which they are subjected in use.

The tips of the central electrode and the insulator are exposed to the combustion gases. This results in the insulators having a tendency to crack from the high thermal and mechanical stresses. Some insulators are also seriously affected by moisture and by abnormal surface deposits. Since, the central electrode and the insulator are subjected to the high temperature of the combustion gases, the heat must flow from the insulator to the steel shell which is in contact with the relatively cool cylinder head in order to cool the electrodes and thereby prevent preignition.

Spark plugs are usually classified as *hot plugs or cold plugs* depending upon the relative operating temperature range of the tip of the high tension electrode. The operating temperature is governed by the amount of heat transferred which in turn depends on the length of the heat transfer path from the tip to the cylinder head and on the amount of surface area exposed to the combustion gases. A *cold plug* has a short heat transfer path and a small area exposed to the

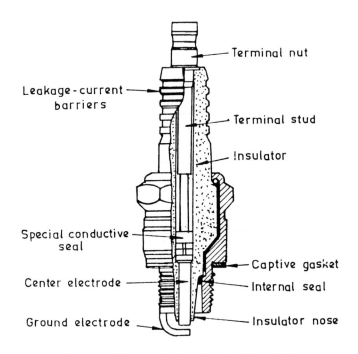

Fig.9.5 Schematic of a Typical Spark Plug

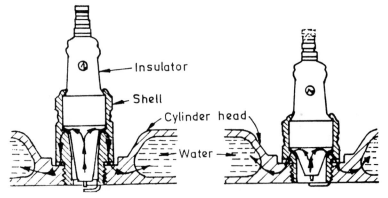

Fig.9.6 Heat Transfer Path of Hot and Cold Spark Plugs

combustion gases as compared to a hot plug as shown in Fig.9.6.

The type of spark plug used in an engine depends on the particular engine requirements. Every engine manufacturer determines the type of plug, cold or hot, that is best suited to his engine. A spark plug which runs at a satisfactory temperature at cruising speeds may run so cool at idling speed that abnormal deposits are likely to foul the electrodes. These deposits may be of soft dull carbon from incomplete combustion or of hard shiny carbon from excess lubricating oil that passes the piston rings and enters the combustion chamber. The carbon deposits from incomplete combustion will burn off at temperatures above 340 °C while the excess oil carbon deposits from lubricating oil require a temperature above 540 °C to burn. If a spark plug runs hot enough at idling speeds to prevent carbon deposits, it may run too hot at high speeds and cause preignition. If a spark plug runs at a temperature above 800 °C preignition usually results. A compromise must be made in order to obtain a proper spark plug which would operate satisfactorily throughout the entire engine operating range. An improper spark plug is always a major source of engine trouble such as misfiring and preignition.

9.7 OPERATION OF A BATTERY IGNITION SYSTEM

The source of the ignition energy in the battery ignition system is the ignition coil. This coil stores the energy in its magnetic field and delivers it at the instant of ignition (firing point) in the form of a surge of high voltage current (ignition pulse) through the high tension ignition cables to the correct spark plug. Storage of energy in the magnetic field is based on an inductive process, as a result of which we also designate the ignition coil as an *inductive storage device*. The schematic diagram of a conventional battery ignition system for a four-cylinder engine is shown in Fig.9.7.

As already explained the ignition coil consists of two coils of wire, one wound around the other, insulated from each other; the primary winding, L_1 with few turns of heavy copper wire and the secondary winding, L_2 with many turns of fine copper wire. The primary and secondary winding are wound around a laminated iron core which has the effect of increasing the strength of the magnetic field and thus of the amount of energy stored.

One end of the primary winding is connected through the ignition switch to the positive terminal post of the storage battery, and the other end is grounded through the contact breaker. The ignition capacitor is connected in parallel with the contact breaker. One end of the secondary winding is also grounded through the contact breaker, and the other end is connected through the distributor and the high-tension ignition cables to the center electrode of the spark plug.

When the ignition switch is closed, the primary winding of the coil is connected to the positive terminal post of the storage battery.

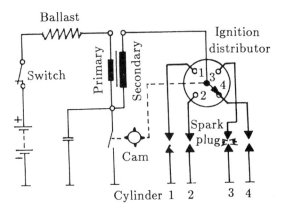

Fig.9.7 Conventional Coil Ignition System for
Four-Cylinder Engines – Schematic Diagram

If the primary circuit is closed through the breaker contacts, a current flows, the so called *primary current*.

This current, flowing through the primary coil, which is wound on a soft iron core, produces a magnetic field in the core. A cam driven by the engine shaft, is arranged to open the breaker points whenever an ignition discharge is required. When the breaker points open, the current which had been flowing through the points now flows into the condenser, which is connected across the points. As the condenser becomes charged, the primary current falls and the magnetic field collapses. The collapse of the field induces a voltage in the primary winding, which charges the condenser to a voltage much higher than battery voltage. The condenser then discharges into the battery, reversing the direction of both the primary current and the magnetic field. The rapid collapse and reversal of the magnetic field in the core induce a very high voltage in the secondary winding of the ignition coil. The secondary winding consists of a large number of turns of very fine wire wound on the same core with the primary. The high secondary voltage is led to the proper spark plug by means of a rotating switch called the distributor, which is located in the secondary or high tension circuit of the ignition system.

If a condenser were not used in the primary circuit, the high primary voltage caused by the collapse of the magnetic field around the primary winding would cause an arc across the breaker points. The arc would burn and destroy the points and would also prevent the rapid drop in primary current and magnetic field which is necessary for the production of the high secondary voltage.

Note that the spark timing is controlled by the crank angle at which the breaker points open, while the distributor merely determines the firing sequence of the spark plugs. Changes in ignition timing may be affected by rotating the plate which holds the breaker points, relative to the cam. Because of this ignition will be delayed if the plate is displaced in the direction in which the camshaft rotates.

9.8 LIMITATIONS

(i) The primary voltage decreases as the engine speed increases due to the limitations in the current switching capability of the breaker system.

(ii) Time available for build-up of the current in the primary coil and the stored energy decrease as the engine speed increases due to the dwell period becoming shorter.

(iii) Because of the high source impedance (about 500 kΩ) the system is sensitive to side-tracking across the spark plug insulator.

(iv) The breaker points are continuously subjected to electrical as well as mechanical wear which results in short maintenance intervals. Increased currents cause a rapid reduction in breaker point life and system reliability. Acceptable life for these systems is obtained with a primary current limited to about 4 amperes.

9.9 DWELL ANGLE

The period, measured in degree of cam rotation, during which the contact points remain *closed* is called the *dwell angle* or the *cam angle*. This is illustrated in Fig.9.8. The dwell angle must be sufficiently large to allow magnetic saturation of the primary coil. Too small a dwell angle will result in lower secondary voltage and hence poor sparks or even misfiring. Too large a dwell angle will lead to burning of condenser and the contact points due to oversaturation of windings.

The magnitude of the dwell angle depends upon the gap between the points and also the angle between the cam lobes. The gap between the points is generally of the order of 0.35 mm to 0.55 mm. The angle between the cam lobes depends upon the number of engine cylinders. If a single contact breaker is used the number of lobes on the cam is the same as the number of cylinders. This means that at any given engine speed the time available to supply energy to the primary winding decreases as the number of cylinders increases, thereby imposing a high-speed limit on an engine having large number of cylinders. The effects of inertia and spring lag further worsen the situation.

As the number of cylinders is increased, the dwell angle must be reduced because more and more closing and opening operations must

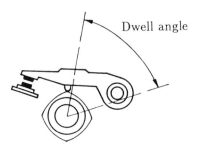

Fig.9.8 Dwell Angle

be accommodated during every rotation of the distributor camshaft; in the four-cylinder engine the dwell angle is about 50°, in the six-cylinder engine it is about 38°, and in the eight-cylinder engine it is about 33°. The following formula shows the way in which the dwell period can be expressed as a function of the dwell angle and the engine speed (in rev/min).

$$\text{Dwell period (milliseconds)} \quad = \quad \frac{1000 \times \text{dwell angle (degrees)}}{6 \times \text{engine speed (rev/min)}}$$

The smaller the gap between the contact points when they are fully open, the larger the dwell angle, and vice versa.

This has led to the use of twin contact breakers connected in parallel between the primary winding and earth and a cam having number of lobes half that of the number of cylinders, to increase the dwell angle and thereby allow sufficient build-up of primary coil energy. Six cylinder engine will, thus have a 60° lobe angle with a twin contact breaker. Such an arrangement however, requires accurate synchronization of the arms of the contact breaker which very often causes problems in service. To avoid this, nowadays a cam having the same number of lobes as there are engine cylinders is used in conjunction with two contact breakers arranged in parallel. One contact breaker always opens the points while the other always closes the points, i.e., the dwell angle is doubled.

9.10 ADVANTAGE OF A 12 V IGNITION SYSTEM

Until about 1950 all car engines had 6-volt ignition systems. The chief advantage of the 6-volt system is that it uses the three cell storage battery which is cheaper, lighter and less bulky than a six-cell battery of the same watt-hour capacity. With the relatively low compression ratios used in those days the 6-volt system gave satisfactory results. As the compression ratios and the engine speeds increased, the voltage required to break down the spark gap rose. Hence, the 12-volt system came to be preferred as considerably higher secondary voltages are obtainable with it.

The other advantages are as follows:

(i) for transmitting equal power without excessive voltage drop, the cables in a 6-volt system need theoretically to be twice the thickness of 12-volt cables.

(ii) starting, in particular, is much improved by the 12-volt system since almost twice the power is available for ignition coil during the starting surge.

(iii) the 12-volt system has adequate electric power to supply the increasing number of electrical accessories used.

9.11 MAGNETO IGNITION SYSTEM

Magneto is a special type of ignition system with its own electric generator to provide the necessary energy for the system. It is mounted on the engine and replaces all the components of the coil ignition system except the spark plug. A magneto when rotated by the engine is capable of producing a very high voltage and does not need a battery as a source of external energy.

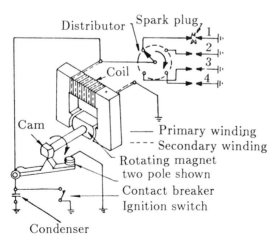

Fig.9.9 High Tension Magneto Ignition System

A schematic diagram of a high tension magneto ignition system is shown in Fig.9.9. The high tension magneto incorporates the windings to generate the primary voltage as well as to step up the voltage and thus does not require a separate coil to boost up the voltage required to operate the spark plug.

Magneto can be either *rotating armature type* or *rotating magnet type*. In the first type, the armature consisting of the primary and secondary windings all rotate between the poles of a stationary magnet, whilst, in the second type the magnet revolves and the windings are kept stationary. A third type of magneto called the *polar inductor type* is also in use. In the polar inductor type *magneto* both the magnet and the windings remain stationary but the voltage is generated by reversing the flux field with the help of soft iron polar projections, called inductors.

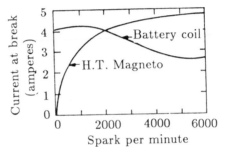

Fig.9.10 Breaker Current Vs Speed in Coil and Magneto Ignition Systems

The working principle of the magneto ignition system is exactly the same as that of the coil ignition system. With the help of a cam, the primary circuit flux is changed and a high voltage is produced in the secondary circuit.

The variation of the breaker current with speed for the coil ignition system and the magneto ignition system is shown in Fig.9.10. It can be seen that since the cranking speed at start is low the current generated by the magneto is quite small. As the engine speed increases the flow of current also increases. Thus, with magneto there is always a starting difficulty and sometimes a separate battery is needed for starting. The magneto is best at high speeds and therefore is widely used for sports and racing cars, aircraft engines, etc.

In comparison, the battery ignition system is more expensive but highly reliable. Because of the poor starting characteristics of the magneto system invariably the battery ignition system is preferred to the magneto system in automobile engines. However, in two wheelers magneto ignition system is favoured due to light weight and less maintenance.

The main disadvantage of the high tension magneto ignition system lies in the fact that the wirings carry a very high voltage and thus there is a strong possibility of causing engine misfire due to leakage. To avoid this the high tension wires must be suitably shielded. The development of the low tension magneto system is an attempt to avoid this trouble. In the low tension magneto system the secondary winding is changed to limit the secondary voltage to a value of about 400 volts and the distributor is replaced by a brush contact. The high voltage is obtained with the help of a step-up transformer. All these changes have the effect of limiting the high voltage current in the small portion of the ignition system wiring and thus avoid the possibilities of leakage etc. Table 9.1 gives the comparison of battery and magneto ignition systems.

Table 9.1 *Comparison between*
Battery Ignition and Magneto Ignition System

Battery Ignition System	Magneto Ignition System
Battery is necessary. Difficult to start the engine when battery is discharged.	No battery is needed and therefore there is no problem of battery discharge.
Maintenance problems are more due to battery.	Maintenance problems are less since there is no battery.
Current for primary circuit is obtained from the battery.	The required electric current is generated by the magneto.
A good spark is available at the spark plug even at low speed.	During starting, quality of spark is poor due to low speed.
Efficiency of the system decreases with the reduction in spark intensity as engine speed rises.	Efficiency of the system improves as the engine speed rises due to high intensity spark.
Occupies more space.	Occupies less space.
Commonly employed in cars and light commercial vehicles.	Mainly used in racing cars and two wheelers.

9.12 MODERN IGNITION SYSTEMS

The major limitations of the breaker-operated ignition systems are the decrease in available voltage as the engine speed increases due to limitations in the current switching capability of the breaker system and the decreasing time available to build up the stored energy in the primary coil. A further disadvantage is that due to their high current load, the breaker points are subjected to electrical wear in addition to mechanical wear which results in short maintenance intervals. The life of the breaker points is dependent on the current they are required to switch.

In order to overcome the above problems in the conventional ignition systems modern ignition systems use electronic circuits. The ability of a transistor to interrupt a circuit carrying a relatively high current makes it an ideal replacement for the breaker points and condenser. One of the pioneering versions was from Ford Motor Company in 1963 and many of the variations are available nowadays.

In modern automobiles the following two types are coming into common use.

(i) Transistorized coil ignition system (TCI system)

(ii) Capacitive discharge ignition system (CDI system)

The details of these ignition systems are explained in the following sections.

9.12.1 Transistorized Coil Ignition (TCI) System

In automotive applications, the transistorized coil ignition systems which provide a higher output voltage and use electronic triggering to maintain the required timing are fast replacing the conventional ignition systems. These systems are also called high energy electronic ignition systems. These have the following advantages:

(i) reduced ignition system maintenance

(ii) reduced wear of the components

(iii) increased reliability

(iv) extended spark plug life

(v) improved ignition of lean mixtures

The circuit diagram of a transistorized coil ignition system is shown in Fig.9.11. The contact breaker and the cam assembly of the conventional ignition system are replaced by a magnetic pulse generating system which detects the distributor shaft position and sends electrical pulse to an electronic control module. The module switches off the flow of current to the primary coil inducing a high voltage in the secondary winding which is distributed to the spark plugs as in the conventional breaker system. The control module contains timing circuit which later closes the primary circuit so that the build up of the primary circuit current can occur for the next cycle. There are many types of pulse generators that could trigger the electronic circuit of the ignition system. A magnetic pulse generator where a gear shaped iron rotor driven by the distributor shaft rotates

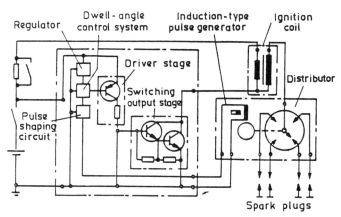

Fig.9.11 Schematic Diagram of Transistorized Coil Ignition (TCI) System with Induction Pulse Generator

past the pole of a stationary magnetic pickup, is generally used. The number of teeth on the rotor is equal to the number of cylinders. The magnetic field is provided by a permanent magnet. As each rotor tooth passes the magnet pole it first increases and then decreases the magnetic field strength ψ linked with the pickup coil wound on the magnet, producing a voltage signal proportional to $\frac{d\psi}{dt}$. In response to this the electronic module switches off the primary circuit coil current to produce the spark as the rotor tooth passes through alignment and the pick up coil voltage abruptly reverses and passes through zero. The increasing portion of the voltage waveform, after this voltage reversal, is used by the electronic module to establish the point at which the primary coil current is switched on for the next ignition pulse.

9.12.2 Capacitive Discharge Ignition (CDI) System

The details of the capacitive discharge ignition system are shown in Fig.9.12. In this system a capacitor rather than an induction coil, is used to store the ignition energy. The capacitance and charging voltage of the capacitor determine the amount of stored energy. The ignition transformer steps up the primary voltage, generated at the time of spark by the discharge of the capacitor through the thyristor, to the high voltage required at the spark plug.

The CDI trigger box contains the capacitor, thyristor power switch, charging device (to convert battery voltage to the charging voltage of 300 to 500 V by means of pulses via a voltage transformer), pulse shaping unit and control unit.

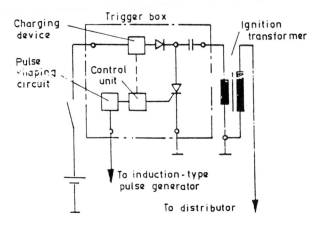

Fig.9.12 Schematic of Capacitive Discharge Ignition (CDI) System

The advantage of using this system is that it is insensitive to electrical shunts resulting from spark plug fouling. Because of the fast capacitive discharge, the spark is strong but short (0.1 to 0.3 ms) which leads to ignition failure during lean mixture operating conditions. This is the main disadvantage of the CDI system.

9.13 FIRING ORDER

Every engine cylinder must fire once in every cycle. This requires that for a four-stroke four-cylinder engine the ignition system must fire for every 180 degrees of crank rotation. For a six-cylinder engine the time available is only 120 degrees of crank rotation.

The order in which various cylinders of a multicylinder engine fire is called the *firing order*. The number of possibilities of firing order depends upon the number of cylinders and throws of the crankshaft. It is desirable to have the power impulses equally spaced and from the point of view of balancing this has led to certain conventional arrangements of crankshaft throws. Further, there are three factors which must be considered before deciding the optimum firing order of an engine. These are:

(i) engine vibrations

(ii) engine cooling and

(iii) development of back pressure

Consider that the first cylinder of the four-cylinder engine shown in Fig.9.13 is fired first. A pressure, p is generated in the cylinder number 1 giving rise to load equal to $p \times a$ and $p \times b$ on the two bearings A and B respectively. The load on bearing A is much less than that on bearing B. Now, if the next cylinder fired is cylinder

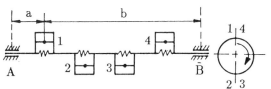

Fig.9.13 Firing Order

number 2, this imbalance in load on the two bearings would be further increased which would result in severe engine vibrations. If we fire the third cylinder after first, the load is equally distributed.

Further, consider the effect of firing sequence on engine cooling. When the first cylinder is fired its temperature increases. If the next cylinder that fires is number 2, the portion of the engine between the cylinder number 1 and 2 gets overheated. If then the third cylinder is fired, overheating is shifted to the portion between the cylinders 2 and 4. Thus we see that the task of the cooling system becomes very difficult because it is then, required to cool more at one place than at other places and this imposes great strain on the cooling system. If the third cylinder is fired after the first the overheating problem can be controlled to a greater extent.

Next, consider the flow of exhaust gases in the exhaust pipe. After firing the first cylinder, exhaust gases flow out to the exhaust pipe. If the next cylinder fired is the cylinder number 2, we find that before the gases exhausted by the first cylinder go out of the exhaust pipe the gases exhausted from the second cylinder try to overtake them. This would require that the exhaust pipe be made bigger. Otherwise the back pressure in it would increase and the possibility of back flow would arise. If instead of firing cylinder number 2, cylinder number 3 is fired then by the time the gases exhausted by the cylinder 3 come into the exhaust pipe, the gases from cylinder 1 would have sufficient time to travel the distance between cylinder 1 and cylinder 3 and thus, the development of a high back pressure is avoided.

It should be noted that to some extent all the above three requirements are conflicting and therefore a trade-off is necessary. For four-cylinder engines the possible firing orders are:

$$1 - 3 - 4 - 2 \quad \text{or} \quad 1 - 2 - 4 - 3$$

The former is more commonly used in the vertical configuration of cylinders. For a six-cylinder engine the firing orders can be:

$$1 - 5 - 3 - 6 - 2 - 4 \quad \text{or} \quad 1 - 5 - 4 - 6 - 2 - 3 \quad \text{or}$$
$$1 - 2 - 4 - 6 - 5 - 3 \quad \text{or} \quad 1 - 2 - 3 - 6 - 5 - 4$$

The first one is more commonly used.

9.14 IGNITION TIMING AND ENGINE PARAMETERS

In order to obtain maximum power from an engine, the compressed mixture must deliver its maximum pressure at a time when the piston is about to commence its downward stroke and is very close to TDC. Since, there is a time lag between occurrence of spark and the burning of the mixture the spark must take place before the piston reaches top dead centre on its compression stroke. Usually the spark should occur at about 15 degrees $bTDC$. If the spark occurs too early the combustion takes place before the compression stroke is completed and the pressure so developed would oppose the piston motion and thereby reduce the engine power. If the spark occurs too late, the piston would have already completed a certain part of the expansion stroke before the pressure rise occurs and a corresponding amount of engine power is lost.

The correct instant for the introduction of a spark is mainly determined by the ignition lag. The ignition lag depends on many factors, such as compression ratio, mixture strength, throttle opening, engine temperature, combustion chamber design and speed. Some of the important engine parameters affecting the ignition timings are discussed in the following sections.

9.14.1 Engine Speed

Suppose an engine has got ignition advance of d degrees and operating speed is N rpm. Then time available for combustion is

$$\frac{d}{360 \ N} \ \text{min}$$

Now if the engine rpm is increased to $2N$, then in order to have the same time available for combustion an ignition advance of $2d$ degrees is required. Thus, if the combustion time for a given mixture strength, compression ratio and volumetric efficiency be assumed constant, then as the engine speed is increased it will be necessary to advance the ignition progressively in order to follow Upton's rule for optimum advance.

The variation of ignition advance required at different speeds is shown in Fig.9.14.

Figure 9.15 shows indicator diagrams taken for a petrol engine running at 1000 and 2000 rpm respectively but with the ignition advance adjusted in each case to give the maximum pressure peak at about 12 degrees after top dead centre. It is seen that at the higher speed a spark advance of 20 degrees is necessary as against 10 degrees at the lower speed. Corresponding delay periods between the passage of the spark and commencement of pressure rise were 20 and 10 degrees respectively, i.e. proportional to the engine speeds.

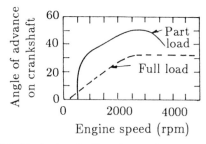

Fig.9.14 Variation in Ignition Advance with Speed

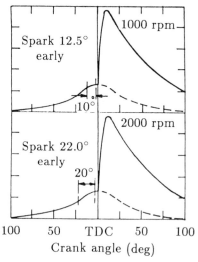

Fig.9.15 Effect of Spark Advance on Pressure-Crank Angle Diagram

9.14.2 Mixture Strength

In general, rich mixtures burn faster. Hence, if the engine is operating with richer mixtures the optimum spark timing must be retarded, i.e., the number of degrees of crank angle before TDC at the time of ignition is decreased and the spark occurs closer to TDC.

9.14.3 Part Load Operation

Part load operation of a spark ignition engine is affected by throttling the incoming charge. Due to throttling a smaller amount of charge enters the cylinder and the dilution due to residual gases is also greater. This results in the combustion process being slower. In order to overcome the problem of exhaust gas dilution and the low charge density at part load operation the spark advance must be increased.

9.14.4 Type of Fuel

Ignition delay will depend upon the type of the fuel used in the engine. For maximum power and economy a slow burning fuel needs a higher spark advance than a fast burning fuel.

9.15 SPARK ADVANCE MECHANISM

It is obvious from the above discussion that the point in the cycle where the spark occurs must be regulated to ensure maximum power and economy at different speeds and loads and this must be done automatically. The purpose of the spark advance mechanism is to assure that under every condition of engine operation, ignition takes place at the most favourable instant in time, i.e., most favourable from a standpoint of engine power, fuel economy, and minimum exhaust dilution. By means of these mechanisms the advance angle is accurately set so that ignition occurs before the top dead-center point of the piston. The engine speed and the engine load are the control quantities required for the automatic adjustment of the ignition timing. Most of the engines are fitted with mechanisms which are integral with the distributor and automatically regulate the optimum spark advance to account for change of speed and load. The two mechanisms used are:

(i) Centrifugal advance mechanism

(ii) Vacuum advance mechanism

These mechanisms are discussed in greater details in the following sections.

9.15.1 Centrifugal Advance Mechanism

The centrifugal advance mechanism controls the ignition timing for full-load operation. The adjustment mechanism is designed so that its operation results in the desired advance of the spark. The cam is mounted, movably, on the distributor shaft so that as the speed increases, the flyweights which are swung farther and farther outward, shift the cam in the direction of shaft rotation. As a result, the cam lobes make contact with the breaker lever rubbing block somewhat earlier, thus shifting the ignition point in the *early* or advance direction. Depending on the speed of the engine, and therefore of the shaft, the weights are swung outward a greater or a lesser distance from the center. They are then held in the extended position, in a state of equilibrium corresponding to the shifted timing angle, by a retaining spring which exactly balances the centrifugal force. The weights shift the cam either on a rolling contact or sliding contact basis; for this reason we distinguish between the rolling contact type and the sliding contact type of centrifugal advance mechanism.

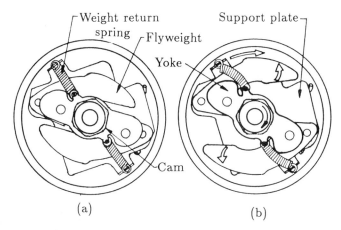

Fig.9.16 Sliding Contact Type Centrifugal Advance Mechanism

The beginning of the timing adjustment in the range of low engine speeds and the continued adjustment based on the full load curve are determined by the size of the weights, by the shape of the contact mechanisms (rolling or sliding contact type), and by the retaining springs, all of which can be widely differing designs. The centrifugal force controlled cam is fitted with a lower limit stop for purposes of setting the beginning of the adjustment, and also with an upper limit stop to restrict the greatest possible full-load adjustment. A typical sliding contact type centrifugal advance mechanism is shown in Fig.9.16(a) and (b).

9.15.2 Vacuum Advance Mechanism

Vacuum advance mechanism shifts the ignition point under partial load operation. The adjustment system is designed so that its operation results in the prescribed partial-load advance curve. In this mechanism the adjustment control quantity is the static vacuum prevailing in the carburettor, a pressure which depends on the position of the throttle valve at any given time and which is at a maximum when this valve is about half open. This explains the vacuum maximum.

The diaphragm of a vacuum unit is moved by changes in gas pressure. The position of this diaphragm is determined by the pressure differential at any given moment between the prevailing vacuum and atmospheric pressure. The beginning of adjustment is set by the pre-established tension on a compression spring. The diaphragm area, the spring force, and the spring rigidity are all selected in accordance with the partial-load advance curve which is to be followed and are

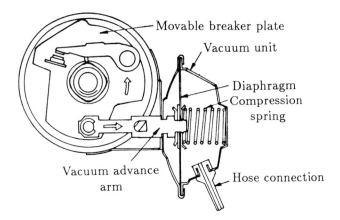

Fig.9.17 Vacuum Advance Mechanism

all balanced with respect to each other. The diaphragm movement is transmitted through a vacuum advance arm connected to the movable breaker plate, and this movement shifts the breaker plate an additional amount under partial-load conditions in a direction opposite to the direction of rotation of the distributor shaft. Limit stops on the vacuum advance arm in the base of the vacuum unit restrict the range of adjustment.

The vacuum advance mechanism operates independent of the centrifugal advance mechanism. The mechanical interplay between the two advance mechanisms, however, permits the total adjustment angle at any given time to be the result of the addition of the shifts provided by the two individual mechanisms. In other words, the vacuum advance mechanism operates in conjunction with the centrifugal advance mechanism to provide the total adjustment required when the engine is operating under partial load. A typical vacuum advance mechanism is shown in Fig.9.17.

9.16 IGNITION TIMING AND EXHAUST EMISSIONS

Idling, deceleration and running rich with closed throttle are some engine operating conditions which produce excessive unburnt hydrocarbons and carbon monoxide in exhaust. The emission quality is greatly affected by the ignition timing as shown in Fig.9.18.

Retarding ignition timing at *idle* tends to reduce exhaust emission in two ways. With retarded timing, exhaust gas temperatures are higher (fuel economy is adversely affected), thereby promoting

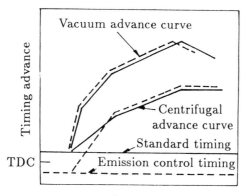

Fig.9.18 Typical Distributor Advance Curve for
Lower HC and CO Exhaust Emission

additional burning of the hydrocarbons in the exhaust manifold. Since
engine efficiency is reduced, retarded timing requires a slightly larger
throttle opening to maintain the same idle speed. The larger throttle
opening means the possibility of better mixing and combustion during
idling. This reduces exhaust emissions appreciably especially during
deceleration.

Questions :-

9.1 What is meant by ignition? What is the interrelation between
ignition and combustion?

9.2 What are various types of ignition system that are commonly
used?

9.3 Explain the basic energy requirements for spark ignition.

9.4 What is capacitance spark and how is it produced?

9.5 With a neat sketch explain an induction coil.

9.6 How does gas movement affect spark-ignition?

9.7 Comment on the spark energy and its duration in the initiation
of combustion.

9.8 What are the important requirements of the high voltage ignition
source for the spark-ignition process?

9.9 What are the two conventional types of ignition systems that are
normally used in automobiles?

9.10 Mention the various important qualities of a good ignition system.

9.11 With a neat sketch explain the battery ignition system.

9.12 Explain TCI ignition system with a sketch.

9.13 Explain CDI ignition system with a suitable diagram.

9.14 With a neat sketch explain the magneto ignition system.

9.15 What is the main function of a spark plug? Draw a neat sketch of a spark plug and explain its various parts.

9.16 Explain the details of firing order.

9.17 Briefly discuss the various factors which affect the ignition timing.

9.18 Why spark advance is required? Explain.

9.19 Briefly explain the centrifugal advance mechanism.

9.20 Explain with a neat sketch the vacuum advance mechanism.

10

COMBUSTION
AND
COMBUSTION CHAMBERS

10.1 INTRODUCTION

Combustion is a chemical reaction in which certain elements of the fuel like hydrogen and carbon combine with oxygen liberating heat energy and causing an increase in temperature of the gases. The conditions necessary for combustion are the presence of combustible mixture and some means of initiating the process. The theory of combustion is a very complex subject and has been a topic of intensive research for many years. In spite of this, not much knowledge is available concerning the phenomenon of combustion.

The process of combustion in engines generally takes place either in a homogeneous or a heterogeneous fuel vapour-air mixture depending on the type of engine.

10.2 HOMOGENEOUS MIXTURE

In spark-ignition engines a nearly homogeneous mixture of air and fuel is formed in the carburettor and burnt in the engine cylinder. Homogeneous mixture is formed outside the engine cylinder and the combustion is initiated inside the cylinder at a particular instant towards the end of the compression stroke. The flame front spreads over a combustible mixture with a certain velocity. In a homogeneous gas mixture the fuel and oxygen molecules are more or less uniformly distributed.

Once the fuel vapour-air mixture is ignited at a point, a flame front appears and rapidly spreads in the mixture. The flame propagation is caused by heat transfer and diffusion of burning fuel molecules from the combustion zone to the adjacent layer of fresh mixture. The flame front is a narrow zone separating the fresh mixture from the combustion products. The velocity with which the flame front moves, with respect to the unburned mixture in a direction normal to its surface is called the normal flame velocity.

In a homogeneous mixture with an equivalence ratio, ϕ, *(the ratio of the actual fuel-air ratio to the stoichiometric fuel-air ratio)* close to 1.0, the flame speed is normally of the order of 40 cm/s. However, in a spark-ignition engine the maximum flame speed is obtained when ϕ is between 1.1 and 1.2, i.e., when the mixture is slightly richer than stoichiometric. If the equivalence ratio is outside this range the flame speed drops rapidly to a low value. When the flame speed drops to a very low value, the heat loss from the combustion zone becomes equal to the amount of heat-release due to combustion and the flame gets extinguished. Therefore, it is quite preferable to operate the engine within an equivalence ratio of 1.1 to 1.2 for proper combustion. However, by introducing turbulence and incorporating proper air movement, the flame speed can be increased in mixtures outside the above range.

10.3 HETEROGENEOUS MIXTURE

In a heterogeneous gas mixture, the rate of combustion is determined by the velocity of mutual diffusion of fuel vapours and air and the rate of chemical reaction is of minor importance. Self-ignition or spontaneous ignition of fuel-air mixture, at the high temperature developed due to high compression, is of primary importance in determining combustion characteristics.

When the mixture is heterogeneous the combustion can take place in an overall lean mixture since, there are always local zones where ϕ varies between 1.1 and 1.2 corresponding to maximum rate of chemical reaction. Ignition starts in this zone and the flame produced helps to burn the fuel in the adjoining zones where the mixture is leaner. Similarly, in the zones where the mixture is rich the combustion occurs because of the high temperature produced due to combustion initiated in the zones where ϕ is 1.1 to 1.2.

A comprehensive study of combustion in both spark-ignition and compression-ignition engines is given in the following sections.

10.4 COMBUSTION IN SPARK–IGNITION ENGINES

As already mentioned, in a conventional spark-ignition engine, the fuel and air are homogeneously mixed together in the intake system, inducted through the intake valve into the cylinder where it mixes with residual gases and is then compressed. Under normal operating conditions, combustion is initiated towards the end of the compression stroke at the spark plug by an electric discharge. A turbulent flame develops following the ignition and propagates through this premixed charge of fuel and air, and also the residual gas in the clearance volume until it reaches the combustion chamber walls. Combustion in the SI engine may be broadly divided into two general types, viz., normal combustion and abnormal combustion.

10.5 STAGES OF COMBUSTION IN SI ENGINES

A typical theoretical pressure-crank angle diagram, during the process of compression (a→b), combustion (b→c) and expansion (c→d) in an ideal four-stroke spark-ignition engine is shown in Fig.10.1. In an ideal engine, as can be seen from the diagram, the entire pressure rise during combustion takes place at constant volume i.e., at TDC. However, in an actual engine this does not happen. The detailed process of combustion in an actual SI engine is described below.

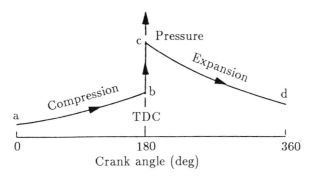

Fig.10.1 Theoretical p-θ Diagram

Sir Ricardo, known as the father of engine research, describes the combustion process in a spark-ignition engine as consisting of three stages:

The pressure variation due to combustion in a practical engine is shown in Fig.10.2. In this figure, A is the point of passage of spark (say $20°bTDC$), B is the point at which the beginning of pressure rise can be detected (say 8° $bTDC$) and C the attainment of peak pressure. Thus AB represents the first stage and BC the second stage and CD the third stage.

The *first stage* (A→B) is referred to as the ignition lag or preparation phase in which growth and development of a self propagating nucleus of flame takes place. This is a chemical process depending upon both temperature and pressure, the nature of the fuel and the proportion of the exhaust residual gas. Further, it also depends upon the relationship between the temperature and the rate of reaction.

The *second stage* (B→C) is a physical one and it is concerned with the spread of the flame throughout the combustion chamber. The starting point of the second stage is where the first measurable rise of pressure is seen on the indicator diagram i.e., the point where the line of combustion departs from the compression line (point B). This can be seen from the deviation from the motoring curve.

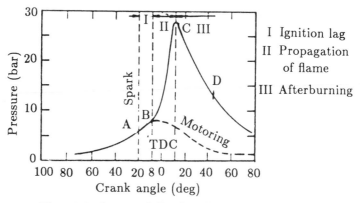

I Ignition lag
II Propagation of flame
III Afterburning

Fig.10.2 Stages of Combustion in an SI Engine

During the second stage the flame propagates practically at a constant velocity. Heat transfer to the cylinder wall is low, because only a small part of the burning mixture comes in contact with the cylinder wall during this period. The rate of heat-release depends largely on the turbulence intensity and also on the reaction rate which is dependent on the mixture composition. The rate of pressure rise is proportional to the rate of heat-release because during this stage, the combustion chamber volume remains practically constant (since piston is near the top dead centre).

The starting point of the *third stage* is usually taken as the instant at which the maximum pressure is reached on the indicator diagram (point C). The flame velocity decreases during this stage. The rate of combustion becomes low due to lower flame velocity and reduced flame front surface. Since the expansion stroke starts before this stage of combustion with the piston moving away from the top dead centre there can be no pressure rise during this stage.

10.6 FLAME FRONT PROPAGATION

For efficient combustion the rate of propagation of the flame front within the cylinder is quite critical. The two important factors which determine the rate of movement of the flame front across the combustion chamber are the *reaction rate* and the *transposition rate*. The *reaction rate* is the result of a purely.chemical combination process in which the flame *eats* its way into the unburned charge. The *transposition rate* is due to the physical movement of the flame front relative to the cylinder wall and is also the result of the pressure differential setup between the burning gases and the unburnt gases in the combustion chamber.

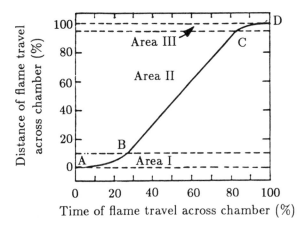

Fig.10.3 Details of Flame Travel

Figure 10.3 shows the rate of flame propagation. In area I, (A→B), the flame front progresses relatively slowly due to a low *transposition rate* and low turbulence. The transposition of the flame front is very little since there is a comparatively small mass of charge burned at the start. The low reaction rate plays a dominant role resulting in a slow advance of the flame. Also, since the spark plug is to be necessarily located in a quiescent layer of gas that is close to the cylinder wall, the lack of turbulence reduces the reaction rate and further lowers the flame speed.

As the flame front leaves the quiescent zone and proceeds into more turbulent areas (area II) where it consumes a greater mass of mixture, it progresses more rapidly and at a constant rate (B→C) as shown in Fig.10.3.

The volume of unburned charge is reduced appreciably towards the end of flame travel and the *transposition rate* again becomes negligible thereby reducing the flame speed. The reaction rate is also reduced again since the flame is entering a zone (area III) of relatively low turbulence (C→D) in Fig.10.3.

10.7 FACTORS INFLUENCING THE FLAME SPEED

The study of factors which affect the velocity of flame propagation is important since the flame velocity influences the rate of pressure rise in the cylinder and it is related to certain types of abnormal combustion that occur in spark-ignition engines. There are several factors which affect the flame speed, to a varying degree, the most important being the turbulence and the fuel-air ratio. Details of various factors that affect the flame speed are discussed below.

Turbulence: The flame speed is quite low in non-turbulent mixtures and increases with increasing turbulence. This is mainly due to the additional physical intermingling of the burning and unburned particles at the flame front which expedites reaction by increasing the rate of contact. The turbulence in the incoming mixture is generated during the admission of fuel-air mixture through comparatively narrow sections of the intake pipe, valve openings etc., in the suction stroke. Turbulence which is supposed to consist of many minute swirls appears to increase the rate of reaction and produce a higher flame speed than that made up of larger and fewer swirls. A suitable design of the combustion chamber which involves the geometry of cylinder head and piston crown increases the turbulence during the compression stroke.

Generally, turbulence increases the heat flow to the cylinder wall. It also accelerates the chemical reaction by intimate mixing of fuel and oxygen so that spark advance may be reduced. This helps in burning lean mixtures also. The increase of flame speed due to turbulence reduces the combustion duration and hence minimizes the tendency of abnormal combustion. However, excessive turbulence may extinguish the flame resulting in rough and noisy operation of the engine.

Fuel-Air Ratio: The fuel-air ratio has a very significant influence on the flame speed. The highest flame velocities (minimum time for complete combustion) are obtained with somewhat richer mixture (point A) as shown in Fig.10.4 which shows the effect of mixture strength on the rate of burning as indicated by the time taken for complete burning in a given engine.

When the mixture is made leaner or richer (with respect to point A in Fig.10.4) the flame speed decreases. Less thermal energy is released in the case of lean mixtures resulting in lower flame temperature. Very rich mixtures lead to incomplete combustion which results again in the release of less thermal energy.

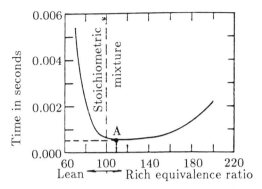

Fig.10.4 Effect of Mixture Strength on the Rate of Burning

Temperature and Pressure: Flame speed increases with an increase in intake temperature and pressure. A higher initial pressure and temperature may help to form a better homogeneous air-vapour mixture which helps in increasing the flame speed. This is possible because of an overall increase in the density of the charge.

Compression Ratio: A higher compression ratio increases the pressure and temperature of the working mixture which reduce the initial *preparation phase* of combustion and hence less ignition advance is needed. High pressures and temperatures of the compressed mixture also speed up the second phase of combustion. Increased compression ratio reduces the clearance volume and therefore increases the density of the cylinder gases during burning. This increases the peak pressure and temperature and the total combustion duration is reduced. Thus engines having higher compression ratios have higher flame speeds.

Engine Output: The cycle pressure increases when the engine output is increased. With the increased throttle opening the cylinder gets filled to a higher density. This results in increased flame speed. When the output is decreased by throttling, the initial and final compression pressures decrease and the dilution of the working mixture increases. The smooth development of self-propagating nucleus of flame becomes unsteady and difficult. The main disadvantages of SI engines are the poor combustion at low loads and the necessity of mixture enrichment (ϕ between 1.2 to 1.3) which causes wastage of fuel and discharge of unburnt hydrocarbon and the products of incomplete combustion like carbon monoxide etc. in the atmosphere.

Engine Speed: The flame speed increases almost linearly with engine speed since the increase in engine speed increases the turbulence inside the cylinder. The time required for the flame to traverse the combustion space would be halved, if the engine speed is doubled.

Double the engine speed and hence half the original time would give the same number of crank degrees for flame propagation. The crank angle required for the flame propagation during the entire phase of combustion, will remain nearly constant at all speeds.

Engine Size: The size of the engine does not have much effect on the rate of flame propagation. In large engines the time required for complete combustion is more because the flame has to travel a longer distance. This requires increased crank angle duration during the combustion. This is one of the reasons why large sized engines are designed to operate at low speeds.

10.8 RATE OF PRESSURE RISE

The rate of pressure rise in an engine combustion chamber exerts a considerable influence on the peak pressure developed, the power produced and the smoothness with which the forces are transmitted to the piston. The rate of pressure rise is mainly dependent upon the mass rate of combustion of mixture in the cylinder.

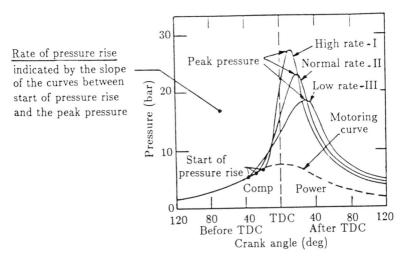

Rate of pressure rise indicated by the slope of the curves between start of pressure rise and the peak pressure

Fig.10.5 Illustrations of Various Combustion Rates

The relationship between the pressure and the crank angle for three different combustion rates is shown in Fig.10.5. Curve I is for a high, curve II for the normal and curve III for a low rate of combustion. It is clear from the figure that with lower rate of combustion longer time is required for the completeness of combustion which necessitates the initiation of burning at an early point on the compression stroke. Also, it may be noted that higher rate of pressure rise, as a result of the higher rate of combustion, generally produce higher peak pressures at a point closer to *TDC*. This generally is a desirable feature because higher peak pressures closer to *TDC* produce a greater force acting through a large part of the power stroke and hence, increase the power output of the engine. The higher rate of pressure rise causes rough running of the engine because of vibrations and jerks produced in the crankshaft rotation and also tend to create a situation conducive to an undesirable occurrence known as knocking. A compromise between these opposing factors is accomplished by designing and operating the engine in such a manner that approximately one-half of the pressure rise takes place by the moment the piston reaches *TDC*. This results in the peak pressure being reasonably close to the beginning of the power stroke, yet maintaining smooth engine operation.

10.9 ABNORMAL COMBUSTION

In normal combustion, the flame initiated by the spark travels across the combustion chamber in a fairly uniform manner. Under certain operating conditions the combustion gets deviated from its normal course leading to loss of performance and possible damage to the engine. This type of combustion may be termed as an abnormal combustion or knocking combustion. The consequences of this abnormal combustion process are the loss of power, recurring preignition and mechanical damage to the engine.

10.10 THE PHENOMENON OF KNOCK IN SI ENGINES

In a spark-ignition engine combustion which is initiated between the spark plug electrodes spreads across the combustible mixture. A definite flame front which separates the fresh mixture from the products of combustion travels from the spark plug to the other end of the combustion chamber. Heat-release due to combustion increases the temperature and consequently the pressure, of the burned part of the mixture above those of the unburned mixture. In order to effect pressure equalization the burned part of the mixture will expand, and compress the unburned mixture adiabatically thereby increasing its pressure and temperature. This process continues as the flame front advances through the mixture and the temperature and pressure of the unburned mixture are increased further.

If the temperature of the unburnt mixture exceeds the self-ignition temperature of the fuel and remains at or above this temperature during the period of preflame reactions (ignition delay),

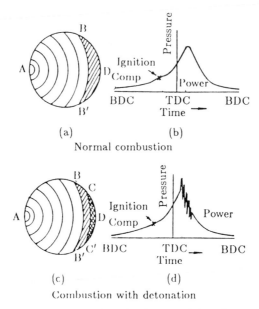

(a)   (b)

Normal combustion

(c)   (d)

Combustion with detonation

Fig.10.6 Normal and Abnormal Combustion

spontaneous ignition or autoignition occurs at various pin-point locations. This phenomenon is called knocking. The process of autoignition leads towards engine knock.

The phenomenon of knock may be explained by referring to Fig.10.6(a) which shows the cross-section of the combustion chamber with flame advancing from the spark plug location A without knock whereas Fig.10.6(c) shows the combustion process with knock. In the normal combustion the flame travels across the combustion chamber from A towards D. The advancing flame front compresses the end charge BB'D farthest from the spark plug, thus raising its temperature. The temperature is also increased due to heat transfer from the hot advancing flame-front. Also some preflame oxidation may take place in the end charge leading to further increase in temperature. Inspite of these factors if the temperature of the end charge had not reached its self-ignition temperature, the charge would not autoignite and the flame will advance further and consume the charge BB'D. This is the normal combustion process which is illustrated by means of the pressure-time diagram, Fig.10.6(b).

However, if the end charge BB'D reaches its autoignition temperature and remains for some length of time equal to the time of preflame reactions the charge will autoignite, leading to knocking combustion. In Fig.10.6(c), it is assumed that when flame has reached the position BB', the charge ahead of it has reached critical autoignition temperature. During the preflame reaction period if the flame front could move from BB' to only CC' then the charge ahead of CC' would autoignite.

Because of the autoignition, another flame front starts traveling in the opposite direction to the main flame front. When the two flame fronts collide, a severe pressure pulse is generated. The gas in the chamber is alternatively compressed and expanded by the pressure pulse until pressure equilibrium is restored. This disturbance can force the walls of the combustion chambers to vibrate at the same frequency as the gas. Gas vibration frequency in automobile engines is of the order of 5000 cps. The pressure-time trace of such a situation is shown in Fig.10.6(d).

It is to be noted that the onset of knocking is very much dependent on the properties of fuel. It is clear from the above description that if the unburned charge does not reach its autoignition temperature there will be no knocking. Further, if the initial phase i.e., ignition delay period, is longer than the time required for the flame front to burn through the unburned charge, there will be no knocking. But, if the critical temperature is reached and maintained, and the ignition delay is shorter than the time it takes for the flame front to burn through the unburned charge then the end charge will detonate. Hence, in order to avoid or inhibit detonation, a high autoignition temperature and a long ignition delay are the desirable qualities for SI engine fuels.

In summary, when autoignition occurs, two different types of vibration may be produced. In one case a large amount of mixture may autoignite giving rise to a very rapid increase in pressure throughout the combustion chamber and there will be a direct blow on the engine structure. The human ear will detect a thudding sound from the impact and consequent noise from free vibrations of the engine parts. In the other case, large pressure differences may exist in the combustion chamber and the resulting gas vibrations can force the walls of the chamber to vibrate at the same frequency as the gas. An audible sound may be evident.

The impact of knock on the engine components and structure can cause engine failure and it may be noted that noise from engine vibration is always objectionable.

The pressure differences in the combustion chamber cause the gas to vibrate and scrub the chamber walls causing increased loss of heat to the coolant.

The presence or absence of knocking in engines is often judged by the appearance of a distinctly audible sound. A scientific method of detecting the knocking phenomenon is by using a pressure break transducer. The output of this transducer is measured, usually, by using a cathode ray oscilloscope. Typical pressure versus time curves which can be obtained in this manner are given in Fig.10.6(b) for normal combustion and in Fig.10.6(d) for knocking combustion.

10.10.1 Knock Limited Parameters

It should be the aim of the designer to reduce the tendency of knock in the engine. In this connection, certain knock limited parameters are explained. They are:

Knock Limited Compression Ratio: The knock limited compression ratio is obtained by increasing the compression ratio on a variable compression ratio engine until incipient knocking is observed. Any change in operating conditions, in fuel consumption or in engine design that increases the knock limited compression ratio is said to reduce the tendency towards knocking.

Knock Limited Inlet Pressure: The inlet pressure can be increased by opening the throttle or increasing supercharger delivery pressure until incipient knock is observed. An increase in knock limited inlet pressure indicates a reduction in the knocking tendency.

Knock Limited Indicated Mean Effective Pressure: The indicated mean effective pressure measured at incipient knock is usually abbreviated as *Klimep*. This parameter and the corresponding fuel consumption are obviously of great practical interest.

An useful measure of knocking tendency called the performance number, has been developed from the concept of knock limited indicated mean effective pressure. This number is defined as the ratio of *Klimep* with the fuel in question to *Klimep* with iso-octane when inlet pressure is used as the dependent parameter and closely related to octane number. One great advantage of performance number is that it can apply to fuel whose knocking characteristics are superior to that of iso-octane, i.e., it extends the octane scale beyond 100.

Further simplification on the use of performance number requirements is done by introducing the concept of relative performance number, *rpn*, which is defined as:

$$rpn = \frac{\text{Actual Performance number}}{\text{Performance number corresponding to the imep of 100}}$$

10.11 EFFECT OF ENGINE VARIABLES ON KNOCK

From the discussion on knock in the previous section, it may be seen that four major factors are involved in either producing or preventing knock. These are the temperature, pressure, density of the unburned charge and the time factors. Since, the effect of temperature, pressure and density are closely interrelated, these three are consolidated into one group and the time factors into another group.

10.11.1 Density Factors

Any factor in the design or operation of an engine which tends to reduce the temperature of the unburned charge should reduce the possibility of knocking by reducing the temperature of the end charge for autoignition. Similarly, any factor which reduces the density of the charge tends to reduce knocking by providing lower energy release. Further, the effect of the following parameters which are directly or indirectly connected with temperature, pressure and density factors on the possibility of knocking is discussed below.

Compression Ratio: Compression ratio of an engine is an important factor which determines both the pressure and temperature at the beginning of the combustion process. Increase in compression ratio increases the pressure and temperature of the gases at the end of the compression stroke. This decreases the delay period of the end gas and thereby increasing the tendency for knocking. The overall increase in the density of the charge due to higher compression ratio increases the preflame reactions in the end charge thereby increasing the knocking tendency of the engine. The increase in the knocking tendency of the engine with increasing compression ratio is the main reason for liming the compression ratio to a lower value.

Mass of Inducted Charge: A reduction in the mass of the inducted charge into the cylinder of an engine by throttling or by reducing the amount of supercharging reduces both temperature and density of the charge at the time of ignition. This decreases the tendency of knocking.

Inlet Temperature of the Mixture: Increase in the inlet temperature of the mixture makes the compression temperature higher and thereby increasing the tendency for the knocking. Further, volumetric efficiency will be sufficiently affected. Hence, a lower inlet temperature is always preferable to reduce knocking. It is to be noted that the temperature should not be so low as to cause starting and vaporization problems in the engine.

Temperature of the Combustion Chamber Walls: Temperature of the combustion chamber walls play a predominant role in knocking. In order to prevent knocking the hot spots in the combustion chamber should be avoided. Since, the spark plug and exhaust valve are two hottest parts in the combustion chamber, the end gas should not be compressed against them.

Retarding the Spark Timing: By retarding the spark timing from the optimized timing, i.e., having the spark closer to TDC, the peak pressures are reached farther down on the power stroke and are thus of lower magnitude. This might reduce the knocking. However, the spark timing will be far away from the MBT timing. This will affect the brake torque and power output of the engine.

Power Output of the Engine: A decrease in the output of the engine decreases the temperature of the cylinder and the combustion chamber walls and also the pressure of the charge thereby lowering mixture and end gas temperatures. This reduces the tendency to knock.

10.11.2 Time Factors

Increasing the flame speed or reducing the duration of the ignition delay period or reducing the time of exposure of the unburned mixture to autoignition condition will tend to reduce knocking. The following factors, in most cases, reduce the possibility of knocking.

Turbulence: Turbulence depends on the design of the combustion chamber and on engine speed. Increasing turbulence increases flame speed and reduces the time available for the end charge to attain autoignition conditions thereby decreasing the tendency to knock.

Engine Speed: An increase in engine speed increases the turbulence of the mixture considerably resulting in increased flame speed, and reduces the time available for preflame reactions. Hence knocking tendency is reduced at higher speeds.

Flame Travel Distance: The knocking tendency is reduced by shortening the time required for the flame front to traverse the combustion chamber. Engine size, combustion chamber size and spark plug position are the three important factors governing the flame travel distance.

Engine Size: The flame requires a longer time to travel across the combustion chamber of a larger engine. Therefore, a larger engine has a greater tendency for knocking than a smaller engine giving more time for the end gas to autoignite. Hence, an SI engine is generally limited to size of about 100 mm bore. For this reason the larger engines are slow speed engines.

Combustion Chamber Shape: Generally, the more compact the combustion chamber is, the shorter is the flame travel and the combustion time and hence better antiknock characteristics. Therefore, the combustion chambers are made as spherical as possible to minimize the length of the flame travel for a given volume. If the turbulence in the combustion chamber is high, the combustion rate is high and consequently combustion time and knocking tendency are reduced. Hence, the combustion chamber is shaped in such a way as to promote turbulence.

Location of Spark Plug: In order to have a minimum flame travel, the spark plug is centrally located in the combustion chamber, resulting in minimum knocking tendency. The flame travel can also be reduced by using two or more spark plugs in case of large engines.

10.11.3 Composition Factors

Once the basic design of the engine is finalized, the fuel-air ratio and the properties of the fuel, particularly the octane rating, play a crucial role in controlling the knock.

Fuel-Air Ratio: The flame speeds are affected by fuel-air ratio. Also the flame temperature and reaction time are different for different fuel-air ratios. Maximum flame temperature is obtained when $\phi \approx 1.1$ to 1.2 whereas $\phi = 1$ gives minimum reaction time for autoignition.

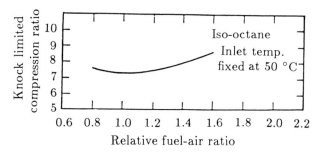

Fig.10.7

Figure 10.7 shows the variation of knock limited compression ratio with respect to relative fuel-air ratio for iso-octane. The maximum tendency to knock takes place for the fuel-air ratio which gives minimum reaction time as discussed earlier. Thus the most predominant factor is the reaction time of the mixture in this case. In general except at rich end, the behaviour in the engine follows the same pattern as the fuel-air ratio versus reaction time discussed earlier. The drop in *Klimep* at very rich end is caused by large drop in thermal efficiency.

Octane Value of the Fuel: A higher self-ignition temperature of the fuel and a low preflame reactivity would reduce the tendency of knocking. In general, paraffin series of hydrocarbon have the maximum and aromatic series the minimum tendency to knock. The naphthene series comes in between the two. Usually, compounds with more compact molecular structure are less prone to knock. In aliphatic hydrocarbons, unsaturated compounds show lesser knocking tendency than saturated hydrocarbon, the exception being ethylene, acetylene and propylene.

Table 10.1 gives the general summary of variables affecting the knock in an SI engine and shows whether the various factors can be controlled by the operator.

Table 10.1 *Summary of Variables Affecting Knock in an SI Engine*

Increase in variable	Major effect on unburned charge	Action to be taken to reduce knocking	Can operator usually control?
Compression ratio	Increases Temperature & pressure	Reduce	No
Mass of charge inducted	Increases pressure	Reduce	Yes
Inlet temperature	Increases temperature	Reduce	In some cases
Chamber wall temperature	Increases temperature	Reduce	Not ordinarily
Spark advance	Increases Temperature & pressure	Retard	In some cases
A/F ratio	Increases Temperature & pressure	Make very rich	In some cases
Turbulence	Decreases time factor	Increase	Somewhat (through engine speed)
Engine speed	Decreases time factor	Increase	Yes
Distance of Flame travel	Increases time factor	Reduce	No

10.12 COMBUSTION CHAMBERS FOR SI ENGINES

The design of the combustion chamber for an SI engine has an important influence on the engine performance and its knocking tendencies. The design involves the shape of the combustion chamber, the location of spark plug and the location of inlet and exhaust valves. Because of this importance, the combustion chamber design has been a subject of considerable amount of research and development in the last fifty years. It has resulted in the raising of the compression ratio of the engine from 4 before the first world war period to 11 in the present times with special combustion chamber designs and suitable antiknock fuels. The important requirements of an SI engine combustion chamber are to provide high power output with minimum octane requirement, high thermal efficiency and smooth engine operation.

Combustion chambers must be designed carefully, keeping in mind the following general objectives.

10.12.1 Smooth Engine Operation

The aim of any engine design is to have a smooth operation and a good economy. These can be achieved by the following:

Moderate Rate of Pressure Rise: The rate of pressure rise can be regulated such that the greatest force is applied to the piston as closely after TDC on the power stroke as possible, with a gradual decrease in the force on the piston during the power stroke. The forces must be applied to the piston smoothly, thus limiting the rate of pressure rise as well as the position of the peak pressure with respect to TDC.

Reducing the Possibility of Knocking: Reducing the possibility of knocking can be achieved by,

(i) Reducing the distance of the flame travel by centrally locating the spark plug and also by avoiding pockets of stagnant charge.

(ii) Satisfactory cooling of the spark plug points and of exhaust valve area which are the source for hot spots in the majority of the combustion chambers.

(iii) Reducing the temperature of the last portion of the charge, through application of a high surface to volume ratio in that part where the last portion of the charge burns. Heat transfer to the combustion chamber walls can be increased by using high surface to volume ratio thereby reducing the temperature.

10.12.2 High Power Output and Thermal Efficiency

The main objective of the design and development of an engine is to obtain high power as well as thermal efficiency. This can be achieved by considering the following factors:

(i) A high degree of turbulence is needed to achieve a high flame front velocity. Turbulence is induced by inlet flow configuration or squish. Squish can be induced in spark-ignition engines by having a bowl in piston or with a dome shaped cylinder head. Squish is the rapid radial movement of the gas trapped in between the piston and the cylinder head into the bowl or the dome.

(ii) High volumetric efficiency, i.e., more charge during the suction stroke, results in an increased power output. This can be achieved by providing ample clearance round the valve heads, large diameter valves and straight passages with minimum pressure drop.

(iii) Any design of the combustion chamber that improves its anti-knock characteristics permits the use of a high compression ratio resulting in high output and efficiency.

(iv) A compact combustion chamber reduces heat loss during combustion and increases the thermal efficiency. In the last two decades many different types of combustion chambers have been developed. Some of them are shown in Fig.10.8. Brief description of these combustion chambers are given below.

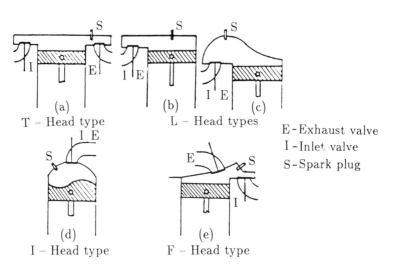

Fig.10.8 Examples of Typical Combustion Chamber

T-Head Type: The T-head combustion chambers [Fig.10.8(a)] were used in the early stage of engine development. Since the distance across the combustion chamber is very long, knocking tendency is high in this type of engines. This configuration provides two valves on either side of the cylinder, requiring two camshafts. From the manufacturing point of view, providing two camshafts is a disadvantage.

L-Head Type: A modification of the T-head type of combustion chamber is the L-head type which provides the two valves on the same side of the cylinder and the valves are operated by a single camshaft. Figures 10.8(b) and (c) show two types of this side valve engine. In these types, it is easy to lubricate the valve mechanism. With the detachable head it may be noted that the cylinder head can be removed without disturbing valve gear etc. In Fig.10.8(b) the air flow has to take two right angle turns to enter the cylinder. This causes a loss of velocity head and a loss in turbulence level resulting in a slow combustion process.

The main objectives of the Ricardo's turbulent head design [Fig.10.8(c)] are to obtain fast flame speed and reduced knock. The main body of the combustion chamber is concentrated over the valves leaving a slightly restricted passage communicating with the cylinder thereby creating additional turbulence during the compression stroke. This design reduces the knocking tendency by shortening the effective flame travel length by bringing that portion of the head which lay over the farther side of the piston into as close a contact as possible with the piston crown, forming a quench space. The thin layer of gas (entrapped between the relatively cool piston and also cooler head) loses its heat rapidly because of large enclosing surface thereby avoiding knocking. By placing the spark plug in the centre of the effective combustion space but with a slight bias towards the hot exhaust valve, the flame travel length is reduced.

I-Head Type or Overhead Valve: The I-head type is also called the overhead valve combustion chamber in which both the valves are located on the cylinder head. The overhead valve engine is superior to a side valve [Fig.10.8(d)] or an L-head engine at high compression ratios. Some of the important characteristics of this type of valve arrangement are:

(i) less surface to volume ratio and therefore less heat loss

(ii) less flame travel length and hence greater freedom from knock

(iii) higher volumetric efficiency from larger valves or valve lifts

(iv) confining the thermal failures to the cylinder head by keeping the hot exhaust valve in the head instead of the cylinder block.

F-Head Type: The F-head type of valve arrangement is a compromise between L-head and I-head types. Combustion chambers in which one valve is in the cylinder head and the other in the cylinder block are known as F-head combustion chambers [Fig.10.8(e)]. Modern F-head engines have exhaust valve in the head and inlet valve in the cylinder block. The main disadvantage of this type is that the inlet valve and the exhaust valve are separately actuated by two cams mounted on two camshafts driven by the crankshaft through gears.

10.13 COMBUSTION IN COMPRESSION–IGNITION ENGINES

There are certain basic differences existing between the combustion process in the SI and CI engines. In the SI engine, a homogeneous carburetted mixture of gasoline vapour and air, in a certain proportion, is compressed (compression ratio 6:1 to 11:1) and the mixture is ignited at one place before the end of the compression stroke by means of an electric spark. A single flame front progresses through the air-fuel mixture after ignition.

In the CI engine, only air is compressed through a high compression ratio (12:1 to 22:1) raising its temperature and pressure to a high value. Fuel is injected through one or more jets into this highly compressed air in the combustion chamber. Here, the fuel jet disintegrates into a core of fuel surrounded by a spray envelope of air and fuel particles [Fig.10.9(a)]. This spray envelope is created both by the atomization and vaporization of the fuel. The turbulence of the air in the combustion chamber passing across the jet separates the fuel particles from the core. A mixture of air and fuel forms at some location in the spray envelope and oxidation starts.

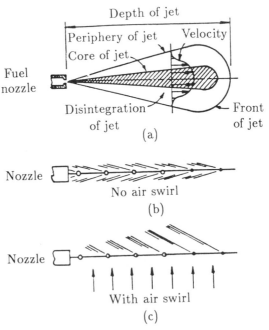

Fig.10.9 Schematic Representation of the Disintegration of a Fuel Jet

The liquid fuel droplets evaporate by absorbing the latent heat of vaporization from the surrounding air which reduces the temperature of a thin layer of air surrounding the droplet and some time elapses before this temperature can be raised again by absorbing heat from the bulk of air. As soon as this vapour and the air reach the level of the autoignition temperature and if the local A/F ratio is within the combustible range, ignition takes place. Thus it is obvious that at first there is a certain delay period before ignition takes place.

Since the fuel droplets cannot be injected and distributed uniformly throughout the combustion space, the fuel-air mixture is essentially heterogeneous. If the air within the cylinder were motionless under these conditions, there will not be enough oxygen in the burning zone and burning of the fuel would be either slow or totally fail as it would be surrounded by its own products of combustion [Fig.10.9(b)]. Hence, an orderly and controlled movement must be imparted to the air and the fuel so that a continuous flow of fresh air is brought to each burning droplet and the products of combustion are swept away. This air motion is called the air swirl and its effect is shown in Fig.10.9(c).

In an SI engine, the turbulence is a disorderly air motion with no general direction of flow. However, the swirl which is required in CI engines, is an orderly movement of the whole body of air with a particular direction of flow and it assists the breaking up of the fuel jet. Intermixing of the burned and unburned portions of the mixture also takes place due to this swirl. In the SI engine, the ignition occurs at one point with a slow rise in pressure whereas in the CI engine, the ignition occurs at many points simultaneously with consequent rapid rise in pressure. In contrast to the process of combustion in SI engines, there is no definite flame front in CI engines.

In an SI engine, the air-fuel ratio remains close to stoichiometric value from no load to full load. But in a CI engine, irrespective of load, at any given speed, an approximately constant supply of air enters the cylinder. With change in load, the quantity of fuel injected is changed varying the air-fuel ratio. The overall air-fuel ratio thus varies from about 18:1 at full load to about 80:1 at no load.

It is the main aim of the CI engine designer that the A/F ratio should be as close to stoichiometric as possible while operating at full load since the mean effective pressure and power output are maximum at that condition. Thermodynamic analysis of the engine cycles has clearly established that operating an engine with a leaner air-fuel ratio always gives a better thermal efficiency but the mean effective pressure and the power output reduce. Therefore, the engine size becomes bigger for a given output if it is operated near the stoichiometric conditions, the A/F ratio in certain regions within the

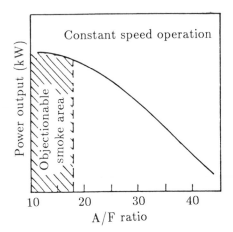

Fig.10.10 Effect of A/F Ratio on Power Output of a CI Engine

chamber is likely to be so rich that some of the fuel molecules will not be able to find the necessary oxygen for combustion and thus produce a noticeably black smoke. Hence the CI engine is always designed to operate with an excess air, of 15 to 40% depending upon the application. The power output curve for a typical CI engine operating at constant speed is shown in Fig.10.10. The approximate region of A/F ratios in which visible black smoke occurs is indicated by the shaded area.

10.14 STAGES OF COMBUSTION IN CI ENGINES

The combustion in a CI engine is considered to be taking place in four phases (Fig.10.11). It is divided into the ignition delay period, the period of rapid combustion, the period of controlled combustion and the period of after-burning. The details are explained below.

10.14.1 Ignition Delay Period

The ignition delay period is also called the preparatory phase during which some fuel has already been admitted but has not yet ignited. This period is counted from the start of injection to the point where the pressure-time curve separates from the motoring curve indicated as start of combustion in Fig.10.11.

The delay period in the CI engine exerts a very great influence on both engine design and performance. It is of extreme importance because of its effect on both the combustion rate and knocking and also its influence on engine starting ability and the presence of smoke in the exhaust.

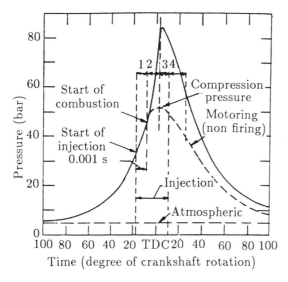

Fig.10.11 Stages of Combustion in a CI Engine

The fuel does not ignite immediately upon injection into the combustion chamber. There is a definite period of inactivity between the time when the first droplet of fuel hits the hot air in the combustion chamber and the time when it starts through the actual burning phase. This period is known as the ignition delay period. In Fig.10.12 the delay period is shown on pressure crank angle (or time) diagram between points a and b. Point a represents the time of injection and point b represents the time at which the pressure curve (caused by combustion) first separates from the compression motoring curve. The ignition delay period can be divided into two parts, the physical delay and the chemical delay.

Physical Delay: The physical delay is the time between the beginning of injection and the attainment of chemical reaction conditions. During this period, the fuel is atomized, vaporized, mixed with air and raised to its self-ignition temperature. This physical delay depends on the type of fuel, i.e., for light fuel the physical delay is small while for heavy viscous fuels the physical delay is high. The physical delay is greatly reduced by using high injection pressures and high turbulence to facilitate breakup of the jet and improving evaporation.

Chemical Delay: During the chemical delay reactions start slowly and then accelerate until inflammation or ignition takes place. Generally, the chemical delay is larger than the physical delay. However, it depends on the temperature of the surroundings and at high temperatures, the chemical reactions are faster and the physical delay

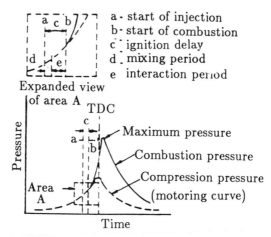

Fig.10.12 Pressure-Time Diagram illustrating Ignition Delay

becomes longer than the chemical delay. It is clear that, the ignition delay in the SI engine is essentially equivalent to the chemical delay for the CI engine. In most CI engines the ignition delay is shorter than the duration of injection.

10.14.2 Period of Rapid Combustion

The period of rapid combustion also called the uncontrolled combustion, is that phase in which the pressure rise is rapid. During the delay period, the droplets have had time to spread over a wide area and fresh air is always available around the droplets. Most of the fuel admitted has evaporated, formed a combustible mixture with air and the preflame reactions would have been completed. The period of rapid combustion is counted from end of delay period or the beginning of the combustion to the point of maximum pressure on the indicator diagram. The rate of heat-release is maximum during this period.

It may be noted that the pressure reached during the period of rapid combustion will depend on the duration of the delay period (the longer the delay the more rapid and higher is the pressure rise since more fuel would have been present in the cylinder before the rate of burning comes under control).

10.14.3 Period of Controlled Combustion

The rapid combustion period is followed by the third stage, the controlled combustion. The temperature and pressure in the second stage is already quite high. Hence the fuel droplets injected during the second stage burn faster with reduced ignition delay as soon as they find the necessary oxygen and any further pressure rise is controlled by the injection rate. The period of controlled combustion is assumed to end at maximum cycle temperature.

10.14.4 Period of After-Burning

Combustion does not cease with the completion of the injection process. The unburnt and partially burnt fuel particles left in the combustion chamber start burning as soon as they come into contact with the oxygen. This process continues for a certain duration called the after-burning period. Usually this period starts from the point of maximum cycle temperature and continues over a part of the expansion stroke. Rate of after-burning depends on the velocity of diffusion and turbulent mixing of unburnt and partially burnt fuel with the air. The duration of the after-burning phase may correspond to 70-80 degrees of crank travel from *TDC*.

The sequence of the events in the entire combustion process in a CI engine including the delay period is shown in Fig.10.13 by means of a block diagram.

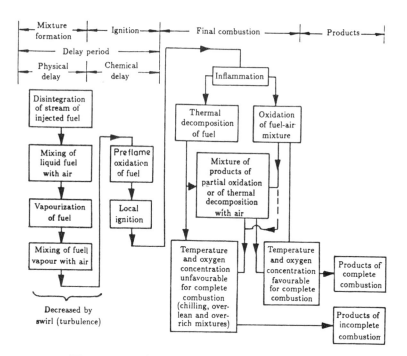

Fig.10.13 Block Diagram illustrating the
Combustion Process in a CI Engine

10.15 FACTORS AFFECTING THE DELAY PERIOD

Many design and operating factors affect the delay period. The important ones are:

  (i) compression ratio
 (ii) engine speed
(iii) output
(iv) atomization of fuel and duration of injection
 (v) injection timing
(vi) quality of the fuel
(vii) intake temperature
(viii) intake pressure

The effect of these factors on the delay period is discussed in detail in the following sections.

10.15.1 Compression Ratio

The increase in the compression temperature of the air with increase in compression ratio evaluated at the end of the compression stroke is shown in Fig.10.14.

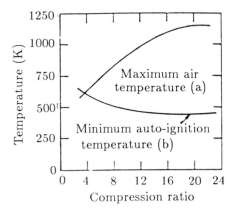

Fig.10.14 Effect of Compression Ratio on Maximum Air Temperature and Minimum Autoignition Temperature

It is also seen from the same figure that the minimum autoignition temperature of a fuel decreases due to increased density of the compressed air. This results in a closer contact between the molecules of fuel and oxygen reducing the time of reaction. The increase in the compression temperature as well as the decrease in the minimum autoignition temperature decrease the delay period. The maximum peak pressure during the combustion process is only marginally affected by the compression ratio (because delay period is shorter with higher compression ratio and hence the pressure rise is lower).

One of the practical disadvantages of using a very high compression ratio is that the mechanical efficiency tends to decrease due to increase in weight of the reciprocating parts. Therefore, in practice the engine designers always try to use a lower compression ratio which helps in easy cold starting and light load running at high speeds.

10.15.2 Engine Speed

The delay period could be given either in terms of absolute time (in milliseconds) or in terms of crank angle degrees. Fig.10.15 shows the decrease in delay period in terms of milliseconds with increase in engine speed in a variable speed operation with a given fuel.

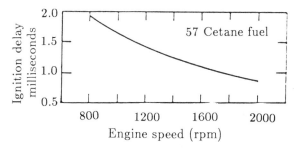

Fig.10.15 Effect of Speed on Ignition Delay in a Diesel Engine

With increase in engine speed, the loss of heat during compression decreases, resulting in the rise of both the temperature and pressure of the compressed air thus reducing the delay period in milliseconds. However, in degrees of crank travel the delay period increases as the engine operates at a higher rpm. The fuel pump is geared to the engine, and hence the amount of fuel injected during the delay period depends on crank degrees and not on absolute time. Hence, at high speeds, there will be more fuel present in the cylinder to take part in the second stage of uncontrolled combustion resulting in high rate of pressure rise.

10.15.3 Output

With an increase in engine output the air-fuel ratio decreases, operating temperatures increase and hence delay period decreases. The rate of pressure rise is unaffected but the peak pressure reached may be high.

10.15.4 Atomization and Duration of Injection

Higher fuel-injection pressures increase the degree of atomization. The fineness of atomization reduces ignition lag, due to higher surface volume ratio. Smaller droplet size will have low depth of penetration due to less momentum of the droplet and less velocity relative to air from where it has to find oxygen after vapourisation. Because of this air utilisation factor will be reduced due to fuel spray path being shorter. Also with smaller droplets, the aggregate area of inflammation will increase after ignition, resulting in higher pressure rise during the second stage of combustion. Thus, lower injection pressure, giving larger droplet size may give lower pressure rise during the second stage of combustion and probably smoother running. Hence, an optimum group mean diameter of the droplet size should be attempted as a compromise. Also the fuel delivery law i.e., change in the quantity of fuel supplied with the crank angle travel will affect the rates of pressure rise during second stage of combustion though ignition delay remains unaffected by the same.

10.15.5 Injection Timing

The effect of injection advance on the pressure variation is shown in Fig.10.16 for three injection advance timings of 9°, 18°, and 27° before TDC. The injected quantity of fuel per cycle is constant. As the pressure and temperature at the beginning of injection are lower for higher ignition advance, the delay period increases with increase in injection advance. The optimum angle of injection advance depends on many factors but generally it is about $20° bTDC$.

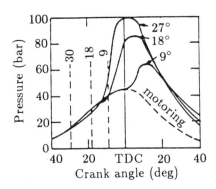

Fig.10.16 Effect of Injection Timing on Indicator Diagram

10.15.6 Quality of Fuel

Self-ignition temperature is the most important property of the fuel which affects the delay period. A lower self-ignition temperature results in a lower delay period. Also, fuels with higher cetane number give lower delay period and smoother engine operation. Other properties of the fuel which affect the delay period are volatility, latent heat, viscosity and surface tension.

10.15.7 Intake Temperature

Increase in intake temperature increases the compressed air temperature resulting in reduced delay period. However, preheating of the charge for this purpose would be undesirable because it would reduce the density of air reducing the volumetric efficiency and power output.

10.15.8 Intake Pressure

Increase in intake pressure or supercharging reduces the autoignition temperature and hence reduces the delay period. The peak pressure will be higher since the compression pressure will increase with intake pressure.

Table 10.2 gives the summary of the factors which influence the delay period in an engine.

10.16 THE PHENOMENON OF KNOCK IN CI ENGINES

In CI engines the injection process takes place over a definite interval of time. Consequently, as the first few droplets to be injected are passing through the ignition delay period, additional droplets are being injected into the chamber. If the ignition delay of the fuel being injected is short, the first few droplets will commence the *actual burning* phase in a relatively short time after injection and a relatively small amount of fuel will be accumulated in the chamber when *actual burning* commences. As a result, the mass rate of mixture burned will be such as to produce a rate of pressure rise that will exert a smooth force on the piston, as shown in Fig.10.17(a). If, on the other hand, the ignition delay is longer, the *actual burning* of the first few droplets is delayed and a greater quantity of fuel droplets gets accumulated in the chamber. When the *actual burning* commences, the additional fuel can cause too rapid a rate of pressure rise as shown in Fig.10.17(b), resulting in a *jamming* of forces against the piston and rough engine operation. If the ignition delay is quite long, so much fuel can accumulate that the rate of pressure rise is almost instantaneous, as shown in Fig.10.17(c). Such a situation produces the extreme pressure differentials and violent gas vibrations known as knocking and is evidenced by audible knock. The phenomenon is similar to that in the SI engine. However, *in the SI engine, knocking occurs near the end of combustion whereas in the CI engine, knocking occurs near the beginning of combustion.*

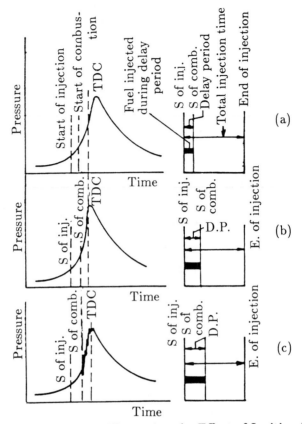

Fig.10.17 Diagrams illustrating the Effect of Ignition Delay on the Rate of Pressure Rise in a CI Engine

In order to decrease the tendency of knock it is necessary to start the *actual burning* as early as possible after the injection begins. In other words, it is necessary to decrease the ignition delay and thus decrease the amount of fuel present when the *actual burning* of the first few droplets start.

10.17 COMPARISON OF KNOCK IN SI AND CI ENGINES

It may be interesting to note that knocking in spark-ignition engines and compression-ignition engines is fundamentally due to the autoignition of the fuel-air mixture. In both the cases, the knocking depends on the autoignition lag of the fuel-air mixture. But careful

Table 10.2 *Effect of Variables on the Delay Period*

Increase in Variable	Effect on Delay Period	Reason
Cetane number of fuel	Reduces	Reduces the self-ignition temperature
Injection pressure	Reduces	Reduces physical delay due to greater surface-volume ratio
Injection timing advance	Increases	Reduces pressures and temperatures when the injection begins
Compression ratio	Reduces	Increases air temperature and pressure and reduces autoignition temperature
Intake temperature	Reduces	Increases air temperature
Jacket water temperature	Reduces	Increases wall and hence air temperature
Fuel temperature	Reduces	Increases chemical reaction due to better vaporization
Intake pressure (supercharging)	Reduces	Increases density and also reduces autoignition temperature
Speed	Increases in terms of crank angle. Reduces in terms of millisecs	Reduces loss of heat
Load (fuel-air ratio)	Decreases	Increases the operating temperature
Engine size	Decreases in terms of crank angle. Little effect in terms of millisecs	Larger engines operate normally at low speeds
Type of combustion chamber	Lower for engines with precombustion chamber	Due to compactness of the chamber

examination of the knocking phenomenon in spark-ignition and the compression-ignition engines reveals the following differences. A comparison of the knocking process in SI and CI engines is shown on the pressure-time diagrams of Fig.10.18.

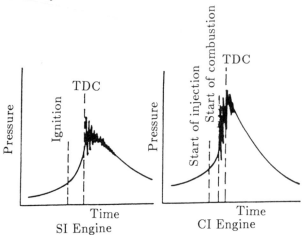

Fig.10.18 Diagrams illustrating Knocking Combustion in SI and CI Engines

(i) In spark-ignition engines, the autoignition of the end gas away from the spark plug, most likely near the end of the combustion causes knocking. But in compression-ignition engines the autoignition of the charge causing knocking is at the start of combustion. It is the first charge that autoignites and causes knocking in the compression-ignition engines. This is illustrated in Fig.10.18. It is clear from Fig.10.18 that explosive auto-ignition is more or less over before the peak pressure for the compression-ignition engines. But for spark-ignition engines, the condition for explosive autoignition of the end charge is more favourable after the peak pressure. In order to avoid knocking in spark-ignition engines, it is necessary to prevent autoignition of the end gas to take place at all. In compression-ignition engine, the earliest possible autoignition is necessary to avoid knocking.

(ii) In spark-ignition engine, the charge that autoignites is homogeneous and therefore intensity of knocking or the rate of pressure rise at explosive autoignition is likely to be more than that in compression-ignition engines where the fuel and air are not homogeneously mixed even when explosive autoignition of the charge occurs.

(iii) In compression-ignition engines, only air is compressed during the compression stroke and the ignition can take place only after fuel is injected just before the top dead centre. Thus there can be no preignition in compression-ignition engines as in spark-ignition engines.

(iv) It has already been pointed out that, the normal process of combustion in compression-ignition engines is by autoignition. And thus normal rate of pressure rise for the first part of the charge for compression-ignition are higher than those for spark-ignition engine, in terms of per degree crank rotation. And normally, audible knock is always present in compression-ignition engine. Thus when the audible noise becomes severe and causes heavy vibrations in the engine, it is said that the engine is knocking. Therefore, it is also a matter of judgement. A definite demarcation between normal combustion and knocking combustion is very difficult. The rate of pressure rise may be as high as 10 bar per degree crank rotation in compression-ignition engines. The factors that tend to increase autoignition reaction time and prevent knock in SI engines promote knock in CI engines. Also, a good fuel for spark-ignition engine is a poor fuel for compression-ignition engine. The spark-ignition fuels have high octane rating 80 to 100 and low cetane rating of about 20, whereas diesel fuels have high cetane rating of about 45 to 65 and low octane rating of about 30.

Table 10.3 gives a comparative statement of various characteristics that reduce knocking in spark-ignition engines and compression-ignition engines.

Table 10.3 *Characteristics Tending to Reduce Detonation or Knock*

S.No.	Characteristics	SI engines	CI engines
1.	Ignition temperature of fuel	High	Low
2.	Autoignition reaction time or ignition lag	Long	Short
3.	Compression ratio	Low	High
4.	Inlet temperature	Low	High
5.	Inlet pressure	Low	High
6.	Combustion wall temperature	Low	High
7.	Speed, rpm	High	Low
8.	Cylinder size	Small	Large

10.18 COMBUSTION CHAMBERS FOR CI ENGINES

The most important function of the CI engine combustion chamber is to provide proper mixing of fuel and air in a short time. In order to achieve this, an organized air movement called the air swirl is provided to produce high relative velocity between the fuel droplets and the air. The effect of swirl has already been discussed in Section 10.13. The fuel is injected into the combustion chamber by an injector having a single or multihole orifices. The increase in the number of jets reduces the intensity of air swirl needed.

When the liquid fuel is injected into the combustion chamber, the spray cone gets disturbed due to the air motion and turbulence inside. The onset of combustion will cause an added turbulence that can be guided by the shape of the combustion chamber. Since the turbulence is necessary for better atomization, and the fact that it can be controlled by the shape of the combustion chamber, makes it necessary to study the combustion chamber design in detail.

CI engine combustion chambers are classified into two categories:

(i) *Direct-Injection (DI) Type*: This type of combustion chamber is also called an open combustion chamber. In this type the entire volume of the combustion chamber is located in the main cylinder and the fuel is injected into this volume.

(ii) *Indirect-Injection (IDI) Type*: In this type of combustion chambers, the combustion space is divided into two parts, one part in the main cylinder and the other part in the cylinder head. The fuel-injection is effected usually into that part of the chamber located in the cylinder head. These chambers are classified further into:

(a) Swirl chamber in which compression swirl is generated.

(b) Precombustion chamber in which combustion swirl is induced.

(c) Air cell chamber in which both compression and combustion swirl are induced.

The details of these chambers are discussed in the following sections.

10.18.1 Direct–Injection Chambers

An open combustion chamber is defined as one in which the combustion space is essentially a single cavity with little restriction from one part of the chamber to the other and hence with no large difference in pressure between parts of the chamber during the combustion process. There are many designs of open chamber some of which are shown in Fig.10.19.

In four-stroke engines with open combustion chambers, induction swirl is obtained either by careful formation of the air intake passages or by masking a portion of the circumference of the inlet valve whereas in two-stroke engines it is created by suitable form for the inlet ports.

These chambers mainly consist of space formed between a flat cylinder head and a cavity in the piston crown in different shapes. The fuel is injected directly into this space. The injector nozzles used for this type of chamber are generally of multihole type working at a relatively high pressure (about 200 bar).

The main advantages of this type of chambers are:

(i) Minimum heat loss during compression because of lower surface area to volume ratio and hence, better efficiency.

(ii) No cold starting problems.

(iii) Fine atomization because of multihole nozzle.

The drawbacks of these combustion chambers are:

(i) High fuel-injection pressure required and hence complex design of fuel-injection pump.

(ii) Necessity of accurate metering of fuel by the injection system, particularly for small engines.

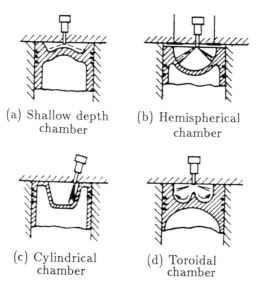

(a) Shallow depth chamber
(b) Hemispherical chamber
(c) Cylindrical chamber
(d) Toroidal chamber

Fig.10.19 Open Combustion Chambers

Shallow Depth Chamber: In shallow depth chamber the depth of the cavity provided in the piston is quite small. This chamber [Fig.10.19(a)] is usually adopted for large engines running at low speeds. Since the cavity diameter is very large, the squish is negligible.

Hemispherical Chamber: This chamber [Fig.10.19(b)] also gives small squish. However, the depth to diameter ratio for a cylindrical chamber can be varied to give any desired squish to give better performance.

Cylindrical Chamber: This design [Fig.10.19(c)] was attempted in recent diesel engines. This is a modification of the cylindrical chamber in the form of a truncated cone with base angle of 30°. The swirl was produced by masking the valve for nearly 180° of circumference. Squish can also be varied by varying the depth.

Toroidal Chamber: The idea behind this shape [Fig.10.19(d)] is to provide a powerful squish along with the air movement, similar to that of the familiar smoke ring, within the toroidal chamber. Due to powerful squish the mask needed on inlet valve is small and there is better utilisation of oxygen. The cone angle of spray for this type of chamber is 150° to 160°.

10.18.2 Indirect–Injection Chambers

A divided combustion chamber is defined as one in which the combustion space is divided into two or more distinct compartments connected by restricted passages. This creates considerable pressure differences between them during the combustion process.

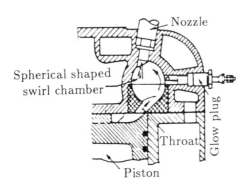

Fig.10.20 Ricardo Swirl Chamber Comet, Mark II

Swirl Chamber: Swirl chamber consists of a spherical-shaped chamber separated from the engine cylinder and located in the cylinder head. Into this chamber, about 50% of the air is transferred during the compression stroke. A throat connects the chamber to the cylinder which enters the chamber in a tangential direction so that the air coming into this chamber is given a strong rotary movement inside the swirl chamber and after combustion, the products rush back into the cylinder through the same throat at much higher velocity. This causes considerable heat loss to the walls of the passage which can be reduced by employing a heat-insulated chamber. However, in this type of combustion chambers even with a heat insulated passage, the heat loss is greater than that in an open combustion chamber which employs induction swirl.

This type of combustion chamber finds application where fuel quality is difficult to control, where reliability under adverse conditions is more important than fuel economy. The use of single hole of larger diameter for the fuel spray nozzle is often important consideration for the choice of swirl chamber engine.

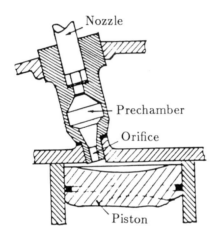

Fig.10.21 Precombustion Chamber

Precombustion Chamber: A typical precombustion chamber (Fig.10.21) consists of an antichamber connected to the main chamber through a number of small holes (compared to a relatively large passage in the swirl chamber). The precombustion chamber is located in the cylinder head and its volume accounts for about 40% of the total combustion space. During the compression stroke the piston forces the air into the precombustion chamber. The fuel is injected into the prechamber and the combustion is initiated. The resulting pressure rise forces the flaming droplets together with some air and their combustion products to rush out into the main cylinder at high velocity through the small holes. Thus it creates both strong secondary turbulence and distributes the flaming fuel droplets throughout the air in the main combustion chamber where bulk of combustion takes place. About 80% of energy is released in main combustion chamber.

The rate of pressure rise and the maximum pressure is lower compared to those of open type chamber. The initial shock of combustion is limited to precombustion chamber only. The precombustion chamber has multi-fuel capability without any modification in the injection system because of the temperature of prechamber. The variation in the optimum injection timing for petrol and diesel operations is only 2° for this chamber compared to 8° to 10° in the other designs.

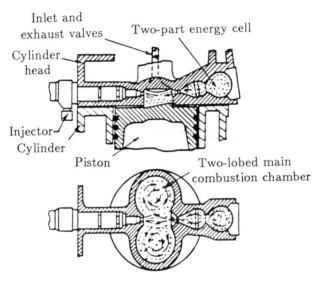

Fig.10.22 Lanova Air-Cell Combustion Chamber

Air-Cell Chamber: In this chamber (Fig.10.22), the clearance volume is divided into two parts, one in the main cylinder and the other called the energy cell. The energy cell is divided into two parts, major and minor, which are separated from each other and from the main chamber by narrow orifices. A pintle type of nozzle injects the fuel across the main combustion chamber space towards the open neck of the air cell.

During compression, the pressure in the main chamber is higher than that inside the energy cell due to restricted passage area between the two. At the TDC, the difference in pressure will be high and air will be forced at high velocity through the opening into the energy cell and this moment the fuel-injection also begins. Combustion starts initially in the main chamber where the temperature is comparatively higher but the rate of burning is very slow due to absence of any air motion. In the energy cell, the fuel is well mixed with air and high pressure is developed due to heat-release and the hot burning gases blow out through the small passage into the main chamber. This high velocity jet produces swirling motion in the main chamber and thereby thoroughly mixes the fuel with air resulting in complete combustion. The design is not suitable for variable speed operation as the combustion induced swirl has no relationship to the speed of the engine. The energy cell is designed to run hot, to reduce ignition lag.

The main advantages of the indirect-injection combustion chambers are:

(i) injection pressure required is low

(ii) direction of spraying is not very important.

These chambers have the following serious drawbacks which have made its application limited.

(i) Poor cold starting performance requiring heater plugs.

(ii) Specific fuel consumption is high because there is a loss of pressure due to air motion through the duct and heat loss due to large heat transfer area.

Questions :-

10.1 What are homogeneous and heterogeneous mixtures? In which engines these mixtures are used? Explain.

10.2 Briefly explain the stages of combustion in SI engines elaborating the flame front propagation.

10.3 Explain the various factors that influence the flame speed.

10.4 What is meant by abnormal combustion? Explain the phenomena of knock in SI engines.

10.5 Explain the effect of various engine variables on SI engine knock.

10.6 What are the various types of combustion chambers used in SI engines? Explain them briefly.

10.7 Bring out clearly the process of combustion in CI engines and also explain the various stages of combustion.

10.8 What is delay period and what are the factors that affect the delay period?

10.9 Explain the phenomenon of knock in CI engines and compare it with SI engine knock.

10.10 Explain with figures various types of combustion chambers used in CI engines.

11

ENGINE FRICTION
AND
LUBRICATION

11.1 INTRODUCTION

Friction generally refers to forces acting between surfaces in relative motion. In engines, frictional losses are mainly due to sliding as well as rotating parts. Normally, engine friction, in its broader sense, is taken as the difference between the indicated power, ip, and the brake power, bp. Usually engine friction is expressed in terms of frictional power, fp. Frictional loss is mainly attributed to the following mechanical losses.

 (i) direct frictional losses
 (ii) pumping losses
 (iii) power loss to drive the components to charge and scavenge
 (iv) power loss to drive other auxiliary components

A good engine design should not allow the total frictional losses to be more than 30% of the energy input in reciprocating engines. It should be the aim of a good designer to reduce friction and wear between the parts having relative motion. This is achieved by proper lubrication. In this section the various losses associated with friction is enumerated.

11.1.1 Direct Frictional Losses

It is the power absorbed due to the relative motion of different bearing surfaces such as piston rings, main bearings, cam shaft bearings etc. Since, there are a number of moving parts, the frictional losses are comparatively higher in reciprocating engines.

11.1.2 Pumping Loss

In case of the four-stroke engines, a considerable amount of energy is spent during intake and exhaust processes. The pumping loss is the net power spent by the engine (piston) on the working medium (gases) during intake and exhaust strokes. In the case of two-stroke engines this is negligible since the incoming fresh mixture is used to scavenge the exhaust gases.

11.1.3 Power Loss to Drive Components to Charge and Scavenge

In certain types of four-stroke engines the intake charge is supplied at a higher pressure than the naturally aspirated engines. For this purpose a mechanically driven compressor or a turbine driven compressor is used. Accordingly the engine is called the supercharged or turbocharged engine. In case of a supercharged engine, the engine itself supplies power to drive the compressor whereas in a turbocharged engine, the turbine is driven by the exhaust gases of the engine. These devices take away a part of the engine output. This loss is considered as negative frictional loss. In case of two-stroke engines with a scavenging pump, the power to drive the pump is supplied by the engine.

11.1.4 Power Loss to Drive the Auxiliaries

A good percentage of the generated power output is spent to drive auxiliaries such as water pump, lubricating oil pump, fuel pump, cooling fan, generator etc. This is considered a loss because the presence of each of these components reduces the net output of the engine.

11.2 MECHANICAL EFFICIENCY

The various losses described above can be clubbed into one heading, viz., the mechanical losses. The mechanical losses can be written in terms of mean effective pressure, that is frictional power divided by engine displacement volume per unit time. Therefore, frictional mean effective pressure, $fmep$, can be expressed as

$$fmep \quad = \quad mmep + pmep + amep + cmep$$

where $mmep$: mean effective pressure required to overcome mechanical friction

$pmep$: Mean effective pressure required for charging and scavenging

$amep$: mean effective pressure required to drive the auxiliary components

$cmep$: mean effective pressure required to drive the compressor or scavenging pump

Because of the various mechanical losses in the engine the term mechanical efficiency is usually associated with the reciprocating internal combustion engine.

As already explained in Chapter 1, mechanical efficiency is defined as the ratio of *bp* to *ip* or *bmep* to *imep*. It is written as

$$\eta_m \;=\; \frac{bp}{ip} \;=\; \frac{bmep}{imep}$$

A knowledge of engine friction is essential for calculating the mechanical efficiency of the engine. Mechanical efficiency indicates how good an engine is, in converting the indicated power to useful power. The value of mechanical efficiency varies widely with the design and operating conditions. It is evidently zero under idling conditions, i.e. all the indicated power developed by the engine is spent in overcoming friction. It should be the aim of any designer to increase the mechanical efficiency to the maximum by reducing the frictional losses to a minimum.

11.3 MECHANICAL FRICTION

As mentioned earlier, friction loss comes into picture in the bearing surfaces of the engine components due to their relative motion. Mechanical friction in engine may be divided into six classes which are discussed in the following sections.

11.3.1 Fluid-film or Hydrodynamic Friction

The hydrodynamic friction is associated with the phenomena when a complete film of lubricant exists between the two bearing surfaces. In this case the friction force entirely depends on the lubricant viscosity. This type of friction is the main mechanical friction loss in the engine.

11.3.2 Partial-film Friction

When rubbing (metal) surfaces are not sufficiently lubricated, there is a contact between the rubbing surfaces in some regions. During normal engine operation there is almost no metallic contact except between the compression (top) piston ring and cylinder walls. This is mainly at the end of each stroke where the piston velocity is nearly zero. During starting of the engine, the journal bearings operate in partial-film friction. Thus, partial-film friction contributes very little to engine friction and compared to total friction this is almost negligible.

11.3.3 Rolling Friction

The rolling friction is due to rolling motion between the two surfaces. Ball and roller bearings and tappet rollers are subjected to rolling friction. Bearings of this type have a coefficient of friction which is nearly independent of load and speed. This friction is partly due to local rubbing from distortion under load and partly due to continuous *climbing* of roller. Rolling friction coefficient is lower than journal bearing friction coefficient during starting and initial running of engine. The reason is that the oil viscosity is high and moreover, partial friction exists in journal bearing during starting where engine uses plain journal bearings on the crankshaft. Rolling friction is negligible compared to total friction.

11.3.4 Dry Friction

Even when an engine is not operated for a long time there is little possibility for direct metal to metal contact. Always some lubricant exists between the rubbing surfaces even after long periods of disuse. One can take the dry friction to be non-existent and hence, this can be safely neglected while considering engine friction.

11.3.5 Journal Bearing Friction

A circular cylindrical shaft called journal rotates against a cylindrical surface called the bearing. Journal bearings are called partial when the bearing surface is less than full circumference. The rotary motion may be either continuous or oscillatory. Much theoretical and experimental work has been done to find the performance of journal bearing under various operating conditions. Engine journal bearing operates under load which varies in magnitude and direction with time. However, the same basic relations obtained for ordinary journal bearing apply to engine journal bearing but the coefficients of friction are usually different.

11.3.6 Friction due to Piston Motion

Friction due to the motion of piston can be divided into

(i) viscous friction due to piston
(ii) non-viscous friction due to piston ring

The non-viscous piston ring friction can be further subdivided into

(i) friction due to ring tension
(ii) friction due to gas pressure behind the ring

Tests show that the quantity of the lubricant between the piston and the cylinder wall is normally insufficient to fill the entire space between the piston and the cylinder wall. The oil film thickness between piston and the cylinder is also affected by the piston side-thrust and the resulting vibrations. Average oil film thickness between the piston and the cylinder wall varies with load and speed. Piston friction also depends upon the viscosity of the oil and the temperature at the various points on the piston.

Piston rings are categorized into compression rings and oil rings. Compression rings are on the top portion of the piston to seal against gas pressure. The pressure exerted by the compression ring on the cylinder wall is partly due to the elasticity of the ring and partly due to the gas pressure which leaks into the space between the ring and the piston. The gas pressure behind the top ring (compression ring) is nearly equal to the cylinder pressure, less than the cylinder pressure in the second ring groove and much less in the third.

Oil rings are designed to scrape some of the oil from the cylinder wall and allow it to return to the oil sump through radial passage in the ring. These grooves for the rings are vented by holes drilled into the piston interior and therefore no gas pressure can act behind it. In this case the pressure of the ring surface on the cylinder wall is entirely due to the elasticity (tension) of the rings. Piston rings press against the cylinder walls at all times because of their spring action. Therefore, the friction due to piston motion always exists.

11.4 BLOWBY LOSSES

It is the phenomenon of leakage of combustion products (gases) from the cylinder to the crankcase past the piston and piston rings. It depends on the compression ratio, inlet pressure and the condition of the piston rings. In case of worn out piston rings this loss is more. This loss is usually accounted in the overall frictional losses.

11.5 PUMPING LOSS

The work spent to charge the cylinder with fresh mixture during the suction stroke and to discharge the combustion products during the exhaust stroke is called the pumping loss. This pumping loss may be reduced by increasing the valve areas. However, this area cannot be greatly increased due to the practical limitation on the availability of space in the cylinder head. Further, engine speed also plays a role on the pumping loss. For four-stroke engines pumping loss may be divided into three parts as shown in the indicator diagram, Fig.11.1.

11.5.1 Exhaust Blowdown Loss

To reduce the work spent by the piston to drive out the exhaust gases the exhaust valve is made to open before piston reaches TDC on its expansion stroke. During this period the combustion gases rush out of cylinder due to pressure difference. Because of this there is a certain loss of power indicated by the area shown in Fig.11.1. This loss is called the blowdown loss. This mainly depends on the exhaust valve timing and its size. With large valve area and earlier exhaust valve opening the blowdown loss will be higher whereas with increase in speed this loss tends to be lower.

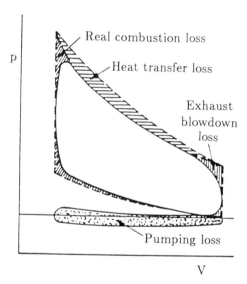

Fig.11.1 Pumping and Blowdown Losses

11.5.2 Exhaust Stroke Loss

The work required to force the products of combustion out of the cylinder after the blowdown process is the exhaust loss. During the starting of blowdown process the gas pressure inside the cylinder is more than three to four times the gas pressure inside the exhaust pipe, therefore, the gases flow out with high velocity. Due to inertia this high velocity of exhaust gases tends to persist even during exhaust stroke and the cylinder pressure may momentarily drop below the gas pressure in the exhaust pipe. When the piston moves up again, pressure rises and gases are pushed into exhaust pipe. Thus the power required to drive the exhaust gases out is called the exhaust stroke loss. This loss is shown in indicator diagram. This loss depends on valve size, valve timing and valve flow coefficient. Increase in valve size, early opening of valve and higher valve flow coefficient may tend to reduce exhaust stroke loss. Increase in speed increases the exhaust stroke loss.

Thus factors like high speed, late opening of exhaust valve, reduced exhaust valve size which tend to increase the exhaust stroke loss may tend to reduce blowdown loss. For getting better performance from the engine the combination of the two losses should be minimum.

11.5.3 Intake Stroke Loss

Energy is supplied to the piston to produce a pressure difference across the inlet valve so that fresh charge could be drawn into the cylinder. This is done to overcome the friction and inertia of the gas in the intake system and the power spent by the piston for doing this is called the intake stroke loss. The combination of intake and exhaust stroke loss is called the pumping loss illustrated in the Fig.11.1.

11.6 FACTORS AFFECTING MECHANICAL FRICTION

Various factors affect the engine friction. In this section, the effect of some of these factors on mechanical friction is discussed.

11.6.1 Engine Design

The design parameters which influence the friction losses are:

(i) *Stroke-bore Ratio :* Lower stroke-bore ratio may tend to slightly decrease the *fmep*. It is mainly due to less frictional area in case of lower stroke to bore ratio.

(ii) *Effect of Engine Size :* Larger engines have more frictional surfaces. Hence, lubrication requirements are more in such engines.

(iii) *Piston Rings :* Reducing the number of piston rings and reducing the contacting surface of the ring with cylinder wall reduces the friction. Light ring pressure also reduces the friction.

(iv) *Compression Ratio :* The friction mean effective pressure increases with increase in compression ratio. But the mechanical efficiency either remains or may improve slightly because of the increase in the *imep*.

(v) *Journal Bearings :* Reducing journal diameter/diametrical clearance ratio in journal bearing reduces the *fmep*. Having short pistons with the non-thrust surfaces cut away and light reciprocating parts which minimize inertia loads which in turn may reduce friction loss.

11.6.2 Engine Speed

Friction increases rapidly with increasing speed. At higher speeds mechanical efficiency starts deteriorating considerably. This is one of the reasons for restricting the speeds of engines.

11.6.3 Engine Load

Increasing the load increases the maximum pressure in the cylinder which results in slight increase in friction values. At the same time increase in load results in increase in temperature inside the cylinder and also temperature of the lubricating oil. The decrease in oil viscosity due to higher temperature slightly reduce the friction.

In gasoline engines the throttling losses reduce as the throttle is opened more so as to supply more fuel. Results of the above may tend to bring down the frictional losses in such engines (SI engine). However, for diesel engines the frictional losses due to engine load are more or less constant since there is no throttling effect.

11.6.4 Cooling Water Temperature

A rise in cooling water temperature slightly reduces engine friction by reducing oil viscosity. However, friction losses are high during starting since temperature of water and oil are low and viscosity is high.

11.6.5 Oil Viscosity

Viscosity and friction loss are (directly) proportional to each other. The viscosity can be reduced by increasing the temperature of the oil. But beyond a certain value of oil temperature, failure of local oil film may occur resulting in partial fluid film friction or even metal to metal contact which is very harmful to the engine.

11.7 LUBRICATION

From the above discussion it can be understood that lubrication is essential to reduce friction and wear between the components in an engine. In the following sections the details of engine lubrication are discussed.

11.7.1 Function of Lubrication

Lubrication is an art of admitting a lubricant (oil, grease, etc.) between two surfaces that are in contact and in relative motion. The purpose of lubrication in an engine is to perform one or several of the following functions.

(i) To reduce friction and wear between the moving parts and thereby the energy loss and to increase the life of the engine.

(ii) To provide sealing action e.g. the lubricating oil helps the piston rings to maintain an effective seal against the high pressure gases in the cylinder from leaking out into the crankcase.

(iii) To cool the surfaces by carrying away the heat generated in engine components.

(iv) To clean the surfaces by washing away carbon and metal particles caused by wear.

Of all these functions, the first function is considered to be the most important one. In internal combustion engines, the problems of lubrication become more difficult because of the high temperatures experienced during the combustion process and by the wide range of temperatures encountered throughout the cycle. The energy loss from the friction between different components of the engine can be minimized by providing proper lubrication.

Moreover, because of the variation in the gas force on the piston and the inertia force of the moving parts, the bearings are subjected to fluctuating loads where effective lubrication at all operating conditions is extremely difficult. The temperature extremes are emphasized for starting during which the flow of lubricating oil is difficult. Therefore, the moving parts may suffer from inadequate supply of lubricating oil and metal to metal contact may result. The basic problems associated with the proper lubrication of the various types of bearings encountered in an internal combustion engine will be discussed in general in this chapter. In addition, the properties of lubricating oils, the effect of engine operation on these properties and the types of lubricating systems will also be discussed.

11.7.2 Mechanism of Lubrication

Consider two solid blocks which are in contact with each other. In order to move the the upper block over the surface of the lower block, a constant tangential force must be applied. The force due to the weight of the upper block acting perpendicular to the surface is called the *normal force*. The ratio of the tangential force to the normal force is known as the *dynamic coefficient of friction or the coefficient of friction, f*.

$$\text{Coefficient of friction, } f = \frac{\text{Tangential force}}{\text{Normal force}}$$

To keep the block in motion, a constant tangential force is required to overcome the frictional resistance between the two surfaces. This frictional resistance arises because the moving surfaces are rough, and hence small irregularities will fit together at the contact area (interface) to give a mechanical lock to the motion. Also, if the moving surfaces are too smooth, the molecular attraction will be more in the interface and will resist the motion. Therefore, friction between dry surfaces may arise either from surface irregularities or from molecular attraction or from both.

The coefficient of friction between two dry surfaces tends to be constant and independent of the load (normal force), relative speed and contact areas of the surfaces but varies with the materials and the surface finish.

The resistance between moving surfaces can be reduced by the introduction of a small film of lubricant between the moving surfaces so that the two surfaces are not in physical contact. Hence, the friction due to surface irregularities and molecular attraction is reduced. The molecules of the lubricant film provide more mobile fluid layer than that of the solid surface and hence less force is required to accomplish a relative motion. The solid friction is replaced with a definitely lesser fluid friction.

Consider the case of two parallel plates filled with viscous oil in between them, of which one is stationary and other is in motion with a constant velocity in the direction as shown in Fig.11.2. It is assumed that

(i) the width of the plates in the direction perpendicular to the motion is large so that the flow of lubricant in this direction is negligibly small

(ii) the fluid is incompressible

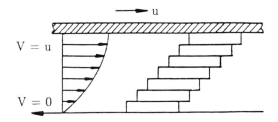

Fig.11.2 Mechanism of Lubrication in Parallel Surfaces

Let us imagine the film as composed of a series of horizontal layers and the force, F causing layers to deform or slide one over another, just like a deck of cards. The first layer clinging to the moving surface will move with the plate because of the adhesive force between the plate and the oil layer while the next layer is moving by at slower pace. The subsequent layers below keep moving at gradually reducing velocities. The layer clinging to the surface of the stationary plate will have zero velocity. The reason is that each layer of the oil is subjected to a shearing stress and the force required to overcome this stress is the fluid friction. Thus the velocity profile across the oil film varies from zero at the stationary plate to the velocity of the plate at the moving surface, as shown in Fig.11.2.

In the above example, the fluid or internal friction arose because of the resistance of the lubricant to shearing stress. A measure of the resistance to shear is a property called *viscosity or coefficient of viscosity*.

Now let us consider that the above mentioned plates are non-parallel and the upper one (A′B′) is moving while the lower one (AB) remains stationary. The cross section of the fluid at the leading edge is less than that of the trailing edge. This arrangement is shown in Fig.11.3. The discussion with regard to Fig.11.2 indicates that the velocity distributions across the oil film at the edges are expected to have the shape shown by dotted line in Fig.11.3.

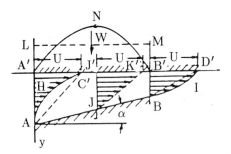

Fig.11.3 Mechanism of Lubrication on Wedge Shaped Surfaces

But as the fluid is incompressible, the actual volume of oil carried into the space at the trailing edge must be equal to the volume discharged from this space at the leading edge. Therefore, the excess volume of oil carried into the space is squeezed out through the sections of either edges producing a constant pressure, which induces a flow through these sections. It is to be remembered that the width of the plates is assumed to be large so that there is no leakage along the side ends of the plate. In addition, the viscosity of the fluid also retards the escape of lubricant from the squeezing action of the normal faces that tends to bring the surface together. This creates a pressure in the lubricant film. Therefore, the actual velocity distribution in the section at the edges is the result of the combined flow of lubricant due to pressure induced flow and due to viscous drag, which is indicated by continuous lines (curved lines) across the section at the edges. This oil film pressure enables to carry the load. This type of lubrication where a wedge shaped oil film is formed between two moving surface is called hydrodynamic lubrication. The important feature of this type of lubrication is that the load carrying capacity of the bearing increases with increase in relative speed of the moving surfaces. This is because at high speed, the available time for the oil to squeeze out is less and hence the pressure rise tends to be high.

It can be seen that with the above mentioned arrangement a positive pressure in the oil film is developed. This condition exists when the thickness of the film decreases in the direction of motion of surface. Such a film is known as convergent film.

If the motion of the upper plate is reversed, with the inclination of the lower plate unchanged, the volume of the lubricant that the moving surface tends to drag into the space becomes less than the volume which tends to discharge from the space and the pressure developed in the oil film tends to be negative (less than atmospheric pressure). The bearing will be unable to support any load by the film. Such a film is known as a diverging film.

11.7.3 Elastohydrodynamic Lubrication

Elastohydrodynamic lubrication is the phenomenon that occurs when the bearing material itself deforms elastically against the pressure built up of the oil film. This type of lubrication occurs between cams and followers, gear teeth and roller bearings where the contact pressures are extremely high. Hydrostatic lubrication is obtained by introducing the lubricant, which is sometimes air or water, into the load-bearing area at a pressure high enough to separate the surfaces with a relatively thick film of lubricant. So, unlike hydrodynamic lubrication, motion of one surface relative to another is not required.

Insufficient surface area, a drop in relative velocity of the moving surface during the period of starting and stopping an inadequate quantity of lubricant, an increase in the bearing load, or a decrease in viscosity of the lubricant due to increase in temperature — anyone of these — may prevent the build up of a film thick enough to full film lubrication. When this happens the highest asperities may be separated by lubricant films only several dimensions in thickness. This is called boundary lubrications. The change from hydrodynamic to boundary lubrication is not an abrupt change. It is probable that a mixed hydrodynamic and boundary type lubrication occurs first, and as the surfaces move closer together, the boundary type lubrication becomes predominant. The viscosity of the lubricant is not as much important with boundary lubrication as the chemical composition of the lubricant. This is illustrated in Fig.11.4.

11.7.4 Journal Bearing Lubrication

Let us examine the formation of a lubricant film in a journal bearing. Fig.11.5 shows a full journal bearing, with the clearance greatly exaggerated. Assume that the space between the journal and bearing is filled with a fluid lubricant.

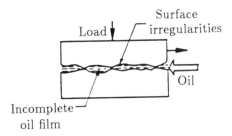

Fig.11.4 Illustration of Boundary Lubrication

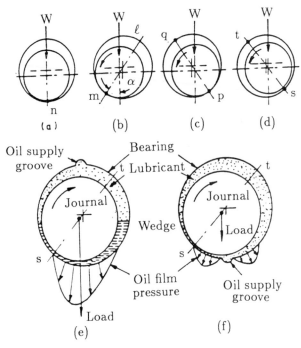

Fig.11.5 Mechanism of Lubrication of Journal Bearings

In Fig.11.5(a), the journal loaded by a vertical load, W, is at rest. In this case the metal to metal contact between the surface of the journal and the bearing is at point n on the line of action of the load. If the journal is just beginning to rotate in an anti-clockwise direction, it tends to climb or roll up the left side of the bearing. Under these conditions, a dry bearing equilibrium is obtained when the friction force is balanced by the tangential component of the bearing load. Therefore, the point of contact will move into position m [Fig.11.5(b)].

It is easy to see that if the speed is quite low, the pressure built in the film can be neglected. The angle α is equal to the angle of sliding friction between the surfaces. With fluid lubricant present, the friction in this case corresponds to the conditions of boundary or extreme boundary conditions.

In this position, the continuous oil film consists of two parts, a convergent part above the line m-ℓ and a divergent part below this line. As explained previously, under such conditions a positive pressure is developed in the converging parts of the oil film. This pressure makes the journal to move to the right. Therefore, with an increase in journal speed, the pressure forces will overcome at the point of contact and will move the journal to the right as shown in Fig.11.5(c).

When the journal reaches a certain speed, the journal is lifted up as shown in Fig.11.5(d) because of the oil pressure. The centre of the journal has moved to such a position that the minimum film thickness is now at point s. From s to t in the direction of motion the oil film is diverging film and t to s it is a converging film and therefore is able to support the load.

An oil film pressure is developed as shown in Fig.11.5(e), the journal is thus operating in the thick film region and the load is carried by the wedge of oil, shown as the shaded area in Fig.11.5(e). To form and to maintain the fluid film, it is necessary to supply a sufficient quantity of lubricant to replace that lost due to end leakage from the bearing and to prevent rupture. This will enable a continuous film around the journal. The proper place where the lubricant must be introduced into the bearing is where the film pressure is low.

The part of the film s to t in the direction of motion is diverging. From the foregoing discussion it is clear that no positive pressure can be developed in this region. On the contrary a negative pressure is expected in this region, which may cause a break in the film. This region can be utilized to supply the lubricant so that a continuous oil film can be maintained. If an oil supply groove is cut in the loaded region as in Fig.11.5(f) the oil film pressure is reduced, resulting in a large reduction in load carrying capacity.

As discussed earlier, if the surface velocity of the journal is high enough and if there is a sufficient oil supply, enough oil pressure may be developed due to the creation of wedge film to raise the shaft off the bearing. The same wedge film can be created if either the journal is held stationary and the bearing rotates or both rotate. The reader should apply the same reasoning for the hydrodynamic pressure development for the journal rotation and stationary bearing as discussed previously. Hence, the relative velocity between journal and the bearing plays a vital role in the hydrodynamic pressure development and thereby the load carrying capacity. In addition, the load may also vary either in magnitude (fluctuating) or in direction (rotating) or both. For example, the bearings of the crankshafts of an internal combustion engine is subjected to a fluctuating and rotating load.

In some other cases, the bearing may still support a load even when the journal and bearing are held stationary but the load is rotating. Also from investigations, one interesting phenomenon to be noted is that the bearing will not support any load when it is stationary and the load rotates at one half of the speed and also in the same direction as the journal.

As regard to the piston pins, the wedge action due to rotation is absent as neither the journal nor the bearing rotates. However, the loads on the piston pins are adequately supported and the piston pin bushings rarely fail in practice. The reason for the above is as discussed below.

Analysis on bearings reveals that a non rotating shaft can support a reciprocating load of considerable magnitude. Physically, it means that as the load is applied in one direction, the journal acts like piston in a hydraulic damping mechanism squeezing out the oil on one side and sucking in on the other side and thereby it can take the load. To explain this, the theory assumes that the journal will move just as slowly away from the bearing as towards it when they are in close proximity due to negative pressures developing in the oil film. Since, cavitation probably occurs long before negative pressures of magnitude, comparable to the positive ones developed during the approach portion of the cycle are reached. The journal can always precede from the bearing wall on the near side much more quickly than it approaches it. In this way it is saved from actual metal to metal contact. Also the relative sliding speeds between the pin and its bushing are generally small, and hence the heat generation due to fluid friction is also minimum. So, one would not expect severe damage in any case. Obviously the higher the frequency of the load cycle, the higher the load that can be carried by the bearing.

11.7.5 Stable Lubrication

As discussed earlier, with lower relative velocity of the moving surfaces, adequate pressure cannot be developed to support the load by the oil film. At this point of time boundary lubrication will exist. This will occur especially during starting and stopping of an engine. As the speed increases, a sufficient film pressure is developed and the load is supported by the oil film. The phenomenon of shift from boundary lubrication to hydrodynamic lubrication is shown in Fig.11.6 for the change in the coefficient of friction, f, versus the characteristic number, $\nu N/p$ of the bearing.

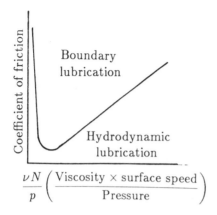

Fig.11.6 Region of Boundary and Hydrodynamic Lubrication

Here ν is the oil viscosity, p is the pressure and N is the speed. To the left of the this point the hydrodynamic pressure developed by the film is too low to support the load and metal to metal contact occurs. This is the zone of boundary lubrication. Even a small reduction in $\nu N/p$ will increase the coefficient of friction drastically and will result in more heat generation which is undesirable for any hydrodynamic bearing.

The bearing operation to the right of this, minimum $\nu N/p$ point is stable. An increase in bearing temperature reduces the oil viscosity and hence, results in lower friction coefficient which gives rise to lower bearing temperature due to reduced heat generation, thereby stabilizing the bearing temperature.

11.8 LUBRICATION OF ENGINE COMPONENTS

In a reciprocating engine there are many surfaces in contact with each other and therefore they should be lubricated to reduce friction. The principal friction surfaces requiring lubrication in an internal combustion engine are

(i) piston and cylinders

(ii) crankshaft and their bearings

(iii) crankpin and their bearings

(iv) wristpin and their bearings

(v) valve gear

11.8.1 Piston

Since the piston and piston rings are exposed to high temperatures, the oil that is supplied to the cylinder walls must provide proper lubrication under extreme conditions. In addition, the lubricant on the cylinder walls must function as a seal to minimise the amount of the gases of combustion that pass the piston rings and enter the crankcase.

Pistons of high speed engines used in automobiles, usually do not need any special lubricating provisions. The oil which flows form the main bearings and crankpin is splashed by the crankwebs and connecting rods and the oil mist which is formed in the enclosed crankcase by the fast moving parts furnish sufficient lubrication for the pistons. These conditions exist not only when the oil in the crankcase is kept at a certain level to allow its pick-up by the connecting rods but even when the oil level is low in the crankcase.

Pistons of low and medium speed engines are lubricated by positive feed mechanical oilers. A positive feed mechanical oiler has a small plunger which discharges oil to the cylinder surface during its forward stroke. On its return stroke a check valve prevents the oil from being sucked back and oil is drawn in through another liner.

Tests indicate that the piston rings and cylinders operate in the thin-film lubrication region a great deal of time. A thick-film lubrication would result in excessive oil consumption. In thin film region, oiliness and surface finish will play a vital role in reducing wear and scuffing. A smooth surface has flat spots with many small indentations interposed among them. These irregularities help the oil to maintain a film on the surface and act as individual oil reservoirs. Surface damage may be further reduced by adding sulphur and chlorine compounds in the lubricant. These compounds convert the material into sulphides and chlorides and prevent welding of stressed points.

11.8.2 Crankshaft Bearings

In small stationary engines, the main bearings of the crankshaft are made with ring oilers. A ring oiler is nothing but a ring which is slipped over the shaft and runs over the journal. The diameter of the ring is bigger than the journal diameters which enables the ring to slip over the shaft when it is rotating. Because of the shaft rotation the ring will also tend to rotate. The oil level in the reservoir, by dipping the ring draws oil with the bottom portion of the ring, and carries it to the top of the journal where it is able to flow into the bearing oil grooves and bearing clearance space, to be distributed to the entire bearing surface.

For large engines, with open or at least not very tight crankcase, as in the case of most horizontal engines, the main bearings are lubricated by positive feed lubricators. Engines with enclosed crankcases as in the case of the most vertical engines, are usually built with pressure lubrication. The oil is drawn by a gear pump and delivered it to the oil gallery in the crankcase which distributes it to the bearings. Oil coming out of the bearing goes to an oil sump where it is picked up by a scavenging pump. Then it is pumped through a filter to an oil cooler and to a supply tank from which it flows back to the pressure pump. Usually, the oil is circulated in large quantities as the oil not only serves as lubricant but also act as a cooling medium to the bearings. In some design of the pistons lubrications helps to remove heat to some extent.

11.8.3 Crankpin Bearings

In small engines, the bearing pressure is low and the crankpins are sometimes lubricated by splashing oil. For this purpose, the bottom of the connecting rod is attached with a dipper which dips into the oil when the connecting rod comes down. This arrangement is illustrated in Fig.11.7(a).

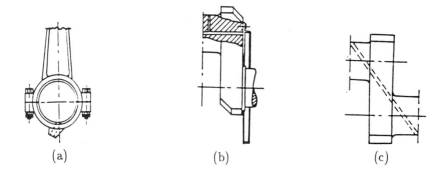

(a) (b) (c)

Fig.11.7 Lubrication of Crankpin Bearings

Small horizontal engines and some two-stroke vertical engines use the centrifugal bango oiler. Fig.11.7(b) shows this kind of arrangement. The oil hole leading to the surface of the crankpin is often drilled to an angle of about 30° in advance of the dead centre, so that the upper shell receives oil before the ignition at a point of relatively low pressure.

For heavy duty engine, the oil is fed from the main bearing to the crankpin by a drilled hole. The details are illustrated in Fig.11.7(c).

11.8.4 Wristpin Bearing

In small horizontal engines the wristpin is lubricated by a sight feed oiler. The details of the arrangements are shown in Fig.11.8(a).

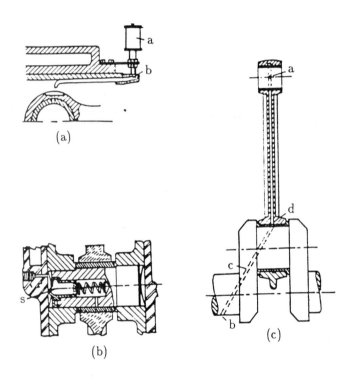

Fig.11.8 Lubrication of Wristpin Bearings

A special scraper scoop is provided to scrap the oil dripping from the cylinder walls and the arrangement is illustrated in Fig.11.8(b).

In other vertical engines where the lubricating oil delivered under pressure to the main and crankpin bearings, the wristpin is lubricated by excess oil from the crankpin bearing. For this purpose, a hole will be drilled through the connecting rod shown. This is illustrated in Fig.11.8(c).

11.9 LUBRICATION SYSTEM

The function of a lubrication system is to provide sufficient quantity of cool, filtered oil to give positive and adequate lubrication to all the moving parts of an engine. The various lubrication systems used for internal combustion engines may be classified as

(i) mist lubrication system

(ii) wet sump lubrication system

(iii) dry sump lubrication system

11.9.1 Mist Lubrication System

This system is used where crankcase lubrication is not suitable. In two-stroke engine, as the charge is compressed in the crankcase, it is not possible to have the lubricating oil in the sump. Hence, mist lubrication is adopted in practice. In such engines, the lubricating oil is mixed with the fuel, the usual ratio being 3% to 6%. The oil and the fuel mixture is inducted through the carburettor. The fuel is vaporized and the oil in the form of mist goes via the crankcase into the cylinder. The oil which strikes the crankcase walls lubricates the main and connecting rod bearings, and the rest of the oil lubricates the piston, piston rings and the cylinder.

The advantage of this system is simplicity and low cost as it does not require an oil pump, filter, etc. However, there are certain disadvantages which are enumerated below.

(i) It causes heavy exhaust smoke due to burning of lubricating oil partially or fully and also forms deposits on piston crown and exhaust ports which affect engine efficiency.

(ii) Since the oil comes in close contact with acidic vapours produced during the combustion process, the bearing surfaces may be corroded.

(iii) This system calls for a thorough mixing for effective lubrication. This requires either separate mixing prior to use or use of some additive to give the oil good mixing characteristics.

(iv) During closed throttle operation as in the case of the vehicle moving down the hill, the engine will suffer from insufficient lubrication as the supply of fuel is less. This is an important limitation of this system.

In some of the modern engines, the lubricating oil is directly injected into the carburettor and the quantity of oil is regulated. Thus the problem of oil deficiency is eliminated to a very great extent. In this system the main bearings also receive oil from a separate pump. For this purpose, they will be located outside the crankcase. With this system, formation of deposits and corrosion of bearings are also eliminated.

11.9.2 Wet Sump Lubrication System

In the wet sump system, the bottom of the crankcase contains an oil pan or sump from which the lubricating oil is pumped to various engine components by a pump. After lubricating these parts, the oil flows back to the sump by gravity. Again it is picked up by a pump and recirculated through the engine lubricating system. There are three varieties in the wet sump lubrication system. They are

(i) the splash system

(ii) the splash and pressure system

(iii) the pressure feed system

Splash System: This type of lubrication system is used in light duty engines. A schematic diagram of this system is shown in Fig.11.9.

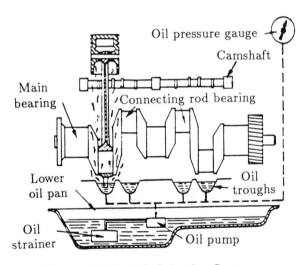

Fig.11.9 Splash Lubrication System

The lubricating oil is charged into in the bottom of the engine crankcase and maintained at a predetermine level. The oil is drawn by a pump and delivered through a distributing pipe extending the length of the crankcase into splash troughs located under the big end of all the connecting rods. These troughs were provided with overflows and the oil in the troughs are therefore kept at a constant level. A splasher or dipper is provided under each connecting rod cap which dips into the oil in the trough at every revolution of the crankshaft and the oil is splashed all over the interior of the crankcase, into the pistons and onto the exposed portions of the cylinder walls. A hole is drilled through the connecting rod cap through which oil will pass to the bearing surface. Oil pockets are also provided to catch the splashing oil over all the main bearings and also over the camshaft bearings. From the pockets the oil will reach the bearings surface through a drilled hole. The oil dripping from the cylinders is collected in the sump where it is cooled by the air flowing around. The cooled oil is then recirculated.

The Splash and Pressure Lubrication System: This system is shown in Fig.11.10, where the lubricating oil is supplied under pressure to main and camshaft bearings. Oil is also supplied under pressure to pipes which direct a stream of oil against the dippers on the big end of connecting rod bearing cup and thus the crankpin bearings are lubricated by the splash or spray of oil thrown up by the dipper.

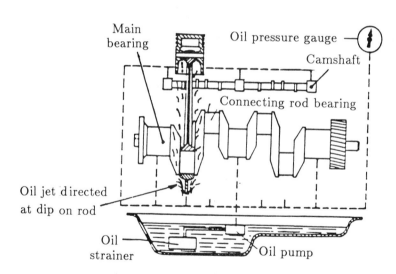

Fig.11.10 Splash and Pressure Lubrication System

Pressure Feed System: The pressure feed system is illustrated in Fig.11.11 in which oil is drawn in from the sump and forced to all the main bearings of the crankshaft through distributing channels. A pressure relief valve will also be fitted near the delivery point of the pump which opens when the pressure in the system attains a predetermined value. An oil hole is drilled in the crankshaft from the centre of each crankpin to the centre of an adjacent main journal, through which oil can pass from the main bearings to the crankpin bearing. From the crankpin it reaches piston pin bearing through a hole drilled in the connecting rod. The cylinder walls, tappet rollers, piston and piston rings are lubricated by oil spray from around the piston pins and the main and connecting rod bearings.

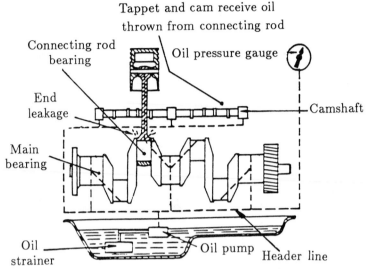

Fig.11.11 Pressure Feed Lubrication System

The basic components of the wet sump lubrication systems are (i) pump (ii) strainer (iii) pressure regulator (iv) filter (v) breather

A typical wet sump and its components are shown in Fig.11.12.

Oil is drawn from the sump by a gear or rotor type of oil pump through an oil strainer. The strainer is a fine mesh screen which prevents foreign particles from entering the oil circulating systems. A pressure relief valve is provided which automatically keeps the delivery pressure constant and can be set to any value. When the oil pressure exceeds that for which the valve is set, the valve opens and allows some of the oil to return to the sump thereby relieving the oil pressure in the systems. Fig.11.13 shows a typical gear pump, pressure relief valve and by-pass. Most of the oil from the pump goes directly to the engine bearings and a portion of the oil passes through a cartridge filter which removes the solid particles from the oil.

This reduces the amount of contamination from carbon dust and other impurities present in the oil. Since all the oil coming from the pump does not pass directly through the filter, this filtering system is called by-pass filtering system. All the oil will pass through the filter over a period of operation. The advantage of this system is that a clogged filter will not restrict the flow of oil to the engine.

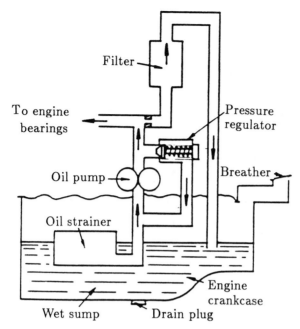

Fig.11.12 Basic Components of Wet sump Lubrication System

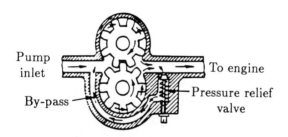

Fig.11.13 Gear Type Lubricating Pump

11.9.3 Dry Sump Lubrication System

A dry sump lubricating system is illustrated in Fig.11.14. In this, the supply of oil is carried in an external tank. Oil dripping from the cylinders and bearings into the sump is removed by a scavenging pump. The lubricating oil passed through a filter returns to the supply tank. The capacity of the scavenging pump is always greater than the oil pump. An oil pump draws oil from the supply tank and circulates it under pressure to the various bearings of the engine. Oil is prevented from accumulating in the base of the engine. In this system a filter with a bypass valve is placed in between the scavenge pump and the supply tank. If the filter is clogged, the pressure relief valve opens permitting oil to by-pass the filter and reaches the supply tank. From the supply tank oil is recirculated to the various bearings of the engine. A separate oil cooler either water or air is usually provided in the dry sump system to remove heat from the oil.

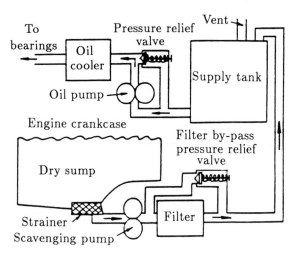

Fig.11.14 Dry Sump Lubrication System

11.10 CRANKCASE VENTILATION

During the compression and the expansion strokes the gas inside the cylinder gets past the piston rings and enters the crankcase which is called the blowby. It contains water vapour and sulphuric acid if either the oil or the fuel contains appreciable amount of sulphur. They might cause corrosion of steel parts in the crankcase. This may also promote sludge formation in the lubricating oil. A considerable amount of water condensed may be collected in the sump, and in cold weather this may freeze and may cause damage to the lubricating oil pump. Hence it has become inevitable to remove the blowby from the crankcase. This removal of the blowby can be achieved effectively by passing a constant stream of fresh air through the crankcase known as crankcase ventilation. By doing so, not only all the water vapour but also a considerable proportion of the fuel in the blowby may be removed from the crankcase. The problem of excessive crankcase dilution and the crankcase corrosion is atleast materially lessened.

The crankcase must have an air inlet and air outlet for the effective crankcase ventilation. The breather and oil filter (if any) forms a suitable inlet placed near the forward end of the case where the fan blows the cooling air and an outlet opening is then provided near the rear end of the engine block and a tube is taken from this outlet to a point below the crankcase where rapid flow of air flows past its outlet when the vehicle is in motion causing an ejector effect. This type of crankcase ventilation is illustrated in Fig.11.15. Air enters through the breather on the right, passes through the crankcase and exits through a pipe down into the air stream at the bottom of the crankcase. The outlet from the crankcase should be located at a point where the air is comparatively quiescent and therefore does not hold much oil in suspension.

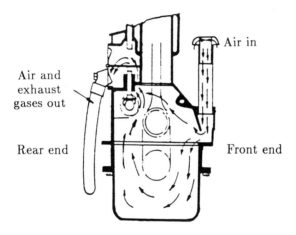

Fig.11.15 Crankcase Ventilation

It is also possible to connect the crankcase outlet to the air cleaner, where the intake suction serves to ventilate the crankcase and the unburned fuel, gases as well as the water vapour are then drawn into the cylinders where the fuel has another chance to burn. This scheme, however has been given up evidently in the belief that corrosion might be promoted by circulating these vapours through the engine.

11.11 PROPERTIES OF LUBRICANTS

The duties of the lubricant in an engine are many and varied in scope. The lubricant is called upon to limit and control the following

(i) friction between the components

(ii) metal to metal contact

(iii) overheating of the components

(iv) wear of the components

 (v) corrosion

(vi) deposits

To accomplish the above functions, the lubricant should have

 (i) suitable viscosity

 (ii) oiliness to ensure adherence to the bearings, and for less friction and wear when the lubrication is in the boundary region, and as a protective covering against corrosion

(iii) high strength to prevent the metal to metal contact and seizure when under heavy load

(iv) should not react with the lubricating surfaces

 (v) a low pour point to allow flow of the lubricant at low temperatures to the oil pump

(vi) no tendency to form deposits by uniting with air, water, fuel or the products of combustion

(vii) cleaning ability

(viii) non-foaming characteristics

(ix) nontoxic and non-inflammable

 (x) low cost

11.11.1 Viscosity

The viscosity of the oil at the temperature and pressure of the operation must be compatible with the load and speed to ensure hydrodynamic lubrication. In general, large clearances and high loads require high-viscosity oils whereas high speeds require low viscosity oils. Hence, the oil supplied must be in a position to meet the variable viscosity requirements.

Viscosity Index: The viscosity index is a measure of change in viscosity of an oil with temperature as compared to two reference oils having the same viscosity at 100 °C. It is an empirical number wherein a typical Pennsylvania (paraffinic base) oil is assigned an index of 100 and a Gulf Coast (naphthenic-base) oil is assigned an index of zero. In general, the high viscosity index number indicates relatively a smaller change in viscosity with temperature. A low index number for a given oil indicates relatively large change of viscosity with temperature.

Thus, the viscosity index of a lubricating oil is an important factor where extreme ranges in temperature are encountered. The oil must necessarily maintain sufficient viscosity at operating temperatures. Yet it should not be too viscous for starting of the engine especially at low temperature. Normally, a high viscosity index is preferred for engine lubrication, because of relatively smaller changes in viscosity of the oil with temperature.

11.11.2 Flash and Fire Points

The flash point of an oil is the minimum temperature at which sufficient flammable vapour is driven off to flash when brought into contact with a flame. The fire point is the minimum temperature at which the inflammable vapours will continue to form and steadily burn once ignited. Flash and fire points may vary with the nature of the original crude oil, the viscosity and the method of refining. For the same viscosities and degree of refinement, the paraffinic oils have higher flash and fire points than naphthenic oils.

11.11.3 Cloud and Pour Points

Petroleum oils when cooled may become plastic solids as a result of either of partial separation of wax or of congealing of the hydrocarbons. With some oils, the separation of wax becomes visible at temperatures slightly above the solidification point and this temperature (under prescribed condition) is known as the cloud point. With oils in which the separation is invisible, the cloud point cannot be determined. That temperature at which the oil will not flow when the test container is tilted under prescribed conditions is known as the pour point.

The pour point indicates the lowest temperature at which an oil stops flowing to the pump, bearings or cylinder walls. It is particularly important for immediate oil circulation in respect of starting of engines in very cold climates with gravity lubricating systems as the fluidity is a factor of pour point and viscosity of the cold oil. Pour point depressants may be added to wax containing oils to lower the pour points instead of dewaxing the oil.

11.11.4 Oiliness and Film Strength

Oiliness or film strength of a lubricant is a measure of the protective film between shaft and bearing. Thus two oils of identical viscosity may show differences in their coefficients of friction in the boundary region because one oil has more oiliness than the other. Most probably, the lubricant reacts with the bearing on shaft to form a protective grease or soap, thus giving rise to the lower coefficient of friction (higher oiliness).

Film strength refers to the ability of the lubricant to resist welding and scuffing. The lubricating oil used must be of enough film strength to take care of welding and scuffing.

11.11.5 Corrosiveness

The oil should be noncorrosive and should protect against corrosion. It is probable that the absorbed film that rise to the concept of oiliness is also related to the protection of the surface against corrosion.

11.11.6 Detergency

An oil has the property of detergency if it acts to clean the engine deposits. A separate property is the dispersing ability which enables the oil to carry small particles uniformly distributed without agglomeration. In general, the term is the name for both detergent and dispersing properties.

11.11.7 Stability

The ability of oil to resist oxidation that would yield acids, lacquers and sludge is called stability. Oil stability demands low-temperature (under 90 °C) operation and the removal of all hot areas from contact with the oil.

11.11.8 Foaming

Foaming describes the condition where minute bubbles of air are held in the oil. This action accelerates oxidation and reduces the mass flow of oil to the bearings thus reducing the hydrodynamic pressure in the bearing and hence the load bearing capacity of the bearing.

11.12 ADDITIVES FOR LUBRICANTS

The lubricating oil should possess all the above properties for the satisfactory engine performance. The modern lubricants for heavy duty engines are highly refined which otherwise may produce sludge or suffer a progressive increase in viscosity. For these reasons the lubricants are tempered or seasoned by the addition of certain oil-soluble organic compounds containing inorganic elements such as phosphorus, sulphur, amine derivatives. Metals are added to the mineral based lubricating oil to exhibit the desired properties. Thus oil soluble organic compounds added to the present day lubricants to impart one or more of the following characteristics.

   (i) anti-oxidant and anticorrosive agents
  (ii) detergent-dispersant
 (iii) extreme pressure additives
  (iv) pour point depressors
   (v) viscosity index improvers
  (vi) antifoam agent
 (vii) oiliness and film-strength agents

11.12.1 Anti-oxidants and Anticorrosive Agents

Oxidation of the lubricating oil is slow at temperatures below 90 °C but increase at an exponential rate when high temperatures are encountered. Oxidation is undesirable, not only because sludge and varnish are created but also because acids are formed which may be corrosive. Thus the additive has the dual purpose of preserving both the lubricant and the components of the engine. To accomplish these purposes, the additive must nullify the action of metals in catalyzing oxidation; copper is especially active as an oxidation catalyst of hydrocarbon. The additives may be alkaline to neutralize acids formed by oxidation, or, it may be non-alkaline and protect the metal by forming a surface film.

Some additives may unite with oxygen, either preferentially to the oil or else with some already oxidized portion of the oil or fuel contaminant. Other additives might act as metal deactivators and as corrosion shields by chemically combining with the metal. Thus a thin sulphide or phosphide coating on the metal deactivates those metals that act as catalysts while protecting other metals from corrosive attack. Zinc ditinophosphate serves as an anti-oxidant and anticorrosive additive.

11.12.2 Detergent-Dispersant

This type of additives improve the detergent action of the lubricating oil. These additives might be metallic salts or organic acids. The mechanism of the additive may arise either from direct chemical reaction or from polar attraction. Thus the additives may chemically unite with the compounds in the oil that would otherwise form sludge and varnish. On the other hand, both the additive and the deposits in the engine are polar compounds, and therefore detergent action may arise from neutralization of the electric moment of one or more deposit molecules by adherence of one or more molecules of additive. In this manner the deposit could be neutralized and would not tend to cling to other molecules or to the surface. Therefore, agglomeration of deposits would be prevented.

11.12.3 Extreme Pressure Additives

At high loads and speeds with high surface temperatures, an extreme pressure additive is necessary. Such additives unite with the metal surface to form a complex inorganic film containing iron, oxygen, carbon and hydrogen. Welding is prevented by the physical presence of the film.

11.12.4 Pour Point Depressors

In order to obtain fluidity or flow of oil at low temperatures, pour point depressants are added to the lubricating oils to lower the pour point. An engine having a lubricant with higher pour point will not get adequate lubrication during starting at low ambient temperatures and excessive wear would result. These additives tend to prevent the formation of wax at the low temperatures encountered in starting.

11.12.5 Viscosity Index Improvers

High molecule polymers are added to the lubricating oils to increase their viscosity index. An increase in the viscosity index increase the resistance of an oil to change in viscosity with a change in temperature. A high viscosity index oil will have good starting characteristics plus satisfactory operation at high speed and heavy load conditions.

11.12.6 Oiliness and Film Strength Agents

Oiliness and high film-strength can be improved by adding organic sulphur, chlorine and phosphorus compounds.

11.12.7 Antifoam Agents

Foaming, to some extent is due to the violent agitation and aeration of the oil that occurs in an operating engine. The minute particles of air in a foaming oil increase oxidation and reduce the mass flow of oil to the bearings. In addition foaming may cause abnormal loss of oil through the crankcase breather.

Antifoam agents are used to reduce the foaming tendencies of the lubricant. Although many chemical compounds are available for this purpose, the most effective are the silicone polymers.

Questions :-

11.1 Explain the various frictional losses in an engine.

11.2 Explain the six classes of mechanical friction and the various factors affecting them.

11.3 Explain the various mechanism of lubrication bringing out their functions.

11.4 What are the various components to be lubricated in an engine and explain how it is accomplished?

11.5 Clearly explain the various wet sump lubrication system. Compare wet sump and dry sump lubrication system.

11.6 What is meant by crankcase ventilation? Explain the details.

11.7 What are the various desired properties of a lubricant and explain how do additives help to achieve the desired properties.

12

HEAT REJECTION
AND
COOLING

12.1 INTRODUCTION

Internal combustion engines at best can transform about 25 to 35 per cent of the chemical energy in the fuel into mechanical energy. About 35 per cent of the heat generated is lost into the surroundings of combustion space, remainder being dissipated through exhaust and radiation from the engine.

During the process of combustion, the cylinder gas temperature often reaches quite a high value. A considerable amount of heat is transferred into the walls of the combustion chamber. Therefore, it is necessary to provide proper cooling especially to the walls of the combustion chamber. Due to the high temperatures, chemical and physical changes in the lubricating oil may also occur. This causes wear and sticking of the piston rings, scoring of cylinder walls or seizure of the piston. Excessive cylinder-wall temperatures will therefore cause the rise in the operating temperature of piston head. This in turn will affect the strength of piston seriously.

In addition, overheated cylinder head may lead to overheated spark plug electrodes causing preignition. The exhaust valve may become hot enough to cause preignition or may fail structurally. Moreover, preignition would increase the cylinder head temperature further until engine failure or complete loss of power results.

As the last part of the charge to burn is in contact with the walls of the combustion space during the burning period, a high cylinder wall or cylinder head temperature will reduce the delay period and may cause knocking.

In view of the above, the inside surface temperature of the cylinder walls should be kept in a range which will ensure correct clearances between parts, promote vaporization of fuel, keep the oil at its best viscosity and prevent the condensation of harmful vapours. Therefore, the heat that is transferred into the walls of the combustion chamber is continuously removed by employing a cooling system. Almost 30 to 35 per cent of the total heat supplied by the fuel is removed by the cooling medium. Heat carried away by lubricating oil and heat lost by radiation amounts to 5 per cent of the total heat supplied. Unless the engine is adequately cooled engine seizure will result. In this chapter the details of engine heat transfer, heat rejection and cooling are considered.

12.2 VARIATION OF GAS TEMPERATURE

There is an appreciable variation in the temperature of the gases inside the engine cylinder during different processes of the cycle. Temperature inside the engine cylinder is almost the lowest at the end of the suction stroke. During combustion there is a rapid rise in temperature to a peak value which again drops during the expansion. This variation of gas temperature is illustrated in Fig.12.1 for various processes in the cycle.

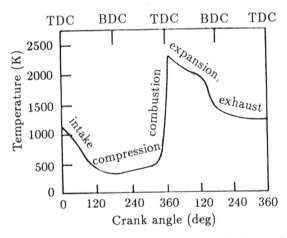

Fig.12.1 Gas Temperature Variation during a Cycle

12.3 PISTON TEMPERATURE DISTRIBUTION

The piston crown is exposed to very high combustion temperatures. Figure 12.2 gives the typical values of temperature at different parts of a cast iron piston. It may be noted that maximum temperature occurs at the centre of the crown and decreases at the outer edge. The temperature is the lowest at the bottom of the skirt. Poor design may result in the thermal overloading of the piston at the centre of the crown. The temperature difference between piston outer edge and the centre of the crown is responsible for the flow of heat to the ring belt through the path offered by metal section of the crown. It is, therefore, necessary to increase the thickness of the crown from the centre to the outer edge in order to make a path of greater cross-section available for the increasing heat quantity. The length of the path should not be too long or the thickness of the crown cross-section too small for the heat to flow. This will cause the temperature at the centre of crown to build up and thereby excessive temperature difference between the crown and the outer edge of the piston will result. This may even lead to cracking of piston during overload operation.

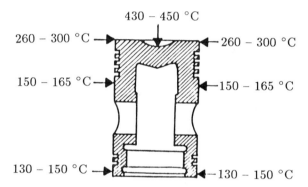

Fig.12.2 Piston Temperature Distribution

12.4 CYLINDER TEMPERATURE DISTRIBUTION

Whenever a moving gas comes into contact with a wall, there exists a relatively stagnant gas layer which acts as a thermal insulator. The resistance of this layer to heat flow is quite high. Heat transfer from the cylinder gases takes place through the gas layer and through the cylinder walls into the cooling medium. A large temperature drop is produced in the stagnant layer adjacent to the walls.

The peak cylinder gas temperature may be 2800 K while the temperature of the cylinder inner wall surface may be only 450 K due to cooling, (Fig.12.3). Heat is transferred from the gases to the cylinder walls when temperature difference exists. The rate of flow varies depending upon the temperature difference. If no cooling is provided, there could be no heat flow, so that the whole cylinder

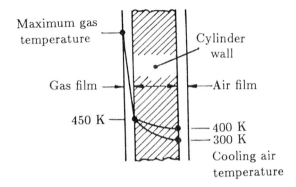

Fig.12.3 Cylinder Wall Temperature Distribution of a
Properly Cooled Cylinder

wall would soon reach an average temperature of the cylinder gases.
By providing adequate cooling, the cylinder wall temperature can be
maintained at optimum level.

12.5 HEAT TRANSFER

Heat transfer occurs when a temperature difference exists. As a
result of combustion, high temperatures are produced, inside the engine cylinder as compared to the surroundings. Considerable heat
flow occurs from the gases to the surrounding metal walls. In addition to this the shearing of the oil film separating the bearing surfaces
transforms available energy into internal energy of the oil film. This
increases the temperature of oil film and results in heat transfer from
the oil to the bearing surfaces. However, the heat transfer on this
account is quite small. Hence, the cylinder walls must be adequately
cooled to maintain safe operating temperatures in order to maintain
the quality of the lubricating oil.

Heat transfer from gases to the cylinder walls may occur by conduction, convection and radiation whereas the heat transfer through
the cylinder wall occurs only by conduction. Heat is ultimately transferred to the cooling medium by all the three modes of heat transfer.
The temperature profiles across the cylinder barrel wall are shown for
both water-cooled and air-cooled engine in Fig.12.4. In this case, T_g,
is the mean gas temperature which may be as high as 650 °C. This
may not be confused with the peak temperature of the cycle which
may be two or three times this value. Largest temperature drop,
however, occurs in the boundary-layer of the gas which lies adjacent
to the cylinder wall. There is a corresponding boundary-layer in the

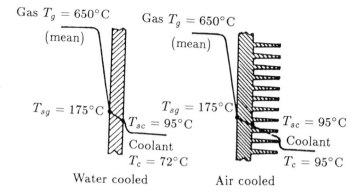

Gas $T_g = 650°C$ (mean)   Gas $T_g = 650°C$ (mean)

$T_{sg} = 175°C$ $T_{sg} = 175°C$ $T_{sc} = 95°C$

$T_{sc} = 95°C$

Coolant   Coolant
$T_c = 72°C$ $T_c = 95°C$

Water cooled   Air cooled

Fig.12.4 Temperature Profile across Cylinder Wall

cooling medium on the outer side of the cylinder. However, because of fins in the air-cooled engines the effect of external boundary-layer is reduced.

The conduction of heat through cylinder walls with corresponding temperature gradients is illustrated in Fig.12.5. The gas film, being of low conductivity, offers a relatively high resistance to the heat flow, whilst on the water jacket side there is usually a layer of corrosion products, scale etc., which are poor heat conductors. The least resistance to the heat flow occurs through the metal cylinder wall, as shown by temperature gradient there.

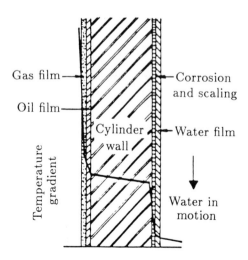

Gas film — — Corrosion and scaling

Oil film —

Cylinder wall — Water film

Temperature gradient

Water in motion

Fig.12.5 Temperature Gradient along Cylinder Wall

In actual practice because of the cyclic operation of engines, there is a cyclic variation of the gas temperature within the cylinder the effect of which is to cause a wave of heat to travel into the metal which gradually dies out and after the warmup period a steady flow condition prevails. It has been experimentally established that in case of small internal combustion engines the cyclic temperature variations die out fast before the fluctuations reach the outside surface of the cylinder. Maximum temperature of the cylinder walls, in a properly designed engine, seldom exceeds 10 °C above the mean temperature.

12.6 THEORY OF ENGINE HEAT TRANSFER

In spite of its high temperature, the cylinder gas is a poor radiator and almost all the heat transfer to the cylinder walls from combustion space is by convection. In order to understand the engine heat transfer, a simple analysis can be followed for the flow of hot gases through a pipe.

For gases in pipes it can be shown by dimensional analysis and also through experiments that

$$\frac{hl}{k} = Z \times \left(\frac{\rho C l}{\mu}\right)^n \left(\frac{C_p \mu}{k}\right)^m \tag{12.1}$$

where | h | = coefficient of heat transfer
| l | = any characteristic length
| k | = thermal conductivity of gases
| Z | = constant
| ρ | = mass density of gases
| C | = velocity of gases
| C_p | = specific heat of gas
| μ | = viscosity of gases
| n, m | = exponents

The term $(\rho C l/\mu)$ can be recognized as Reynolds number for cylinder gases. The term $(C_p \mu/k)$ is called the Prandtl number and is nearly constant for gases. Therefore, Prandtl number can be absorbed in the constant Z in Eqn.12.1 so that

$$\frac{hl'}{k} = Z \times \left(\frac{\rho C l}{\mu}\right)^n \tag{12.2}$$

Since Prandtl number is constant, $k \propto \mu C_p$ and substituting μC_p for k in Eqn. (12.2),

$$\frac{hl}{\mu C_p} = Z \times \left(\frac{\rho C l}{\mu}\right)^n \tag{12.3}$$

$$h = Z \times C_p(\rho C)^n \left(\frac{l}{\mu}\right)^{n-1} \tag{12.4}$$

The rate of heat transfer can be written as

$$\dot{q} = h \times A \times \Delta T$$

where ΔT is the temperature difference between the gas and the wall. Substituting the value of h from Eqn.12.4, we get

$$\dot{q} = Z \times C_p(\rho C)^n \left(\frac{l}{\mu}\right)^{n-1} A \Delta T \tag{12.5}$$

In the above expression, A is the area of heat transfer which is proportional to l^2 and S is the mean piston speed which is proportional to gas velocity, C. When the average gas temperature is considered C_p and μ can be assumed to have constant values. Then,

$$\dot{q} = Z(\rho S)^n l^{n+1} \Delta T \tag{12.6}$$

Piston speed is proportional to the product of l and N where N is the rpm of the engine. Volumetric efficiency, η_v is proportional to the density of the charge, ρ. Then

$$\dot{q} = Z(\eta_v N)^n l^{2n+1} \Delta T$$

The average temperature of the cooling medium, the fuel-air ratio of the mixture and the compression ratio of the engine directly influence the value of ΔT. The density is mainly affected by the intake pressure, compression ratio and the volumetric efficiency. Those engines which have nearly equal value of ΔT, the heat transfer rate depends on the product of $\eta_v N$ and the size of the engine.

For engines when ΔT is assumed to be invariant

$$\dot{q} = Z(\eta_v N)^n (l)^{2n+1} \tag{12.7}$$

The values of Z and n are determined from experiments on a particular type of engine under various operating conditions. The constants so obtained can be used for calculating heat transfer rate for other operating conditions of the same engine or for geometrically similar engines.

12.7 PARAMETERS AFFECTING ENGINE HEAT TRANSFER

From the above discussion, it may be noted that the engine heat transfer depends upon many parameters. Unless the effect of these parameters is known, the design of a proper cooling system will be difficult. In this section, the effect of various parameters on engine heat transfer is briefly discussed.

12.7.1 Fuel-Air Ratio

A change in fuel-air ratio will change the temperature of the cylinder gases and affect the flame speed. The maximum gas temperature will occur at an equivalence ratio of about 1.12 i.e., at a fuel-air ratio about 0.075. At this fuel-air ratio ΔT will be maximum. However, the maximum heat rejection usually occurs for a mixture slightly leaner than this value.

12.7.2 Compression Ratio

An increase in compression ratio causes only a slight increase in gas temperature near the top dead centre; but, because of greater expansion of the gases, there will be a considerable reduction in gas temperature near bottom dead centre where a large cylinder wall is exposed. The exhaust gas temperature will also be much lower because of greater expansion so that the heat rejected during blowdown will be less. In general, as compression ratio increases there tend to be a marginal reduction in heat rejection.

12.7.3 Spark Advance

A spark advance more than the optimum as well as less than the optimum will result in increased heat rejection to the cooling system. This is mainly due to the fact that the spark timing other than MBT value (Minimum spark advance for Best Torque) will reduce the power output and thereby more heat is rejected.

12.7.4 Preignition and Knocking

Effect of preignition is the same as advancing the ignition timing. Large spark advance might lead to erratic running and knocking. Though knocking causes large changes in local heat transfer conditions, the over-all effect on heat transfer due to knocking appears to be negligible. However, no authentic information is available regarding the effect of preignition and knocking on engine heat transfer.

12.7.5 Engine Output

Engines which are designed for high mean effective pressures or high piston speeds, heat rejection will be less. Less heat will be lost for the same indicated power in large engines.

12.7.6 Cylinder Wall Temperature

The average cylinder gas temperature is much higher in comparison to the wall temperature. Hence, any change in combustion chamber wall temperature will have very little effect on the temperature difference and thus on heat rejection.

12.8 POWER REQUIRED TO COOL THE ENGINE

Theoretically the thermal efficiency of the engine will improve if there is no cooling system, but actually the engine will cease to operate. It is mainly because of high temperature, the metals will loose their characteristics and piston will expand considerably and cease the movement. In order to avoid seizure, cooling of the engine is a must. However, the power required to run the cooling system should come from the engine. In this section, the power required to run a cooling system is briefly examined.

In order to simplify the calculations and obtain an expression for the power required for cooling an engine, some assumptions have to be made. The important ones are:

(i) Engine is assumed to operate at constant fuel-air ratio.

(ii) Changes in fin or coolant temperature are assumed to be small enough so that ΔT, the temperature difference between cylinder gases and the cylinder wall may be assumed to be constant.

(iii) The density and temperature of cooling media are assumed to be unaffected by its flow through the engine or radiator fins.

With the above assumption, it can be shown that variation of cooling power required in terms of the indicated power of the engine can be shown to be

$$P_c = \frac{A_e(ip)^{2.25}}{(A_f \Delta T_f)^{3.75}\rho_a^2} \tag{12.8}$$

Equation 12.8 shows that, for a given power, a great reduction in required cooling power can be effected by increasing the fin area and the temperature difference as much as possible. In the above equation, A_e is the effective area, ip is the indicated power, A_f is the fin area and ΔT_f is the temperature difference between the air and the fin and ρ_a is the density of air. It may be noted that power required for the cooling can be reduced by reducing the air velocity. It can be shown that $P_c \propto C_a^3$, where C_a is the velocity of air. In order to reduce the air velocity, adjustable flaps are usually provided in the cooling air duct of air-cooled engines.

12.9 NEED FOR COOLING SYSTEM

From the discussion on heat rejection in the previous sections, it may be noted that during the process of converting thermal energy to mechanical energy, high temperatures are produced in the cylinders of the engine as a result of the combustion process. A large portion of the heat from the gases of combustion is transferred to the cylinder head and walls, piston and valves. Unless this excess heat is carried away and these parts are adequately cooled, the engine will be damaged. A cooling system must be provided not only to prevent damage to the vital parts of the engine, but the temperature of these components must be maintained within certain limits in order to obtain maximum performance from the engine. Adequate cooling is then a fundamental requirement associated with reciprocating internal combustion engines. Hence, a cooling system is needed to keep the engine from not getting so hot as to cause problems and yet to permit it to run hot enough to insure maximum efficiency operation. The duty of cooling system, in other words, is to keep the engine from getting not too hot and at the same time not to keep it cool!

12.10 CHARACTERISTICS OF AN EFFICIENT COOLING SYSTEM

The following are the two main characteristics desired of an efficient cooling system:

(i) It should be capable of removing about 30% of heat generated in the combustion chamber while maintaining the optimum temperature of the engine under all operating conditions of the engine.

(ii) It should remove heat at a fast rate when engine is hot. However, during starting of the engine the cooling should be very slow, so that the working parts of the engine reach their operating temperatures in a short time.

12.11 TYPES OF COOLING SYSTEMS

In order to cool the engine a cooling medium is required. This can be either air or a liquid. Accordingly there are two types of systems in general use for cooling the IC engines. They are

(i) liquid or indirect cooling system

(ii) air or direct cooling system

12.12 LIQUID COOLED SYSTEMS

In this system mainly water is used and made to circulate through the jackets provided around the cylinder, cylinder-head, valve ports and seats and other hot spots where it extracts most of the heat.

The diagrammatic sketch of water circulating passage, viz., water jacket is shown in Fig.12.6. It consists of a long flat, thin-walled tube with an opening, facing the water pump outlet and a number of small openings along its length that direct the water against the exhaust valves. The tube fits in the water jacket and can be removed from the front end of the block.

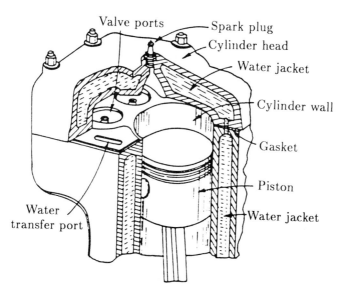

Fig.12.6 Cooling Water Passages

The heat is transferred from the cylinder walls and other parts by convection and conduction. The liquid becomes heated in its passage through the jackets and is in turn cooled by means of an air-cooled radiator system. The heat from liquid in turn is transferred to air. Hence it is called the indirect cooling system.

Water-cooling can be carried out by any one of the following five methods

(i) Direct or non-return system

(ii) Thermosyphon system

(iii) Forced circulation cooling system

(iv) Evaporative cooling system

(v) Pressure cooling system

12.12.1 Direct or Non-return System

This system is useful for large installations where plenty of water is available. The water from a storage tank is directly supplied through an inlet valve to the engine cooling water jacket. The hot water is not cooled for reuse but simply discharged.

12.12.2 Thermosyphon System

The basic principle of thermosyphon can be explained with respect to Fig.12.7. Heat is supplied to the fluid in the tank A. The hot fluid travels up its place being taken up by comparatively cold fluid from the tank B through the pipe P_2. The hot fluid flows through the pipe P_1 to the tank B where it gets cooled. Thus the fluid circulates through the system in the form of convection currents.

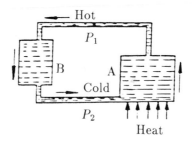

Fig.12.7 Principle of Thermosyphon System

For engine application, tank A represents the cylinder jackets while tank B represents a radiator and water acts as the circulating fluid. In order to ensure that coolest water is always made available to cylinder jackets, the water jackets are located at a lower level than the radiator.

The main advantages of the system are its simplicity and automatic circulation of cooling water. The main limitation of the system is its inability to meet the requirement of large flow rate of water, particularly for high output engines.

12.12.3 Forced Circulation Cooling System

This system is used in a large number of automobiles like cars, buses and even heavy trucks. Here, flow of water from radiators to water jackets is by convection assisted by a pump.

The main principle of this system is explained with the help of a block diagram shown in Fig.12.8.

The water or coolant is circulated through jackets around the parts of the engine to be cooled, and is kept in motion by a centrifugal

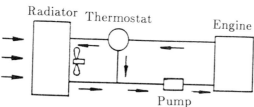

Fig.12.8 Principle of Forced Circulation Cooling System using the Thermostat

pump which is driven from the engine. The water is passed through the radiator where it is cooled by air drawn through the radiator by a fan and by the air flow developed by the forward motion of the vehicle. A thermostat is used to control the water temperature required for cooling.

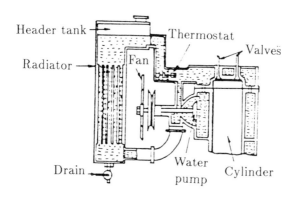

Fig.12.9 Cooling of an Automobile

This system mainly consists of four components, viz., a radiator, fan, water pump and a thermostat. The details of these components are shown in Fig.12.9.

Radiator: The purpose of a radiator is to provide a large amount of cooling surface area to the air so that the water passing downward through it in thin streams is cooled efficiently. To accomplish this, there are many arrangements that are possible. One such arrangement is shown in Fig.12.10.

The radiator consists essentially of an upper tank (header tank) and a lower tank. The upper tank in some design may contain a removable filter mesh to avoid dust particles going in into the radiator while filling water in the radiator. Between the two tanks is the core or

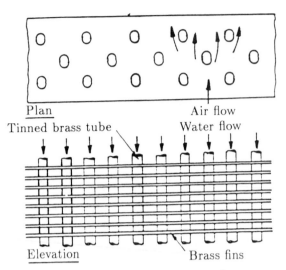

Fig.12.10 Radiator Construction

radiating element. The upper tank is connected to the water outlets from the engine jacket by rubber hose, and the lower tank is connected by another rubber hose to the jacket inlet through the pump.

Radiator cores are classified as tubular or cellular. A tubular radiator, consists of a large number of elliptical or circular brass tubes pressed into a number of suitable punched brass fins. The tubes are finned to guard against corrosion and are staggered as shown in Fig.12.10. The main disadvantage is the great inconvenience to repair any of the damaged tubes. But initial cost of the system is comparatively less.

The other type of radiator core arrangement, called *honey comb* or *cellular radiator core* is shown in Fig.12.11.

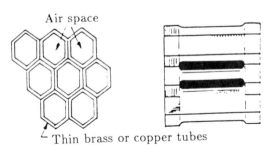

Fig.12.11 Honey Comb Radiator Core

The water used for cooling should be soft. If hard water is used, it forms sediments on water jackets and tubes, which acts as insulator and make the cooling inefficient. If soft water is not available, 30 g of sodium bichromate should be added for every 10 litres of water.

Fan: The fan mounted on the impeller spindle driven by a suitable belt pulley arrangement as shown in Fig.12.9 draws air through the spaces between the radiator tubes thus bringing down the temperature of the water appreciably.

Pump: The pump maintains the circulation of the water through the system. The bottom of radiator is connected to the suction side of the pump. The power is transmitted to the pump spindle from a pulley mounted on the end of the camshaft or crankshaft.

A positive supply of water is achieved in all conditions by centrifugal pump placed in this system (Fig.12.9). This ensures good velocity of water circulation. Consequently less quantity of water and a smaller radiator would suit the purpose.

A pump is mounted conveniently on the engine and driven by the crankshaft with a fan belt. Adjustable packing glands are provided on the driving shaft to prevent water leakage. Lubrication of bearings is done by using high melting point grease. In certain cases special bushes are used which do not require lubrication. In case of multi-cylinder engines an header is usually employed to provide equal distribution of water to all the cylinders. The header is supplemented by tubes or ducts which give high rate of flow around critical sections of the engine such as the exhaust valve seats. This system is employed on most diesel and automotive spark-ignition engines. The rate of circulation is usually 3 to 4 litres per minute per kilowatt.

In some illustrations the pump is included in the line from the bottom of the radiator to the engine block and forces cool water from the radiator into the engine jacket. On automobiles, however, this arrangement would result in such a low location of the pump that the fan could not be well placed on the pump shaft. A disadvantage of this installation would seem to be that in case of loss of water, circulation stops as soon as the level drops to the bottom of the cylinder head jacket while with the pump in the supply line continues as long as there is any water left in the system.

Thermostat: Whenever the engine is started from cold, some heat has to be supplied to the large quantity of water surrounding the engine in order to acquire good working temperature. This may take an appreciable time. A thermostat fitted in the system initially prevents the circulation of water below a certain temperature through the radiator so that the water gets heated up quickly. When the preset temperature is reached the thermostat allows the water to flow

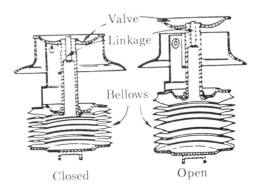

Fig **12.12** Bellows Type Thermostat

through the radiator. Usually a Bellow type thermostat is used, the details of which are shown in Fig.12.12.

The unit consists of a closed bellows with volatile liquid under reduced pressure. When the bellows is heated the liquid vaporizes and creates enough pressure to expand the bellows. The movement of bellows operates a linkage which opens the valve. When the unit is cooled, the gas condenses, the pressure is reduced and the bellows collapses to close the valve.

12.12.4 Evaporative Cooling System

In this system, the engine will be cooled because of the evaporation of the water in the cylinder jackets into steam. Here, the advantage is taken from the high latent heat of vapourization of water by allowing it to evaporate in the cylinder jackets. If the steam is formed at a pressure above atmospheric the temperature will be above the normal permissible temperature.

Figure 12.13 illustrates evaporative cooling with air-cooled condenser. In this case water is circulated by the pump A and when delivered to the overhead tank B part of it boils out. The tank has a partition C. The vapour rises above the partition C and because of the condensing action of the radiator tubes D, condensate flows into the lower tank E from which it is picked up and returned to the tank B by the small pump F. The vertical pipe G is in communication with the outside atmosphere to prevent the collapsing of the tanks B and E when the pressure inside them due to condensation falls below atmospheric.

Figure 12.14 illustrates evaporate cooling with the water-cooled condenser. In this case condensation of the vapour formed in the overhead tank B occurs in the heat exchanger C cooled by a secondary

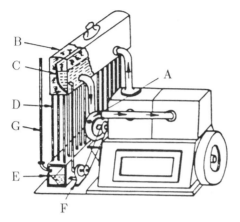

Fig.12.13 Evaporative Cooling with Air-Cooled Condenser

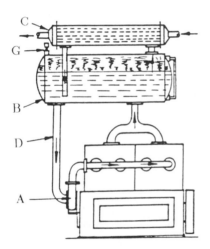

Fig.12.14 Evaporative Cooling with Water-Cooled Condenser

water circuit and the water returns to B by gravity. The pump A circulates the cooling water to the engine and the heated water from the engine is delivered to tank B thereby the circulation is maintained.

12.12.5 Pressure Cooling System

As already mentioned, the rate of heat transfer depends upon the temperature difference between the two mediums, the area of exposed surface and conductivity of materials. In case of radiators, in order to reduce the size of radiator, it is proposed to seal the cooling system from the atmosphere and to allow a certain amount of pressure to occur in the system, so that the advantage may be taken of the fact that the temperature of the boiling point of water increases as the pressure on it is raised. Boiling point of water at various pressures is shown in Table 12.1.

Table 12.1 *Boiling Point of Water at Various Pressures*

Pressure (bar)	1.0	2.0	5.0	10.0
Temperature (°C)	100	121	153	180

In pressure cooling system moderate pressures, say upto 2 bar, are commonly used. As shown in Fig.12.15, a cap is fitted with two valves, a safety valve which is loaded by a compression spring and a vacuum valve. When the coolant is cold both valves are shut but as the engine warms up the coolant temperature rises until it reaches a certain preset value corresponding to the desired pressure when the safety valve opens; but if the coolant temperature falls during the engine operation the valve will close again until the temperature again rises to the equivalent pressure value.

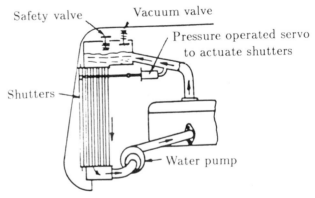

Fig.12.15 Pressure Cooling

When the engine is switched off and the coolant cools down vacuum begins to form in the cooling system but when the internal pressure falls below atmospheric the vacuum valve is opened by the higher outside pressure and the cooling system then attains atmospheric pressure.

A safety device is incorporated in the filler cap so that if an attempt is made to unscrew it while the system is under pressure, the first movement of the cap at once relieves the pressure and thus prevents the emission of scalding steam or the blowing off the cap due to higher internal pressure.

12.13 AIR–COOLED SYSTEM

In an air-cooled system a current of air is made to blow past the outside of the cylinder barrel whose outer surface area has been considerably increased by providing cooling fins as shown in Fig.12.16. This method will increase the rate of cooling.

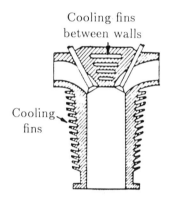

Cooling fins
between walls

Cooling
fins

Fig.12.16 Cooling Fins on an Engine Cylinder increase the Surface Area of Cooling

Application : This method is mainly applicable to engines in motor cycles, small cars, airplanes and combat tanks where motion of vehicle gives a good velocity to cool the engine. In bigger units a circulating fan is also used. In addition to these engines, air-cooling is also used in some small stationary engines.

The value of heat transfer coefficient between metal and air is appreciably low. As a result of this the cylinder wall temperatures of the air-cooled cylinders are considerably higher than those of water-cooled type. In order to lower the cylinder wall temperature the area of the outside surface which directly dissipates heat to the atmosphere must be sufficiently high. This is usually done by providing the fins.

12.13.1 Cooling Fins

Cooling fins are either cast integral with the cylinder and cylinder head or can be fixed with the cylinder block separately. Various shapes of cooling fins are shown in Fig.12.17. The heat dissipating capacity of fins depends upon their cross-section and length. At the same time as heat is gradually dissipated from the fin surface, the temperature of the fin decreases from its root to its tip. Hence, the fin surface nearer to the tip dissipates heat at a lower rate and is less efficient. On the other hand as the quantity of heat flowing towards the tip gradually decreases, the thickness of the fin can be decreased. The material of the fin is used most efficiently if the drop in temperature from the root to the tip is constant per unit length. A comparison of fins of different cross-sections is shown in Fig.12.17 with drop in temperature from root to tip. The rectangular section has least temperature drop whereas the maximum temperature drop is for the fin marked 'a'.

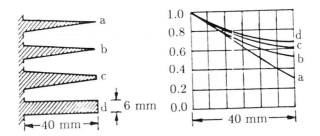

Fig.12.17 Types of Cooling Systems

Fins are usually given a taper of 3 to 5 degrees in order to give sufficient draft to the pattern. The tip is made 0.5 to 1.25 mm thick and a clearance of 2.5 to 5 mm is allowed at the root. The fins are made 25 to 50 mm long. Too close spacing of the fins is undesirable as mutual interference of the boundary-layers of adjacent layers restricts the air flow and results in small quantity of heat dissipated.

12.13.2 Baffles

The rate of heat transfer from the cylinder walls can be substantially increased by using baffles which force the air through the space between the fins. Figure 12.18 shows various types of baffles commonly used on engines. The arrangement at 'a' has got the highest pressure drop. It is always desired to have negligible kinetic energy loss between the entrance and the exit. Usually the normal type of baffle, 'b', is used on petrol engines. The arrangement 'c' for minimising the kinetic energy loss is shown with a well rounded entrance to reduce the entrance loss and an exit section that will transform the velocity head into pressure head and thus decrease the pressure drop. Arrangement 'd' is adopted for diesel engines.

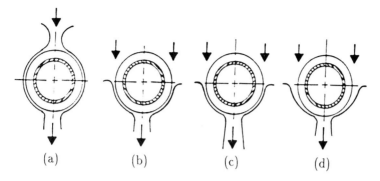

(a)　　　　(b)　　　　(c)　　　　(d)

Fig.12.18 Types of Baffles in Air-Cooled Engines

12.14 COMPARISON OF LIQUID AND AIR–COOLING SYSTEMS

In view of the wide spread use of these two alternative cooling systems for petrol as well as diesel engines it is of interest to summarize the respective advantages and limitations of these systems.

12.14.1 Advantages of Liquid-Cooling System

(i) Compact design of engines with appreciably smaller frontal area is possible.

(ii) The fuel consumption of high compression liquid-cooled engines are rather lower than for air-cooled ones.

(iii) Because of the even cooling of cylinder barrel and head due to jacketing makes it possible to reduce the cylinder head and valve seat temperatures.

(iv) In case of water-cooled engines, installation is not necessarily at the front of the mobile vehicles, aircraft etc. as the cooling system can be conveniently located wherever required. This is not possible in case of air-cooled engines.

(v) The size of engine does not involve serious problems as far as the design of cooling systems is concerned. In case of air-cooled engines particularly in high horsepower range difficulty is encountered in the circulation of requisite quantity of air for cooling purposes.

12.14.2 Limitations

(i) This is dependent system in which supply of water for circulation in the jackets is required.

(ii) Power absorbed by the pump for water circulation is considerable and this affects the power output of the engine.

(iii) In the event of failure of the cooling system serious damage may be caused to the engine.

(iv) Cost of the system is considerably high.

(v) System requires considerable maintenance of various parts of the system.

12.14.3 Advantages of Air-Cooling System

(i) The design of the engine becomes simpler as no water jackets are required. The cylinder can be identical dimensions and individually detachable and therefore cheaper to renew in case of accident etc.

(ii) Absence of cooling pipes, radiator, etc. makes the cooling system simpler thereby has minimum maintenance problems.

(iii) No danger of coolant leakage etc.

(iv) The engine is not subject to freezing troubles etc., usually encountered in case of water cooled engines.

(v) The weight per kW of the air-cooled engine in less than that of water-cooled engine.

(vi) In this case the engine is rather a self-contained unit as it requires no external components like radiator, header, tank etc.

(vii) Installation of air-cooled engines is easier.

12.14.4 Limitations

(i) Can be applied only to small and medium sized engines

(ii) Useful mainly to medium sized engines

(iii) Cooling is not uniform

(iv) Higher working temperatures compared to water-cooling

(v) Produce more aerodynamic noise

(vi) Specific fuel consumption is slightly higher

(vii) Lower maximum allowable compression ratios

(viii) The fan, if used absorbs as much as 5% of the power developed by the engine

Questions :-

12.1 With a suitable figure explain the variation of gas temperature during a cycle.

12.2 Explain with sketches piston and cylinder temperature distribution.

12.3 Explain briefly the engine heat transfer. Develop the necessary equation for the rate of heat transfer.

12.4 Mention the various parameters which affect the engine heat transfer and explain their effect.

12.5 Explain the reasons for cooling an engine.

12.6 What are the various characteristics of an efficient cooling system?

12.7 Explain the two types of cooling systems and compare them.

12.8 Explain the following:
   (i) thermosyphon cooling system
   (ii) forced circulation cooling system
   (iii) evaporative cooling system
   (iv) pressure cooling system

12.9 What is an air-cooling system and in which type of engine it is normally used?

12.10 Why fins and baffles are required in an air-cooled engine? Explain.

13

MEASUREMENTS
AND
TESTING

13.1 INTRODUCTION

The basic task in the design and development of engines is to reduce the cost and improve the efficiency and power output. In order to achieve the above task, the development engineer has to compare the engine developed with other engines in terms of its output and efficiency. Towards this end he has to test the engine and make measurements of relevant parameters that reflect the performance of the engine. In general, he has to conduct a wide variety of engine tests. The nature and the type of the tests to be conducted depend upon a large number of factors. It is beyond the scope of this book to discuss all the factors. In this chapter certain basic important measurements and tests are considered.

   (i) Friction power
  (ii) Indicated power
 (iii) Brake power
 (iv) Fuel consumption
  (v) Air-fuel ratio
 (vi) Speed
(vii) Exhaust and coolant temperature
(viii) Emissions
 (ix) Noise
  (x) Combustion phenomenon

     The details of measurement of these parameters are discussed in the following sections.

13.2 FRICTION POWER

The difference between the indicated and the brake power of an engine is known as friction power. The internal losses in an engine are essentially of two kinds, viz., pumping losses and friction losses. During the inlet and exhaust stroke the gaseous pressure on the piston is greater on its forward side (on the under side during the inlet and on the upper side during the exhaust stroke), hence during both strokes the piston must be moved against a gaseous pressure, and this causes the so called pumping loss. The friction loss is made up of the friction between the piston and cylinder walls, piston rings and cylinder walls, and between the crankshaft and camshaft and their bearings, as well as by the loss incurred by driving the essential accessories, such as the water pump, ignition unit etc.

It should be the aim of the designer to have minimum loss of power in friction. Friction power is used for the evaluation of indicated power and mechanical efficiency. Following methods are used to find the friction power to estimate the performance of the engine.

(i) Willan's line method

(ii) Morse test

(iii) Motoring test

(iv) From the measurement of indicated and brake power

(v) Retardation test

13.2.1 Willan's Line Method

This method is also known as fuel rate extrapolation method. A graph connecting fuel consumption (y-axis) and brake power (x-axis) at constant speed is drawn and it is extrapolated on the negative axis of brake power. The intercept of the negative axis is taken as the friction power of the engine at that speed. The method of extrapolation is shown in Fig.13.1 (dotted lines). As shown in the Fig.13.1, since, in most of the power range the relation between the fuel consumption and brake power is linear which permits extrapolation. Further, when the engine does not develop any power, i.e., $bp = 0$, it consumes a certain amount of fuel. The energy would have been spent in overcoming the friction. Hence, the extrapolated negative intercept of the x-axis will be the work representing the combined losses due to mechanical friction, pumping and blowby and as a whole it is termed the frictional loss of the engine. It should be noted that the measured frictional power by this method will hold good only for a particular speed and is applicable mainly to CI engines.

The main drawback of this method is the long distance to be extrapolated from data obtained between 5 and 40% load towards

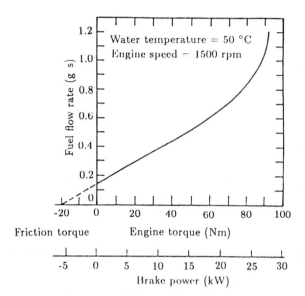

Fig.13.1 Willan's Line Method

the zero line of fuel input. The directional margin of error is rather wide because the graph is not exactly linear. The changing slope along the curve indicates the effect of part load efficiency of the engine. The pronounced change in the slope of this line near full load reflects the limiting influence of the air-fuel ratio and of the quality of combustion. Similarly, there may be slight curvature at light loads. This is perhaps due to the difficulty in injecting accurately and consistently very small quantities of fuel per cycle. Therefore, it is essential that great care should be taken in extrapolating the line and as many readings as possible should be taken at light loads to establish the true nature of the curve. The accuracy obtained in this method is reasonably good and compares favourably with other methods if extrapolation is carefully done.

13.2.2 Morse Test

The Morse test consists of obtaining indicated power of the engine without any elaborate equipment. The test consists of making inoperative, in turn, each cylinder of the engine and noting the reduction in brake power developed. With a gasoline engine each cylinder is rendered inoperative by *shorting* the spark plug of the cylinder; with a diesel engine by cutting off the supply of fuel to each cylinder. It is assumed that pumping and friction losses are the same when the cylinder is inoperative as well as during firing. This test is applicable only to multi cylinder engines. Referring to the Fig.13.2, the unshaded area of the indicator diagram is a measure of the gross power, *gp* developed by the engine, the dotted area being the pumping power, *pp*.

Net indicated power per cylinder   =   *gp* − *pp*

In this test the engine is first run at the required speed by adjusting the throttle in SI engine or the pump rack in CI engine and the output is measured. The throttle rack is locked in this position. Then, one cylinder is cut out by short circuiting the spark plug in the SI engine or by disconnecting the injector in the CI engine. Under this condition all the other cylinders will *motor* the cut out cylinder and the speed and output drop. The engine speed is brought to its original value by reducing the load. This will ensure that the frictional power is the same while the brake power of the engine will be with one cylinder less.

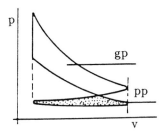

Fig.13.2 *p-V* Diagram of an Otto Engine

If there are k cylinders, then

$$ip_1 + ip_2 + ip_3 + ip_4 + \ldots + ip_k = \sum_1^k bp_k + fp_k \quad (13.1)$$

where ip, bp and fp are respectively indicated, brake and frictional power and the suffix k stands for the cylinder number.

If the first cylinder is cut-off, it will not produce any power but it will have friction, then

$$ip_2 + ip_3 + ip_4 + \ldots + ip_k = \sum_2^k bp_k + fp_k \quad (13.2)$$

Subtracting Eqn.13.2 from Eqn.13.1

$$ip_1 \quad = \quad \sum_{1}^{k} bp_k - \sum_{2}^{k} bp_k$$

Similarly we can find the indicated power of the cylinders, viz., $ip_2, ip_3, ip_4, \ldots, ip_k$.

The total indicated power developed by the engine, ip_k, is given by

$$ip_k \quad = \quad \sum_{1}^{k} ip_k \qquad (13.3)$$

when all the k cylinders are working, it is possible to find the brake power, bp_k, of the engine.

The frictional power of the engine is given by

$$fp_k \quad = \quad ip_k - bp_k \qquad (13.4)$$

For a better understanding of this test refer to the worked out examples 14.24 and 14.25.

13.2.3 Motoring Test

In Motoring test the engine is steadily operated at the rated speed by its own power and allowed to remain under the given speed and load conditions for sufficient time so that the temperature of the engine components, lubricating oil and cooling water reaches a steady state. A swinging field type electric dynamometer is used to absorb the power during this period which is most suitable for this test. The ignition is then cut-off and by suitable electric switching devices the dynamometer is converted to run as a motor so as to crank the engine at the same speed at which it was previously operating. The power supply from the above dynamometer is measured which is a measure of the frictional power of the engine at that speed. The water supply is also cut-off during the motoring test so that the actual operating temperatures are maintained to the extent possible.

This method though determines the fp at conditions very near to the actual operating temperatures at the test speed and load, it does not give the true losses occurring under firing conditions due to following reasons:

(i) The temperatures in the motored engine are different from those in a firing engine.

(ii) The pressure on the bearings and piston rings is lower than in the firing engine.

(iii) The clearance between piston and cylinder wall is more (due to cooling) and this reduces the piston friction.

(iv) The air is drawn at a temperature much lower than when the engine is firing because it does not get heat from the cylinder (rather losses heat to the cylinder).

Motoring method, however, gives reasonably good results and is very suitable for finding the losses imparted by various engine components. This insight on the losses caused by various components and other parameters is obtained by progressive stripping off of the engine. First the full engine is motored, then the test is conducted under progressive dismantling conditions keeping water and oil circulation intact. Then the cylinder head can be removed to evaluate by difference, the compression loss. In this manner, piston rings, pistons, etc. can be removed and evaluated for their effect on overall friction.

13.2.4 From the Measurement of Indicated and Brake Power

This is an ideal method by which fp is obtained by computing the difference between indicated power obtained from an indicator diagram and brake power obtained by a dynamometer. This method is mostly used only in research laboratories as it is necessary to have elaborate equipment to obtain accurate indicator diagrams at high speeds.

13.2.5 Retardation Test

This test involves the method of retarding the engine by cutting the fuel supply. The engine is made to run at no load and rated speed taking into all usual precautions. When the engine is running under steady operating conditions the supply of fuel is cut-off and simultaneously the time of fall in speeds by say 20%, 40%, 60%, 80% of the rated speed is recorded. The tests are repeated once again with 50% load on the engine. The values are usually tabulated in an appropriate table.

A graph connecting time for fall in speed (x-axis) and speed (y-axis) at no load as well as 50% load conditions is drawn as shown in Fig.13.3. From the graph the time required to fall through the same range (say 100 rpm) in both, no load and load conditions are found. Let t_2 and t_3 be the time of fall at no load and load conditions respectively. The frictional torque and hence frictional power are calculated as shown below.

Moment of inertia of the rotating parts is constant throughout the test.

$$\text{Torque} \quad = \quad \text{Moment of Inertia} \times \text{Angular Acceleration}$$

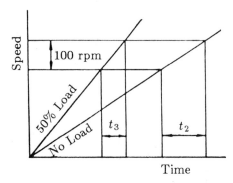

Fig.13.3 Graph for Retardation Test

Let ω be the angular velocity and $\frac{d\omega}{dt}$ be the angular acceleration.

$$T = I\frac{d\omega}{dt} \tag{13.5}$$

But,

$$I = MK^2 \tag{13.6}$$

therefore,

$$T = MK^2\frac{d\omega}{dt} \tag{13.7}$$

$$d\omega = \frac{T}{MK^2}dt \tag{13.8}$$

Now integrating between the limits ω_1 and ω_2 for time t_1 and t_2,

$$\int_{\omega_1}^{\omega_2} d\omega = \frac{T}{MK^2}\int_{t_1}^{t_2} dt \tag{13.9}$$

therefore,

$$(\omega_2 - \omega_1) = \frac{T}{MK^2}(t_2 - t_1) \tag{13.10}$$

Let T_f be the friction torque and T_l the load torque. At no load the torque is only friction torque, T_f, and at load the torque is $T_l + T_f$.

Hence at no load

$$\omega_0 - \omega_1 = \frac{T_f}{MK^2}(t_2 - 0) \tag{13.11}$$

The reference angular velocity, ω_0 is that at, say 1000 rpm, the time of fall for the same range at load

$$\omega_0 - \omega_1 = \frac{T_f + T_l}{MK^2}(t_3 - 0) \tag{13.12}$$

therefore,

$$(T_f + T_l)t_3 = T_f \, t_2$$

$$\frac{t_2}{t_3} = \frac{T_l + T_f}{T_f} = 1 + \frac{T_l}{T_f} \tag{13.13}$$

$$\frac{T_l}{T_f} = \frac{t_2}{t_3} - 1 = \frac{t_2 - t_3}{t_3} \tag{13.14}$$

therefore,

$$T_f = T_l \left(\frac{t_3}{t_2 - t_3} \right) \tag{13.15}$$

T_l is the load torque which can be measured from the loading, t_2 and t_3 are observed values. From the above T_f can be calculated and thereby the frictional power. For a better understanding of this test refer the worked out example 14.26.

13.2.6 Comparison of Various Methods

The Willan's line method and Morse tests are comparatively easy to conduct. However, both these tests give only an overall idea of the losses whereas motoring test gives a very good insight into the various causes of losses and is a much more powerful tool. As far as accuracy is concerned, the $ip - bp$ method is the most accurate if carefully done. Motoring method usually gives a higher value for fp as compared to that given by the Willan's line method. Retardation method, though simple, requires accurate determination of the load torque and the time for the fall in speed for the same range.

13.3 INDICATED POWER

Indicated power of an engine tells about the health of the engine and also gives an indication regarding the conversion of chemical energy in the fuel into heat energy. Indicated power is an important variable because it is the potential output of the cycle. Therefore, to justify the measurement of indicated power, it must be more accurate than motoring and other indirect methods of measuring frictional power. For obtaining indicated power the cycle pressure must be determined as a function of cylinder volume. It may be noted that it is of no use to determine pressure accurately unless volume or crank angle can be accurately measured.

In order to estimate the indicated power of an engine the following methods are usually followed.

(i) using the indicator diagram

(ii) by adding two measured quantities viz. brake power and friction power

13.3.1 Method using the Indicator Diagram

The device which measures the variation of the pressure in the cylinder over a part or full cycle is called an *indicator* and the plot of such information obtained is called an *indicator diagram*. Indicator diagram is the only intermediate record available in the account of total liberated energy before it is measured at the output shaft. Thus an indicator diagram gives a very good indication of the process of combustion and in the associated factors such as rate of pressure rise, ignition lag, etc. by its analysis. Also the losses occurring in the suction and exhaust strokes can be studied. It is very rare that an indicator diagram is taken to find indicated power only. It is almost invariably used to study engine combustion, knocking, tuning of inlet and exhaust manifolds, etc.

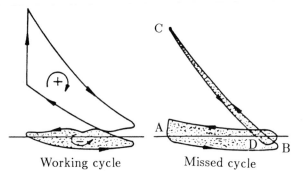

Working cycle   Missed cycle

Fig.13.4 Actual Indicator Diagram of an Otto Engine

Pressure-volume, *p-V* and pressure-crank angle, *p-θ*, are the two types of indicator diagrams that can be obtained from an engine. Both these indicator diagrams are mutually convertible. An actual indicator diagram is shown in Fig.13.4(a) for a working cycle whereas Fig.13.4(b) is for a missed cycle. During a missed cycle of operation there is no power developed and therefore the entire area is shaded. The direction of the arrows show the path to be followed in the diagram. The sign of an area depends upon the direction in which it is traced and since the shaded area is traced in the reverse direction compared to the unshaded area, which has the opposite sign. The shaded area represents the work done in charging and discharging the cylinder. The elasticity of the column of exhaust gas results in a wavy line on the exhaust stroke. The unshaded area represents the gross power, *gp*, developed and the shaded one represents the pumping power, *pp*. Therefore, the indicated power, $ip = (gp - pp)$. In practice *pp* is generally ignored since it is very small. Thus the area of the indicator diagram if accurately measured will represent the indicated power of the engine.

13.3.2 Engine Indicators

Basically an engine indicating device consists of

  (i) a pressure sensing device

 (ii) a device for sensing the piston displacement or the angular position of the crankpin over the complete cycle and

(iii) a display device which can depict both pressure and piston displacement on paper or screen.

    Some indicators also need additional equipment such as preamplifier to amplify the pressure signal before it can be displayed.

    The main types of engine indicators are

  (i) Piston indicator

 (ii) Balanced diaphragm type indicator

(iii) Electronic indicator

    In addition to this, optical indicators are also used. Thus, a large number of indicators are in use. Of the above types, only electronic indicators are in common use at present.

13.3.3 Electronic Indicators

In order to investigate the individual cycles or a part of the cycle, electronic indicators have been developed which consist of four main parts.

  (i) a pressure pick-up

 (ii) a preamplifying device

(iii) a time-base recording device

(iv) a display unit

    The pressure pick-up shown in Fig.13.5 is usually called a pressure transducer. It generates an electric signal in proportion to the pressure to which it is subjected. The transducer is usually fitted in the cylinder head just like a spark plug without projecting into the combustion space. The pressure is picked-up with respect to displacement of a diaphragm. This diaphragm is made very stiff in order to reduce the displacement and hence the inertia effects are reduced to minimum. The displacement of the diaphragm is transmitted to the transducer element which may be any one of the following.

  (i) piezo electric crystal

 (ii) electromagnetic type

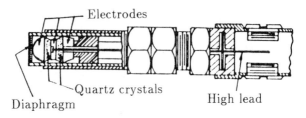

Fig.13.5 Pressure Transducer

(iii) capacitance type

(iv) strain gauge-type element

The detailed description of these transducer elements and their characteristics is beyond the scope of this book and is not discussed. The transducer element produces an electrical signal which is usually too small to be displayed on any device. Therefore, a preamplifying device is used to augment the signal so that it can be displayed on the oscilloscope or recorded in a recorder.

The pick-up used for measuring the cylinder pressure must have a linear output and a good frequency response. Its temperature behaviour should also be satisfactory, i.e., the effect of heat should not affect its performance. Also it must have a low acceleration sensitivity. Of all the pressure transducers currently available the piezo electric quartz transducers are most satisfactory for all normal uses in internal combustion engine measurements as far as the sensitivity is concerned. The quartz transducer, which has a natural frequency greater than 50 kHz, has a sensitivity of about one-tenth of the other types of pressure transducers in which inductive, capacitive or strain-gauge principles are applied. The temperature effect is about 0.005 per cent per degree centigrade change in temperature of the crystal.

The use of electrical pressure pickups avoid almost all the difficulties of a mechanical indicator and gives inertia free operation. However, the greatest problem when using these transducers is the difficulty in calibrating them. A notable exception is strain gauge type transducer which can be satisfactorily calibrated.

Usually the device used to display the *p-θ* diagram is the storage type cathode ray oscilloscope, CRO. The CRO provides almost inertia free recording and displaying of the pressure signal. The principle of CRO is that a cathode ray can be deflected by the variation of an electric current. It can be of electromagnetic or electrostatic type. Usually the pressure signal from the pressure transducer and the time signal from the time-base pick-up are applied to the two beams of dual beam oscilloscope. This produces a *p-θ* or *p-t* diagram on the screen of the oscilloscope which can be observed or photographed. A typical time base unit consists of a permanent magnet with coil and a *V*-shaped pole piece and a rotating disc having slots in which a magnetic material is fixed. When the disc rotates and a slot passes the permanent magnet it generates a voltage according to its depth and produces a peak on the oscilloscope screen. Usually slots are 1° apart with deeper slots at 10° intervals and still deeper at 90° interval so that a complete degree timing diagram is produced. The disc is so adjusted on the engine shaft that when the deepest slot is against the magnet poles it shows the top dead centre. This type of time base device does not work below about 150 rpm because of weak impulse signal. Hence an oscilloscope triggering circuit is used. A typical *p-θ* diagram obtained from the electronic indicator is shown in Fig.13.6(a) for 0 to 360° and 13.6(b) for 360 to 720°.

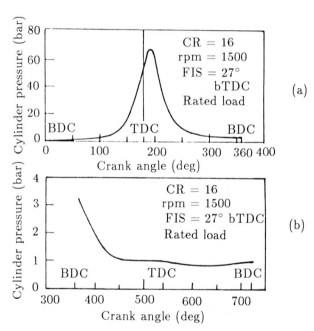

Fig.13.6 A Typical *p-θ* Diagram

13.4 BRAKE POWER

Measurement of brake power is one of the most important measurements in the test schedule of an engine. It involves the determination of the torque and the angular speed of the engine output shaft. The torque measuring device is called a dynamometer.

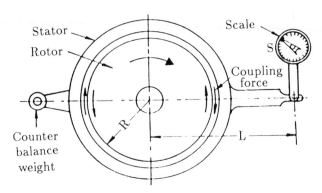

Fig.13.7 Principle of a Dynamometer

Figure 13.7 shows the basic principle of a dynamometer. A rotor driven by the engine under test, is mechanically, hydraulically or electromagnetically coupled to a stator. For every revolution of the shaft, the rotor periphery moves through a distance $2\pi R$ against the coupling force, F. Hence the work done per revolution is

$$W \;=\; 2\pi RF \qquad\qquad (13.16)$$

The external moment or torque is equal to $S \times L$, where S is the scale reading and L is the arm length. This moment balances the turning moment $R \times F$, i.e.,

$$S \times L \;=\; R \times F \qquad\qquad (13.17)$$

Therefore

$$\text{Work done/revolution} \;=\; 2\pi SL$$

$$\text{Work done/minute} \;=\; 2\pi SLN$$

Hence brake power is given by

$$bp \;=\; 2\pi NT \text{ Watts} \qquad (13.18)$$

where T is the torque and N is rpm

Dynamometers can be broadly classified into two main types:

(i) *Absorption Dynamometers*: These dynamometers measure and absorb the power output of the engine to which they are coupled.

The power absorbed is usually dissipated as heat by some means. Examples of such dynamometers are prony brake, rope brake, hydraulic, eddy current dynamometers, etc.

(ii) *Transmission Dynamometer*: In transmission dynamometers the power is transmitted to the load coupled to the engine after it is indicated on some type of scale. These are also called torquemeters.

The terms brake and dynamometer mean the same. A dynamometer is also a brake except the measuring devices are included to indicate the amount of force required in attempting to stop the engine.

13.4.1 Prony Brake

One of the simplest methods of measuring power output of an engine is to attempt to stop the engine by means of a mechanical brake on the flywheel and measure the weight which an arm attached to the brake will support, as it tries to rotate with the flywheel. This system is known as the *prony brake* and from its use, the expression brake power has come.

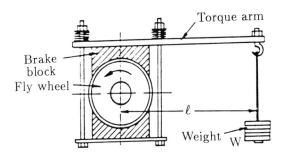

Fig.13.8 Prony Brake

The prony brake consists of a frame with two brake shoes gripping the flywheel (see Fig.13.8). The pressure of the brake shoes on the fly wheel can be varied by the spring loaded using nuts on the top of the frame. The wooden block when pressed into contact with the rotating drum opposes the engine torque and the power is dissipated in overcoming frictional resistance. The power absorbed is converted into heat and hence this type of dynamometer must be cooled. The brake power is given by

$$bp = 2\pi NT \qquad (13.19)$$

where

$$T = Wl \qquad (13.20)$$

W being the weight applied at a distance l.

13.4.2 Rope Brake

The rope brake as shown in Fig.13.9 is another simple device for measuring *bp* of an engine. It consists of a number of turns of rope wound around the rotating drum attached to the output shaft. One side of the rope is connected to a spring balance and the other to a loading device. The power absorbed is due to friction between the rope and the drum. The drum therefore requires cooling.

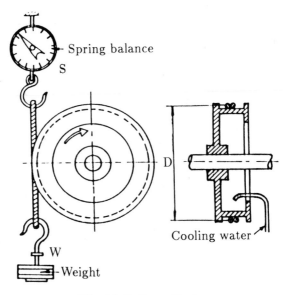

Fig.13.9 Rope Brake

Rope brake is quite cheaper and can be easily fabricated but not very accurate because of changes in the friction coefficient of the rope with temperature.

The *bp* is given by

$$bp = \pi DN(W - S) \qquad (13.21)$$

where D is the brake drum diameter, W is the weight and S is the spring scale reading.

13.4.3 Hydraulic Dynamometer

The hydraulic dynamometer (Fig.13.10) works on the principle of dissipating the power in fluid friction rather than in dry friction. In principle its construction is similar to that of a fluid flywheel. It consists of an inner rotating member or impeller coupled to the output shaft of the engine. This impeller rotates in a casing filled with some hydraulic fluid. The outer casing, due to the centrifugal force developed, tends to revolve with the impeller, but is resisted by a torque arm supporting the balance weight. The frictional forces between the impeller and the fluid is measured by the spring-balance fitted on the casing. The heat developed due to dissipation of power is carried away by a continuous supply of the working fluid, usually water. The output can be controlled by regulating the sluice gates which can be moved in and out to partially or wholly obstruct the flow of water between impeller and the casing.

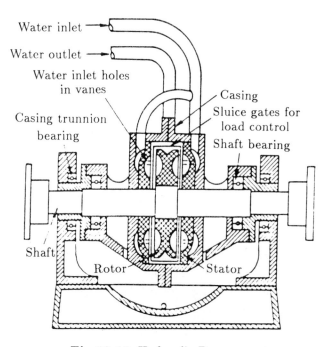

Fig.13.10 Hydraulic Dynamometer

13.4.4 Eddy Current Dynamometer

The details of eddy current dynamometer is shown in Fig.13.11. It consists of a stator on which are fitted a number of electromagnets and a rotor disc made of copper or steel and coupled to the output shaft of the engine. When the rotor rotates eddy currents are produced in the stator due to magnetic flux set up by the passage of field current in the electromagnets. These eddy currents oppose the rotor motion, thus loading the engine. These eddy currents are dissipated in producing heat so that this type of dynamometer also requires some cooling arrangement. The torque is measured exactly as in other types of absorption dynamometers i.e., with the help of a moment arm. The load is controlled by regulating the current in the electromagnets.

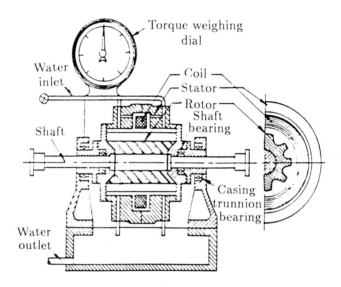

Fig.13.11 Eddy Current Dynamometer

The following are the main advantages of eddy current dynamometers:

(i) Capable of measuring high power per unit weight of the dynamometer.

(ii) They offer the highest ratio of constant brake power to speed range (up to 5:1).

(iii) Level of field excitation is below 1 per cent of total power being handled by dynamometer, thus easy to control and operate.

(iv) Development of eddy current is smooth hence the torque is also smooth and continuous under almost all conditions.

(v) Relatively higher torque under low speed conditions.

(vi) Has no intricate rotating parts except shaft bearing.

(vii) No natural limit to size, either small or large.

13.4.5 Swinging Field dc Dynamometer

Basically a swinging field dc dynamometer is a dc shunt motor so supported in *trunnion* bearings to measure the reaction torque that the outer case and field coils tend to rotate due to the magnetic drag. Hence the name *swinging field*. The torque is measured with an arm and weighing equipment in the usual manner.

These dynamometers are provided with suitable electric connections to run as motor also. Then the dynamometer is reversible, i.e., works as a motoring as well as power absorbing device. When used as an absorption dynamometer it works as a dc generator and converts mechanical energy into electric energy which is dissipated in an external resistor or fed back to the mains. When used as a motoring device an external source of dc voltage is needed to drive the motor. The load is controlled by changing the field current.

13.4.6 Fan Dynamometer

A fan dynamometer is shown in Fig.13.12. This dynamometer is also an absorption type of dynamometer in that when driven by the engine it absorbs the engine power. Such dynamometers are useful mainly for rough testing and running in. The accuracy of the fan dynamometer is very poor. It is also quite difficult to adjust the load. The power absorbed is determined by using previous calibration of the fan brake.

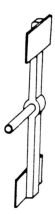

Fig.13.12 Fan Dynamometer

13.4.7 Transmission Dynamometer

Transmission dynamometers, also called torquemeters, mostly consist of a set of strain gauges fixed on the rotating shaft and the torque is measured by the angular deformation of the shaft which is indicated as strain of the strain gauge. Usually a four-arm bridge is used to reduce the effect of temperature to minimum and the gauges are arranged in pairs such that the effect of axial or transverse load on the strain gauges is avoided.

Figure 13.13 shows a transmission dynamometer which employs beams and strain-gauges for a sensing torque.

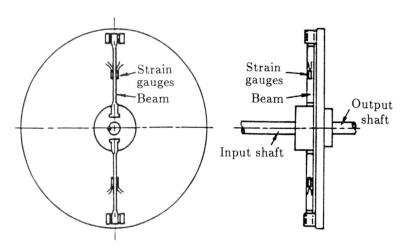

Fig.13.13 Transmission Dynamometer

Transmission dynamometers are very accurate and are used where continuous transmission of load is necessary. These are used mainly in automatic units.

13.4.8 Chassis Dynamometer

In order to test the complete vehicle a *chassis dynamometer* is used. Use of chassis dynamometer allows the test engineer to simulate road load conditions in a laboratory and it can be termed as *all weather road* for the test vehicle. The *road load* on the vehicle can be put on the engine on a chassis dynamometer and quick results can be obtained. However, this dynamometer is quite expensive to install.

13.5 FUEL CONSUMPTION

There are two ways of expressing fuel consumption viz. by volume or by weight during a specified time. For automobiles it is expressed in terms of kilometers per litre.

Accurate measurement of fuel consumption is very important in engine testing work. Though this seems to be a simple matter, it is by no means so as apparent from the occurrence of the following phenomena:

(i) Due to engine heat, vapour bubbles are formed in the fuel line. When the bubble grows the fuel volume increases and back flow of fuel take place. Some fuel flowmeters measure this backflow as if it was forward flow. Some meters do not count backward flow but when the bubble collapses a forward flow takes place which is counted.

(ii) If bubbles are formed before or inside the flowmeter the measured flow can be much higher than actual.

(iii) If there is any swirl in the fuel flow especially in the case of turbine type flowmeter it may register a higher flow rate.

(iv) The density of the fuel is dependent on temperature which can vary over a wide range (-10 °C to 70 °C) giving rise to a error in measurement.

(v) Some flowmeters which use a light beam, the measurement may be affected by the colour of the fuel.

(vi) The needle valve in the float bowl of the carburettor opens and closes periodically allowing fuel to surge into the float bowl. This may cause water hammer type effect making the turbine type flowmeter to continue to rotate even when fuel flow has stopped, thereby producing errors in flow measurements.

As already mentioned two basic type of fuel measurement methods are

(i) Volumetric type

(ii) Gravimetric type

13.5.1 Volumetric Type Flowmeter

The simplest method of measuring volumetric fuel consumption is using glass bulbs of known volume and having a mark on each side of the bulb. Time taken by the engine to consume this volume is measured by a stop watch. Volume divided by time will give the volumetric flow rate.

Burette Method: It consists of two spherical glass bulbs having 100 cc and 200 cc capacity respectively (Fig.13.14). They are connected by three way cocks so that one may feed the engine while the other is being filled. The glass bulbs are of different capacities so as to make the duration of the tests approximately constant irrespective of the engine load whilst the spherical form combines strength with a small variation of fuel head which is most important particularly in case of carburettor engines.

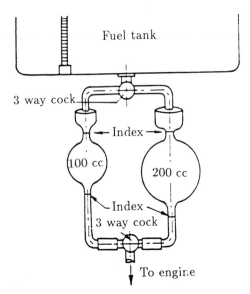

Fig.13.14 Burette Method of Measuring Fuel Consumption

In order to avoid the error in sighting the fuel level against the mark on the burette photocells are used. Figure 13.14 shows such an arrangement in which the measurement is made automatic.

Automatic Burette Flowmeter: Figure 13.15 shows an automatic volumetric type fuel flow measuring system which is commercially available. It consists of a measuring volume (A) which has a photocell (B) and a light source (C) fitted in tubular housings. These housings are put opposite to each other at an angle such that a point of light is formed on the axis of the measuring volume as shown in Fig.13.15 and one each is put on lower and upper portions of the measuring cylinder. An equalization chamber (D) is connected to the measuring tube via the air tube (E) and magnetic valve (F) and equalization pipe (G) to provide an air cushion at supply line pressure and to store fuel during measurement.

On pressing the start button the lamps in the two photoelectric systems light up and the magnetic valve stops the flow through the instrument. The fuel level in the measuring volume starts falling at a rate depending upon the engine consumption. At the same time an equal amount of the flows through the equalization tube to the equalization chamber. When the fuel level reaches the upper measuring level (H), the focussed beam of light from the lamp is reflected on to the opposite photocell and converted into an electrical signal to start a timer counter. When the fuel level has fallen further to

reach the lower measuring level the new signal generated stops the
timer counter. The lamps are automatically switched off and the
valve opened and the regular flow is restored. The time period for the
consumption of the chosen volume of fuel is thus recorded.

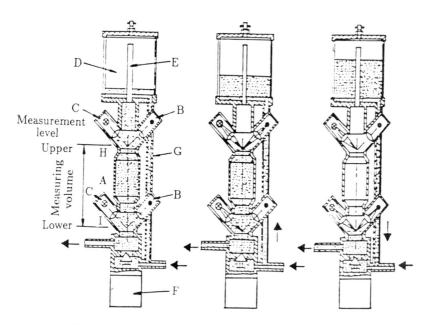

Fig.13.15 Automatic Volumetric Fuel Flowmeter

Orifice Flowmeters: Sometimes flowmeters are also used for this
purpose. Flowmeters depend on the pressure drop across an orifice.
Two orifices, X and Y are shown in Fig.13.16. These orifices are pre-
calibrated in terms of the fuel being used and direct observations in
terms of volume of fuel supplied per hour may be recorded. One or
two orifices may be used at a time. Orifices can also be changed for
different rates of flow. In that case calibration scales have also got to
be changed.

13.5.2 Gravimetric Fuel Flow Measurement

The efficiency of an engine is generally related to the kilograms of fuel
which are consumed and not to the number of litres. The method
of measuring volume flow and then correcting it for specific gravity
variations is quite inconvenient and inherently limited in accuracy.
Instead if the weight of the fuel consumed is directly measured a
great improvement in accuracy and cost can be obtained.

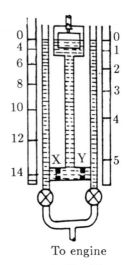

Fig.13.16 Orifice Flowmeter

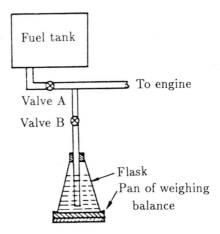

Fig.13.17 Gravimetric Measurement of Fuel Flow

The method involves weighing the fuel supplied to the engine by an arrangement as shown in Fig.13.17. In this method the valve **A** is opened whenever the engine is to be run without measuring the rate of fuel supply and valve **B** is closed so that fuel from tank directly flows to the engine. The fuel from the tank is supplied to the flask by opening valves **A** and **B** whenever measurement of the fuel is to be done. On the balance the amount of fuel is weighed. Keeping the valve

B open the valve A is closed so that the fuel from flask is syphoned off to the engine. This method avoids separate determination of the specific gravity of the fuel. The time taken to syphon off the weighed fuel completely is noted by means of a stop watch. Thus the fuel consumption in gravimetric units are obtained.

13.5.3 Fuel Consumption Measurement in Vehicles

The third way of expressing fuel consumption is by measuring the fuel consumption in kilometers per litre. The instrument as shown in Fig.13.18 is usually used which accurately checks the amount of fuel used by any vehicle under test. A glass burette of 1 litre capacity is connected by tubes and control valves so that the precise number of km/litre is observed on an actual road test. The instrument shows the effect of speed, traffic, loading, driver's habits and engine conditions on fuel consumption. The tester is hung on the top edge of the right front door glass or similarly suitable location and connection is made by special plastic tubes and the adapters provided. The surplus capacity of the fuel pump fills the glass burette when the needle valve is opened, and when a test run is made the control tap directs the fuel to the intake side of the fuel pump. Thus the carburettor is supplied at normal working pressure. Speedometer readings in kilometers and tenths for each 0.5 litres used, make calculations of actual consumption simple and accurate.

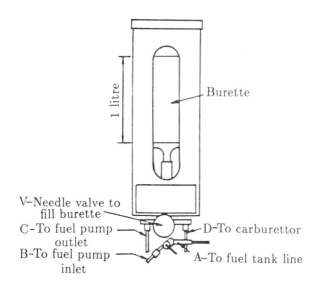

Fig.13.18 Fuel Consumption Measurements in Vehicles

13.6 AIR CONSUMPTION

The diet of an engine consists of air and fuel. For finding out the performance of the engine accurate measurement of both the quantities is essential.

In IC engines the satisfactory measurement of air consumption is quite difficult because *the flow is pulsating due to the cyclic nature of the engine* and because the air is a compressible fluid. Therefore, the simple method of using an orifice in the induction pipe is not satisfactory since the reading will be pulsating and unreliable.

All kinetic flow inferring systems such as nozzles, orifices and ventures have a square law relationship between flow rate and differential pressure which gives rise to severe errors on unsteady flow. Pulsation produced errors are roughly inversely proportional to the pressure across the orifice for a given set of flow conditions.

13.6.1 Air Box Method

The orifice method can be used if pressure pulsations could be damped out by some means. The usual method of damping out pulsations is to fit an air box of suitable volume (500 to 600 times the swept volume in single cylinder engines and less in the case of multi-cylinder engines) to the engine with an orifice placed in the side of the box remote from the engine (Fig.13.19).

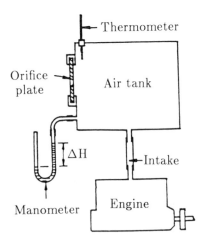

Fig.13.19 Measurement of Air by Air Box Method

13.6.2 Viscous-Flow Air Meter

The use (Fig.13.20) of viscous flow airmeter gives accurate reading for pulsating flows. This meter uses an element where viscous resistance is the principal source of pressure loss and kinetic effects are small. This gives a linear relationship between pressure difference and flow instead of a square-law. From this it follows that a true mean-flow indication is obtained under pulsating flow conditions.

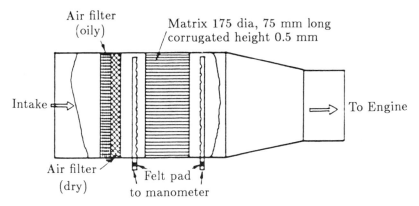

Fig.13.20 Alcock Viscous-Flow Air Meter

The viscous element is in the form of a honeycomb passage (a very large number of passages, Reynolds number being less than 200). The passages are triangular of the size approximately 0.5 × 0.5 × 75 mm.

The chief source of error in viscous meters arises from surface contamination of the small triangular passages. However, by ensuring good filtration at the entry to the meter, and not passing air through the meter unless readings are required, this trouble can be minimized.

An advantage of viscous-flow meter is that larger range of flow can be measured without pressure head being too small. Nowadays positive displacement type of flowmeters are also used for the measurement of air consumption.

13.7 SPEED

Speed measurement is an art. Speed of the engine is widely used in the computation of power, design and development. Measurement of speed is accomplished by instruments like mechanical counters and timers, mechanical tachometers, stroboscope, electric counters, tachometers, electric generators, electronic pulse counters etc.

The best method of measuring speed is to count the number of revolution in a given time. This gives an accurate measurement of speed. Many engines are fitted with such revolution counters. A mechanical or electrical tachometer can also be used for measuring speed. Both these types are affected by temperature variation and are not very accurate. For accurate and continuous measurements of speed a magnetic pick-up placed near a toothed wheel coupled to the engine shaft can be used. The magnetic pick-up will produce a pulse for every revolution and a pulse counter will accurately measure the speed.

13.8 EXHAUST AND COOLANT TEMPERATURE

Simplest way of measuring the exhaust temperature is by means of a thermocouple. Nowadays electronic temperature sensitive transducers are available which can be used for temperature measurements. Ultra violet radiation analyzers also have come into use for measuring temperatures. Coolant temperatures are normally measured using suitable thermometers.

13.9 EMISSION

From the point of view of pollution control, measurement of emissions from engines is very important. Emissions may be divided into two groups, viz., invisible emissions and visible emissions. The exhaust of an engine may contain one or more of the following:

(i) carbon dioxide
(ii) water vapour
(iii) oxides of nitrogen
(iv) unburnt hydrocarbons
(v) carbon monoxide
(vi) aldehydes
(vii) smoke and
(viii) particulate

Out of the eight the first six may be grouped as invisible emissions and the last two as visible emissions. Out of the various invisible emissions carbon dioxide and water vapour are considered harmless compared to others. Hence, their measurements are not discussed. We will briefly describe the measurement of other invisible emissions.

13.9.1 Oxides of Nitrogen

Oxides of nitrogen which also occur only in the engine exhaust are a combination of nitric oxide (NO) and nitrogen dioxide (NO_2). Nitrogen and oxygen react at relatively high temperatures. Therefore, high temperatures and availability of oxygen are the two main reasons for the formation of NO_x. When the proper amount of oxygen is available, the higher the peak combustion temperature the more is the NO formed.

The NO_x concentration in exhaust is affected by engine design and the mode of vehicle operation. Air-fuel ratio and the spark advance are the two important factors which significantly affect NO_x emissions. The maximum NO_x is formed at ratios between 14:1 and 16:1. At lean and rich air-fuel mixtures the NO_x concentration is comparatively low.

Increasing the ignition advance will result in lower peak combustion temperatures and higher exhaust temperatures. This will result in high NO_x concentration in the exhaust.

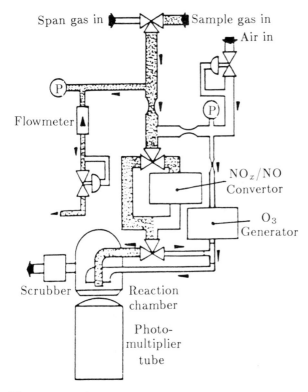

Fig.13.21 Chemiluminescence Method of Measuring Oxides of Nitrogen

Internationally accepted method for measuring oxides of nitrogen is by chemiluminescence analyzer. The details are shown in Fig.13.21. The principle of measurement is based on chemiluminescence reaction between ozone and NO resulting in the formation of excited NO_2.

This excited NO_2 emits light whose intensity is proportional to NO concentration. The light intensity is measured by a photo multiplier tube. The analyzer measures only nitric oxide, NO, and not NO_2. To analyze all the oxides of nitrogen a converter is usually fitted ahead of the reaction chamber to convert all the oxides of nitrogen into nitric oxide. Thereby, the light intensity can be taken to be proportional to the oxides of nitrogen concentration in the sample.

13.9.2 Carbon Monoxide

Carbon monoxide occurs only in engine exhaust. It is a product of incomplete combustion due to insufficient amount of air in the air-fuel mixture or insufficient time in the cycle for completion of combustion.

Theoretically, the gasoline engine exhaust, can be made free of CO by operating it at air-fuel mixture ratios greater than 16:1. However, that some CO is always present in the exhaust even at lean mixtures.

The percentage of CO decreases with speed. In passenger cars CO percentage has been found to be as high as 5 per cent with rich mixtures and 1 per cent with near stoichiometric mixtures. The complete elimination of CO is not possible and 0.5 per cent CO should be considered a reasonable goal.

Carbon monoxide emissions are high when the engine is idling and reach a minimum value during deceleration. They are the lowest during acceleration and at steady speeds. Closing of the throttle which reduces the oxygen supply to engine is the main cause of CO production, so deceleration from high speed will produce highest CO in exhaust gases.

Non Dispersive Infra-Red analyzer (NDIR) as shown in Fig.13.22 is the widely accepted instrument for measuring CO. This instrument is presently used for the testing and legal certification of some automotive exhaust emissions. In the NDIR analyzer the exhaust gas species being measured are used to detect themselves. The method of detection is based on the principle of selective absorption of the infrared energy of a particular wave length peculiar to a certain gas which will be absorbed by that gas.

13.9.3 Unburned Hydrocarbons

Unburnt hydrocarbons emissions are the direct result of incomplete combustion. The pattern of hydrocarbon emissions is closely related to many design and operating variables. Two of the important design variables are induction system design and combustion chamber design, while main operating variables are air-fuel ratio, speed, load and mode of operation. Maintenance is also an important factor.

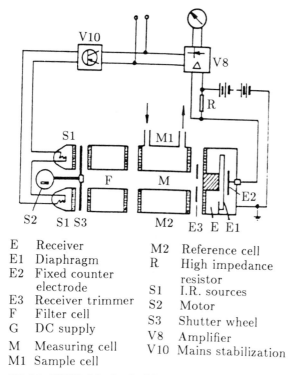

E	Receiver	M2	Reference cell
E1	Diaphragm	R	High impedance
E2	Fixed counter		resistor
	electrode	S1	I.R. sources
E3	Receiver trimmer	S2	Motor
F	Filter cell	S3	Shutter wheel
G	DC supply	V8	Amplifier
M	Measuring cell	V10	Mains stabilization
M1	Sample cell		

Fig.13.22 NDIR Method of Measuring Carbon Monoxide

Induction system design and engine maintenance affect the operating air-fuel ratio of the engine and hence the emission of hydrocarbons and carbon monoxide. Induction system determines fuel distribution of cylinders, fuel economy, available power etc. And the quantity of engine maintenance determines whether the engine will operate at the designed air-fuel ratio and for how long. This will include piston ring wear, lubrication, cooling, deposits and other factors which are likely to affect the air-fuel ratio supplied or its combustion in the combustion chamber.

The design of the combustion chamber is important in that in the combustion chamber portions of the fuel-air mixture which come in direct contact with the chamber walls are quenched and do not burn. Some of this quenched fuel-air mixture is forced out of the chamber during the exhaust stroke and, because of the high local concentration of hydrocarbon in this mixture, contributes to the high hydrocarbon exhaust from the engine. A small displacement engine will have a higher surface-to-volume ratio than an engine with a large displacement. Factors like combustion chamber shape, bore diameter, stroke and compression ratio affect the surface-to-volume ratio and hence the hydrocarbon emission. Lower compression ratio, higher stroke to bore ratio, larger displacement per cylinder and fewer cylinders, all lower the surface-to-volume ratio and hence the hydrocarbons.

The effect of air-fuel ratio on the HC emission is exactly like that on carbon monoxide. At near stoichiometric fuel-air mixtures both hydrocarbon and carbon monoxide (HC/CO) emissions are higher and lean fuel mixtures have substantially low HC/CO emission.

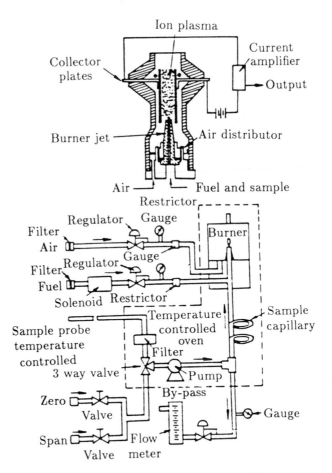

Fig.13.23 Flame Ionization Method for Measuring Unburnt Hydrocarbons

Flame Ionization Detector (FID) is used for measuring hydrocarbons. This instrument as shown in Fig.13.23 is a well established and accepted method for measuring HC. Ionization is a characteristic of HC compound. This principle is employed in the FID detector.

Formation of electrically charged particles of ionized carbon atoms from the hydrocarbons in a hydrogen-oxygen flame is achieved in the FID analyzer. Current flow in micro amperes (amplified for measurements) is a measure of the concentration of hydrocarbons.

13.9.4 Aldehydes

The emission of odourous oxygenated hydrocarbons from the engines is generally carcinogenic. The use of alcohol based fuels can lead to higher levels of oxygenated hydrocarbon emissions. These aldehydes are responsible for the pungent smell of the engine exhaust and a trained human personnel specifies the odour ratings for the emission sample by comparison.

The methods of measurement for aldehydes are based on wet chemical methods (analytical methods) like iodine titration technique (ITT), chromotropic acid (CA) method, 3-methyl 2-benzothiazolene hydrazone (MBTH) method and 2,4 Dinitrophenyl hydrazine (DNPH) method. While the first two methods are meant to measure total aldehydes, the DNPH method is for individual aldehydes and the CA method is only for the formaldehyde.

All the above described wet chemical methods excepting the improved DNPH method, even though time consuming and cumbersome, are economical. Various other methods available are Fourier Transform-Infrared Spectroscopy (FT-IR), Derivative Spectrophotometry and Portable Polarigraph.

13.10 VISIBLE EMISSIONS

Compared to harmful and invisible emissions, visible emissions are more irritating and cause nuisance. Especially in diesel engines, smoke is one of visible emissions. The details are discussed in the following sections.

13.10.1 Smoke

The smoke of the engine exhaust is a visible indicator of the combustion process in the engine. Smoke is due to incomplete combustion. Smoke in diesel engine can be divided into three categories viz., blue, white and black. Visible method of analysis is used for quantifying the above three types of smokes. Smoke measurements can be broadly classified into two groups viz. comparison method and obscuration method.

Comparison Method: Most ordinances regulating smoke emissions are based on estimation of the density of the smoke as it emerges from the exhaust. Of the several available methods, the one of the commonly used method is the Ringelmann Chart. The chart shows four shades of gray, as well as a pure white and an all-black section. To overcome the difficulty of reproducing various shades of gray, the intermediate shades are built from black lines of various widths (Fig.13.24). The four intermediate charts are printed by the United States Bureau of Mines on a single 70 cm × 25 cm sheet. They may be reproduced as follows:

0. All white
1. Black lines 1 mm thick, 10 mm apart, leaving white spaces 9 mm square.
2. Lines 2.3 mm thick, spaces 7.7 mm square.
3. Lines 3.7 mm thick, spaces 6.3 mm square.
4. Lines 5.5 mm thick, spaces 4.5 mm square.
5. All black.

In use, the chart is set up at eye level in line with the stack at such distances (10 m or more) that the sections appear to be different degrees of uniform gray shades. The appearance of the smoke at the top of the stack is matched against one of the shades on the card and reported as a specific *Ringelmann number* ranging from 0 (no smoke) to No.5 (dense black smoke). With practice, an observer can estimate smoke density to half a number, particularly in the Nos.2 to 4 range. Readings below No.2 Ringelmann are subject to considerable error.

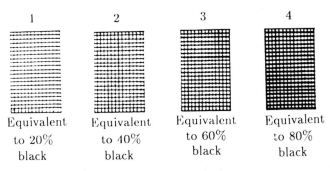

Fig.13.24 Ringelmann Chart

Obscuration Method: It is basically divided into light extinction type, continuous filtering type and spot filtering type.

(i) *Light Extinction Type:* In this method of testing, the intensity of a light beam is reduced by smoke which is a measure of smoke intensity. Schematically this method is shown in Fig.13.25. A continuously taken exhaust sample is passed through a tube of about 45 cm length which has light source at one end and photocell at the other end. The amount of light passed through this column is used as an indication of smoke level or smoke density. *The smoke level or smoke density is defined as the ratio of electric output from photocell when sample is passed through the column to the electric output when clean air is passed through it.*

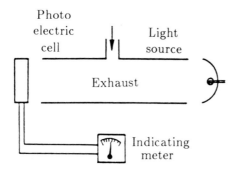

Photo electric cell

Light source

Exhaust

Indicating meter

Fig.13.25 Obscuration Method for Measuring Smoke

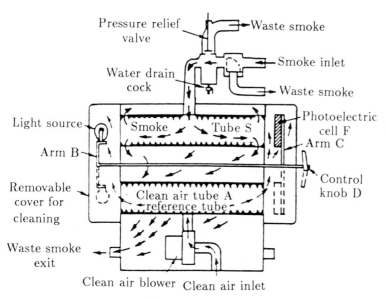

Pressure relief valve

Waste smoke

Smoke inlet

Water drain cock

Waste smoke

Light source

Smoke — Tube S

Photoelectric cell F

Arm B

Arm C

Removable cover for cleaning

Clean air tube A reference tube

Control knob D

Waste smoke exit

Clean air blower Clean air inlet

Fig.13.26 Hartridge Smokemeter

The Hartridge smokemeter is a very commonly used instrument based on this principle. Schematically it is represented in Fig.13.26.

Light from a source is passed through a standard length tube containing the exhaust gas sample from the engine and at its other end the transmitted light is measured by a suitable device. The fraction of the light transmitted through the smoke, (T) and the length of the light path (L_ℓ) are related by the Beer-Lambert law.

$$T = e^{-K_{ac}L_\ell} \tag{13.22}$$

where $K_{ac} = nA\theta$.

K_{ac} is called the optical absorption coefficient of the obscuring matter per unit length, n the number of soot particles per unit volume, A the average projected area of each particles and θ the specific absorbance per particle.

(ii) *Continuous Filtering Type*: In this method provision is made for continuous reading and observing transient conditions. Measurement of smoke intensity is achieved by continuously passing exhaust gas through a moving strip of filter paper and collecting particles. The instrument is schematically represented in Fig.13.27. Van Brand smokemeter is the popular instrument based on this principle.

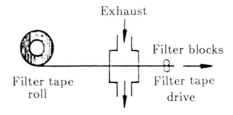

Fig.13.27 Continuous Filter Type Smokemeter

Van Brand smokemeter is also filter darkening type. The exhaust sample is passed at a constant rate through a strip of filter paper moving at a preset speed. A stain is imparted to the paper. The intensity of the stain is measured by the amount of light which passes through the filter and is indication of the smoke density of exhaust. In Van Brand smokemeter the amount of light passing through the filter is used to indicate smoke level.

(iii) *Spot Filtering Type*: A smoke stain obtained by filtering a given quantity of exhaust gas through a fixed filter paper is used for the measure of smoke intensity. The instrument is schematically shown in Fig.13.28. Bosch smokemeter is the popular instrument based on this principle and the details are illustrated in Fig.13.29.

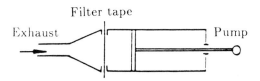

Fig.13.28 Spot Filtering Type Smokemeter

A fixed quantity of exhaust gas is passed through a fixed filter paper and the density of the smoke stains on the paper are evaluated optically. In a recent modification of this type of smokemeter, a pneumatically operated sampling pump and a photo-electric unit are used for the measurement of the intensity of smoke stain on filter paper.

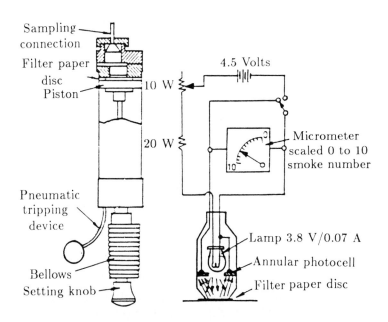

Fig.13.29 Bosch Smokemeter

13.11 NOISE

Noise is a mixture of various sounds which is a source of irritation for the listener. Sound is created by a vibrating object. The vibrations are transmitted to the surrounding air in the form of pressure waves. If the frequency and intensity of the pressure waves are within specified range they produce the sensation of sound (15 to 15000 Hz and 0 to 120 dB intensity).

Sound level meter consists of a microphone, calibrated attenuator, an electronic amplifier and an indicator meter which reads in decibels (dB) as shown in Fig.13.30.

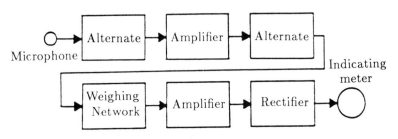

Fig.13.30 Sound Level Meter

Octave band frequency analyzer is suitable for obtaining the frequency distribution in light bands in frequency region between 20 and 10,000 Hz.

The measurement of the noise emitted by motor vehicles is based on a moving vehicle, since it is the total noise emitted by motor vehicle including gear box and transmission.

13.12 COMBUSTION PHENOMENON

Combustion in internal combustion engine is very complex and still it is not fully understood. In order to have an insight into the combustion it is necessary to measure flame temperature, flame propagation, details of combustion and knock.

13.12.1 Flame Temperature Measurement

Flame temperature measurement is a difficult task especially inside the cylinder. Infrared and spectrographic method are some of the methods currently employed. Figure 13.31 shows a schematic of the infrared method. Principle of the method is based on the fact that if infrared waves from a black body source pass through a cylinder containing gas at a temperature T, then the black body temperature T_b equals the cylinder temperature when the rate of absorption of the infrared waves by the gas in the cylinder equals the rate of emission from the gas. Water vapour is taken as the sensing element (that is the detector and optical system is set to record the emissions at one of the wavelengths for water vapour – 2.6 μm).

The black body source for an optical filter is kept ahead of the detector to absorb the unwanted wave lengths. The cylinder consists

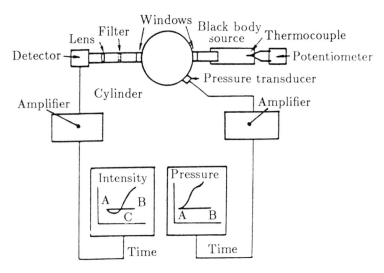

Fig.13.31 Infrared Spectrographic Method

of two quartz windows so that there is a direct optical path from the infrared source to the detector. During the suction stroke there is minimum infrared absorption by the cylinder gases so that infrared waves transmitted from the black body source will pass through the cylinder with almost no change in intensity.

As shown in Fig.13.31 a horizontal line AB will be recorded. The intensity, I will correspond to the black body temperature, T_b measured by the thermocouple. It should be noted that this will not be the same as the gas temperature. Suppose if T_b is set above the suction temperature, then at the beginning of the compression stroke as the temperature raises there will be some absorption of the infrared by the cylinder gas and the emission level will be less than from the black body. The detector will record a decrease in intensity. Further increase in temperature of the cylinder gas will produce a further decrease in recorded intensity until a minimum is reached. As the stroke continues the intensity will now increase until at C the rate of absorption of the infrared waves from the black body source equals the rate of emission from the cylinder gases. At this point the cylinder temperature equals the black body temperature T_b. The crank angle at which this occurs can be measured in the normal manner. As series of black body temperatures T_b is taken and the procedure is repeated, a temperature-crank angle diagram can be obtained. Cleanliness of the quartz windows, window radiation, absorption and homogeneity of the cylinder temperature, cycle-to-cycle variations are factors upon which the accuracy of the method depends.

Another technique is the spectrographic method which is not only used for measuring the local gas temperature but also for studying the chemical reactions in the cylinder. Basis of this method depends on the measurement of the light intensities associated with reacting species at defined wavelengths at which the intensities are at a maximum. One method which has been successfully used to study the formation of oxides of nitrogen during combustion is described. Figures 13.32(a) and 13.32(b) show the details.

A number of quartz windows, W are fitted in the cylinder head. The light source is of 2×2 mm size fitted in the combustion chamber. The core of light emitted from the cylinder is split into four separate beams in mirror M_2, each of which is brought to a separate focus by mirrors M_3 and M_4. Three beams (2, 3 and 4) are focused onto a photo-multiplier with an interference filter to pass wave lengths 0.38 μm and 0.61 μm with a band pass of $100°$A and a wave length of 0.75 μm at a band pass of $300°$A. The fourth channel is monitored with a monochromator and photomultiplier. The light intensities I at the wave length 0.38 μm, 0.68 μm and 0.61 μm are recorded on a oscillograph relative to the crank angle as shown in Fig.13.32(c). The 0.38 μm wavelength corresponds to the reaction

$$CO + O \rightarrow CO_2 + h\nu \qquad (13.23)$$

where the emitted light is mainly associated with the discreet energy pocket $h\nu$. The 0.61 μm and 0.68 μm wave lengths are associated with the reaction

$$NO + O \rightarrow NO_2 + h\nu \qquad (13.24)$$

For the above two reactions the intensities of emitted light I associated with the energy packets $h\nu$ are related to the concentrations by

$$I_{CO} = K_1 [CO] [O] \qquad (13.25)$$

$$I_{NO} = K_2 [NO] [O] \qquad (13.26)$$

K_1 and K_2 parameters are functions of the instantaneous temperature T and the wave length λ. It is assumed that the carbon-hydrogen-oxygen reactions are in equilibrium. Thus for each temperature T there is a known value for $[CO][O]$ from the equilibrium constant. Also for each temperature and wave length there is a known value for K_1. It follows therefore that at a given wave length

$$K_1 [CO] [O] = f(T) \qquad (13.27)$$

when this function equals the numerical value of the emitted light I_{CO} we know the temperature and the oxygen concentration. In this

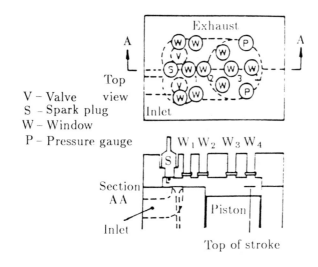

V – Valve
S – Spark plug
W – Window
P – Pressure gauge

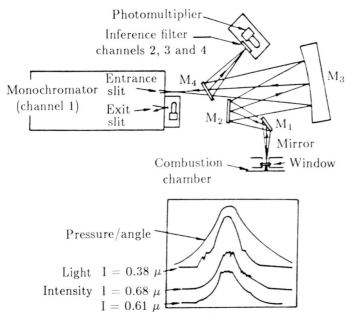

Fig.13.32 Spectrographic Method to Study the
Formation of Oxygen

way by measuring the intensity I_{CO} at the wave length 0.38 μm we
can determine the local temperature T. For the nitric oxide reaction,
if I_{NO} is known at the appropriate wave length, then, since we know
the temperature T and the oxygen concentration [O] and K_2 is a

known function of the wave length and temperature, the nitric oxide concentration can be calculated from

$$[NO] = \frac{I_{NO}}{K_2[O]} \tag{13.28}$$

13.12.2 Flame Propagation

Slow combustion and the propagation of a combustion wave are the two phenomena that occur whenever ignition of combustible mixture takes place. During slow combustion, the fuel molecules that are already burning raise the temperature of adjacent molecules by conduction and radiation causing them to ignite. At the same time the temperature rise of the gas molecules increases their velocity. This raises the pressure in that point and results in an expansion that assists in propagating the ignition. Turbulence created in the charge before ignition increase materially the velocity of flame propagation which in this case occurs not only through conduction and radiation but also through convection. It is not easy to segregate all these processes. Whenever a compression wave occurs its velocity may reach several metres per second and passing through the explosive mixture accompanied by almost instantaneous generation of very high pressures it may cause serious trouble in the engine. That is why the measurement of flame propagation is necessary in the evaluation of the performance of IC engine.

Flame location with time is directly measured by the use of ionization gauges. The gauges are made up of a wire in an annulus. The wire is insulated from the body of the gauge. Whenever flame reaches the annulus it becomes conducting since an electrical circuit is completed between the wire and the gauge body. A dc voltage is normally applied to the wire and an electronic timing circuit connected to the gauge body as well as some reference time location. When the flame arrives at the gauge the timing circuit is triggered and the time from the reference time is measured. By locating a number of gauges in the combustion chamber a quantitative picture of the flame propagation may be made. Allowance must be made for expansion of the hot gas zone which is superimposed on the flame speed during the analysis of the results. Air-fuel ratio affects the minimum voltage.

13.12.3 Combustion Process

High speed colour photography has been successfully used for qualitative combustion analysis notably by Ricardo (Fig.13.33). The cylinder has either a quartz or perspex window. A high speed rotating prism camera, operating up to 16,000 frames per second is mounted on a rigid support and mirrors are arranged to record the combustion and crankshaft simultaneously. The fuel is normally doped with copper

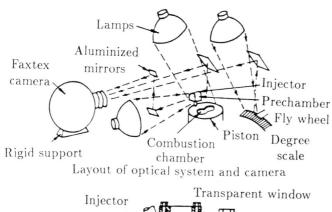

Layout of optical system and camera

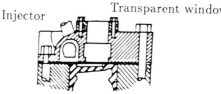

Location of transparent windows in comet
V combustion chamber

Fig.13.33 High Speed Photographic Technique for
Combustion Analysis

oleate to render low luminosity flames visible. A better understanding
of the combustion process can be obtained from combustion films.

Questions :-

13.1 List various methods available for finding friction power of an
engine.

13.2 State the limitations experienced in the evaluation of friction
power using Willan's line method.

13.3 Why Morse test is not suitable for single cylinder engine?
Describe the method of finding friction power using Morse test.

13.4 Explain the method of motoring test for obtaining friction power
of an engine.

13.5 Why the method of obtaining friction power by computing the
difference between indicated power and brake power is mostly
used in research laboratories?

13.6 Explain the principle involved in the measurement of brake
power.

13.7 Explain the use of prony brake and rope brake in measuring the power output of an engine.

13.8 Explain the basic principle and working of hydraulic dynamometer.

13.9 With a neat sketch explain an Eddy current dynamometer.

13.10 What is transmission dynamometer? Explain.

13.11 Enumerate the advantage of gravimetric fuel flow measurement over volumetric fuel flow measurement.

13.12 Explain an automatic fuel flow meter.

13.13 Classify the meters used for measuring air flow and explain.

13.14 "Air flow to engines can not be measured with the same precision as can fuel flow", explain the significance of the above statement.

13.15 What is the need for measurement of speed of an I.C. engine?

13.16 List the types of exhaust temperature measurement.

13.17 List the emissions that are considered significant for measurement and performance study.

13.18 Explain the internationally accepted methods of measuring the following invisible emission.
  (i)   Oxides of nitrogen   (ii)   Carbon monoxide
  (iii) Unburned hydrocarbons (iv) Aldehydes

13.19 What is smoke and classify the measurement of smoke?

13.20 Explain the method of measurement of smoke by comparison method.

13.21 In obscuration method of measuring smoke, list the types and give the name of popular instruments with the principle of working.

13.22 Define noise and explain the method of measurement of noise using sound level meter.

13.23 What parameters are necessary for measurement in order to have an insight into combustion?

13.24 Explain the method of determining the local temperature (flame temperature) by spectrographic method.

13.25 Explain how the flame propagation during the combustion process can be measured using high speed photographic techniques.

14

PERFORMANCE PARAMETERS AND CHARACTERISTICS

14.1 INTRODUCTION

Internal combustion engine generally operates within a useful range of speed. Some engines are made to run at fixed speed by means of a speed governor which is its rated speed. At each speed within the useful range the power output varies and it has a maximum usable value. The ratio of power developed to the maximum usable power at the same speed is called the load. The specific fuel consumption varies with load and speed. The performance of the engine depends on inter-relationship between power developed, speed and the specific fuel consumption at each operating condition within the useful range of speed and load.

The following factors are to be considered in evaluating the performance of an engine:

(i) Maximum power or torque available at each speed within the useful range of speed.

(ii) The range of power output at constant speed for stable operation of the engine. The different speeds should be selected at equal intervals within the useful speed range.

(iii) Brake specific fuel consumption at each operating condition within the useful range of operation.

(iv) Reliability and durability of the engine for the given range of operation.

Engine performance characteristics can be determined by the following two methods.

(i) By using experimental results obtained from engine tests.

(ii) By analytical calculation based on theoretical data

Engine performance is really a relative term. It is represented by typical characteristic curves which are functions of engine operating parameters. The term performance usually means how well an engine is doing its job in relation to the input energy or how effectively it provides useful energy in relation to some other comparable engines.

Some of the important parameters are speed, inlet pressure and temperature, output, air-fuel ratio etc. The useful range of all these parameters is limited by various factors, like mechanical stresses, knocking, over-heating etc. Due to this, there is a practical limit of maximum power and efficiency obtainable from an engine. The performance of an engine is judged from the point of view of the two main factors, viz., engine power and engine efficiency. Besides the overall efficiency, various other efficiencies are encountered when dealing with the theory, design and operation of engines. These factors are discussed in more detail in the following two sections.

14.2 ENGINE POWER

In general, as indicated in section 1.7, the energy flow through the engine is expressed in three distinct terms. They are indicated power, ip, friction power fp and brake power, bp. Indicated power can be computed from the measurement of forces in the cylinder and brake power may be computed from the measurement of forces at the crankshaft of the engine. The friction power can be estimated by motoring the engine or other methods discussed in Chapter 13. It can also be calculated as the difference between the ip and bp if these two are known, then,

$$ip \quad = \quad bp + fp \qquad (14.1)$$

$$fp \quad = \quad ip - bp \qquad (14.2)$$

In the following sections, the usually employed formulae for the computation of power are discussed.

14.2.1 Indicated Mean Effective Pressure (p_{im})

It has been stated in section 14.2 that ip can be computed from the measurement of forces developed in the cylinder, viz., the pressure of the expanding gases.

As already described, in chapters dealing with cycles, the pressure in the cylinder varies throughout the cycle and the variation can be expressed with respect to volume or crank angle rotation to obtain

p-V or p-θ diagrams respectively. However, such a continuous variation does not readily lend itself to simple mathematical analysis in the computation of ip. If an average pressure for one cycle can be used, then the computations becomes far less difficult.

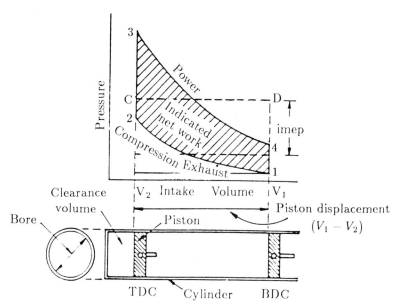

Fig.14.1 p-V Diagram for an Ideal Four-Stroke Cycle Engine

As the piston moves back and forth between TDC and BDC (Fig.14.1), the process lines on the p-V diagram indicate the successive states of the working fluid through the cycle. The indicated net work of the cycle is represented by the area 1234 enclosed by the process lines for that cycle. If the area of rectangle $ABCD$ equals area 1234, the vertical distance between the horizontal lines AB and CD represents the *indicated mean effective pressure, imep*. It is a mean value expressed in N/m^2, which, when multiplied by the displacement volume, V_s, gives the same indicated net work as is actually produced with the varying pressures.

$$p_{im} \times (V_1 - V_2) \quad = \quad \text{Net work of cycle} \qquad (14.3)$$

$$p_{im} \quad = \quad \frac{\text{Net work of cycle}}{V_1 - V_2} \qquad (14.4)$$

$$= \quad \frac{\text{Area of the indicator diagram}}{\text{Length of the indicator diagram}} \qquad (14.5)$$

On an actual engine, the *p-V* diagram (called the *indicator diagram*) is obtained by a mechanical or electrical instrument attached to the cylinder. The area enclosed by the actual cycle on the indicator card may be measured by a planimeter. The value of the area measured, when divided by the piston displacement, results in the mean ordinate, or indicated mean effective pressure, p_{im}.

14.2.2 Indicated Power (ip)

Power is defined as the rate of doing work. In the analysis of cycles the net work is expressed in kJ/kg of air. This may be converted to power by multiplying by the mass flow rate of air through the engine in kg per unit time. Since, the net work obtained from the *p-V* diagram is the net work produced in the cylinder as measured by an indicator diagram, the power based there on is termed indicated power, *ip*.

$$ip \quad = \quad \dot{m}_a \times \text{net work} \qquad (14.6)$$

where $\dot{m}_a$ is in kg/s, network is in kJ/kg of air and *ip* is in kW.

In working with actual engines, it is often desirable to compute *ip* from a given p_{im} and given engine operating conditions. The necessary formula may be developed from the equation of net work based on the mean effective pressure and piston displacement. From Eqn.14.3,

$$\text{Indicated net work/cycle} \quad = \quad p_{im} \, V_s \qquad (14.7)$$

By definition

$$\text{Indicated power} \quad = \quad \text{Indicated net work} \times \text{cycles/s}$$

$$ip \quad = \quad \frac{p_{im} V_s n K}{1000 \times 60}$$

$$= \quad \frac{p_{im} L A n K}{60000} \text{ kW} \qquad (14.8)$$

where ip =   indicated power (kW)

p_{im} =   indicated mean effective pressure (N/m^2)

L =   length of the stroke (m)

A =   area of the piston (m^2)

N =   speed in revolutions per minute

n =   number of power strokes per minute

      $N/2$ for a four-stroke engine

      N for a two-stroke engine

K =   number of cylinders

14.2.3 Brake Power (bp)

Indicated power is based on indicated net work and is thus a measure of the forces developed within the cylinder. More practical interest is the rotational force available at the delivery point, at the engine crankshaft (termed the drive-shaft), and the power corresponding to it. This power is interchangeably referred to as brake power, shaft power or delivered power. In general, only the term *brake power, bp*, has been used in this book to indicate the power actually delivered by the engine.

The bp is usually measured by attaching a power absorption device to the drive-shaft of the engine. Such a device sets up measurable forces counteracting the forces delivered by the engine, and the determined value of these measured forces is indicative of the forces being delivered.

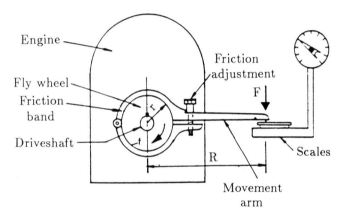

Engine

Friction adjustment

Fly wheel

F

Friction band

Driveshaft

R

Scales

Movement arm

Fig.14.2 Adaptation of Prony Brake for Power Measurement

By using the geometry of a simple prony brake as the basis, a formula can now be developed for computing the bp delivered by an engine. Work has been defined as the product of a force and the distance through which the point of application of force moves. Then the drive-shaft of the engine turns through one revolution, any point on the periphery of the rigidly attached wheel moves through a distance equal to $2\pi r$ (Fig.14.2). During this movement, a friction force, f is acting against the wheel. The force, f is thus acting through the distance $2\pi r$, and producing work. Thus,

$$\text{Work during one revolution} \quad = \quad \text{Distance} \times \text{Force}$$

$$= \quad (2\pi r) \times f \qquad (14.9)$$

The torque, rf, produced by the drive-shaft is opposed by a turning moment equal to the product of the length of the moment arm R and the force F measured by the scale

$$T \;\; = \;\; rf \;\; = \;\; RF \qquad (14.10)$$

$$\text{Work during one revolution} \;\; = \;\; 2\pi RF$$

$$\text{Power} \;\; = \;\; \frac{\text{Work}}{\text{Time}}$$

$$= \;\; 2\pi RF \frac{N}{60} \qquad (14.11)$$

where N = revolutions per minute of the drive-shaft. Therefore,

$$bp \;\; = \;\; \frac{2\pi RFN}{60 \times 1000} \;\; kW \quad (14.12)$$

It should be noted that N is the rpm of the engine. The friction force is acting during every revolution of the crankshaft, regardless of whether or not that revolution contains a power stroke.

The product of the moment arm R and the measured force, F is termed the torque of the engine and is usually expressed in Nm. *Torque*, T is the uniform or fluctuating turning moment, or twist, exerted by a tangential force acting at a distance from the axis of rotation. For an engine operating at a given speed and delivering a given power, the torque must be a fixed amount, or the product of F and R must be constant $(T = FR)$. In such a case, if R is decreased, F will increase proportionately and vice versa.

The brake power, bp, can also be written as

$$bp \;\; = \;\; \frac{2\pi NT}{60000} \;\; kW \qquad (14.13)$$

In practice, the length of the moment arm R of the measuring equipment is so designed that the value of the constants 2π and the constant R and 60000 combine to give a convenient number (i.e., in thousands and ten-thousands) in order to simplify computations.

$$bp \;\; = \;\; \frac{FN}{60000/2\pi R}$$

$$= \;\; \frac{FN}{C_1} \qquad (14.14)$$

In order to have $C_1 = 10000$, R should be 0.955 m.

The reader may recollect that the *torque is the capacity of an engine to do work while power is the rate at which an engine does work*. A simple example is that a tractor pulling a given load. The torque developed will determine whether or not the tractor is capable of pulling the load, and the power delivered will determine how fast the load can be pulled.

14.2.4 Brake Mean Effective Pressure (p_{bm})

Indicated mean effective pressure may be considered to consist of *fmep* and *bmep*, two hypothetical pressures. Friction mean effective pressure is that portion of *imep* which is required to overcome friction losses, and brake mean effective pressure is the portion which produces the useful power delivered by the engine.

$$imep \quad = \quad bmep + fmep$$

Since *bmep* is that portion of *imep* which goes into the development of useful power, it has the same relationship to *bp* as *imep* has to *ip*, or

$$\frac{bmep}{imep} \quad = \quad \frac{bp}{ip} \tag{14.15}$$

Equation 14.8 was developed as a means of computing *ip* when *imep* is determined from an engine indicator diagram

$$ip \quad = \quad \frac{p_{im} L An K}{60000}$$

For a given engine, L, A, n and K are constants. Since *bp* and *bmep* have the same relationship to one another as do *ip* and *imep*, *bp* can similarly expressed as

$$bp \quad = \quad \frac{p_{bm} L An K}{60000} \tag{14.16}$$

where p_{bm} is brake mean effective pressure (N/m^2).

And due to the same relationship, the mechanical efficiency, η_m of the engine can be expressed as the ratio of *bmep* to *imep*.

$$\eta_m \quad = \quad \frac{bp}{ip} \quad = \quad \frac{bmep}{imep} \tag{14.17}$$

It should be noted that for a given engine operating under given conditions, the torque developed is proportional to the *bmep*. This relationship, may be obtained by equating formulae 14.13 and 14.16.

Brake mean effective pressure is very useful in comparing engines or in establishing engine operating limits.

14.3 ENGINE EFFICIENCIES

Apart from expressing engine performance in terms of power, it is also essential to express in terms of efficiencies. Various engine efficiencies are:

  (i) Air-standard efficiency
 (ii) Brake thermal efficiency
(iii) Indicated thermal efficiency
 (iv) Mechanical efficiency
  (v) Relative efficiency
 (vi) Volumetric efficiency
(vii) Scavenging efficiency
(viii) Charge efficiency
 (ix) Combustion efficiency

Most of these efficiencies have already been discussed in various chapters. Generally these efficiencies are expressed in percentage or decimal fractions. We will briefly review them again here.

14.3.1 Air-Standard Efficiency

The air-standard efficiency is also known as thermodynamic efficiency. It is mainly a function of compression ratio and other parameters. It gives the upper limit of the efficiency obtainable from an engine.

14.3.2 Indicated and Brake Thermal Efficiencies

The indicated and brake thermal efficiencies are based on the *ip* and *bp* of the engine respectively. These efficiencies give an idea of the output generated by the engine with respect to heat supplied in the form of fuel. In modern engines an indicated thermal efficiency of almost 28 per cent is obtainable with gas and gasoline spark-ignition engines having a moderate compression ratio and as high as 36 per cent or even more with high compression ratio oil engines.

14.3.3 Mechanical Efficiency

Mechanical efficiency takes into account the mechanical losses in an engine. Mechanical losses of an engine may be further subdivided into the following groups:

  (i) Friction losses as in case of pistons, bearings, gears, valve mechanisms. With the development in bearing design and materials, improvements in gears etc., these losses are usually limited from 7 to 9 per cent of the indicated output.

 (ii) Power is absorbed by engine auxiliaries such as fuel pump, lubricating oil pump, water circulating pump, radiator, magneto and distributor, electric generator for battery charging, radiator fan etc. These losses may account for 3 to 8 per cent of the indicated output.

(iii) Ventilating action of the flywheel. This loss is usually below 4 per cent of the indicated output.

(iv) Work of charging the cylinder with fresh charge and discharging the exhaust gases during the exhaust stroke. In case of two-stroke engines the power absorbed by the scavenging pump etc. These losses may account for 2 to 6 per cent of the indicated output.

In general, mechanical efficiency of engines varies from 65 to 85%.

14.3.4 Relative Efficiency

The relative efficiency or efficiency ratio as it is sometimes called is the ratio of the actual efficiency obtained from an engine to the theoretical efficiency of the engine cycle. Hence,

$$\text{Relative efficiency} \quad = \quad \frac{\text{Actual thermal efficiency}}{\text{Air-standard efficiency}}$$

Relative efficiency for most of the engines varies from 75 to 95% with theoretical air and decreases rapidly with insufficient air to about 75% with 90% air.

14.3.5 Volumetric Efficiency

Volumetric efficiency is a measure of the success with which the air supply, and thus the charge, is inducted into the engine. It is a very important parameter, since it indicates the breathing capacity of the engine. It has been discussed in detail in section 5.5.2. Volumetric efficiency is defined as *the ratio of the actual mass of air drawn into the engine during a given period of time to the theoretical mass which should have been drawn in during that same period of time, based upon the total piston displacement of the engine, and the temperature and pressure of the surrounding atmosphere.*

$$\eta_v \quad = \quad \frac{\dot{m}_{act}}{\dot{m}_{th}} \tag{14.18}$$

where,

$$\dot{m}_{th} \quad = \quad \rho_a n V_s$$

where n is the number of intake strokes per minute. For a four-stroke engine $n = N/2$ and for a two-stroke engine $n = N$, where N is the speed of the engine in rev/min. The actual mass is a measured quantity. The theoretical mass is computed from the geometry of the cylinder, the number of cylinders, and the speed of the engine, in conjunction with the density of the surrounding atmosphere.

14.3.6 Scavenging Efficiency

In case of two-stroke engines (discussed in detail in the next chapter) scavenging efficiency is defined as the ratio of the amount of air or gas-air mixture, which remains in the cylinder, at the actual beginning of the compression to the product of the total volume and air density of the inlet. Scavenging efficiency for most of the two-stroke engines varies from 40 to 95 per cent depending upon the type of scavenging provided.

14.3.7 Charge Efficiency

The charge efficiency shows how well the piston displacement of a four-stroke engine is utilized. Various factors affecting charge efficiency are:

(i) the compression ratio.
(ii) the amount of heat picked up during passage of the charge through intake manifold.
(iii) the valve timing of the engine.
(iv) the resistance offered to air-fuel charge during its passage through induction manifold.

14.3.8 Combustion Efficiency

Combustion efficiency is the ratio of heat liberated to the theoretical heat in the fuel. The amount of heat liberated is less than the theoretical value because of incomplete combustion either due to dissociation or due to lack of available oxygen. Combustion efficiency in a well adjusted engine varies from 92% to 97%.

14.4 ENGINE PERFORMANCE CHARACTERISTICS

Engine performance characteristics are a convenient graphical presentation of an engine performance. They are constructed from the data obtained during actual test runs of the engine and are particularly useful in comparing the performance of one engine with that of another. In this section some of the important performance characteristics of the SI engines are discussed.

It is to be noted that there is a certain speed, within the speed range of a particular engine, at which the charge induced per cylinder per cycle will be the maximum. At this point, the maximum force can therefore be exerted on the piston. For all practical purposes, the torque, or engine capacity to do work, will also be maximum at this point. Thus, *there is a particular engine speed at which the charge per cylinder per cycle is a maximum, and at approximately this same speed, the torque of the engine will be a maximum.*

As the speed of the engine is increased above this speed the quantity of the indicated charge will decrease. However, the power output of the engine increases with speed due to more number of cycles are executed per unit time. It should be noted that the air consumption will continue to increase with increased engine speed until some point is reached where the charge per cylinder per stroke decreases very rapidly than the number of strokes per unit time is increasing. Engines are so designed that the maximum air consumption point is not reached within the operating speed of the engine. Increase in air consumption means that increased quantities of fuel can be added per unit time increasing the power output. In fact the *ip* produced in the cylinder is almost directly proportional to the engine air consumption.

The relationship between air charge per cylinder per cycle and torque, as well as air consumption and *ip* is illustrated in Fig.14.3. Note that the maximum torque occurs at a lower speed than the maximum *ip*.

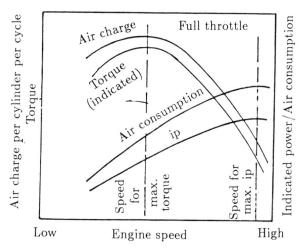

Fig.14.3 Typical Performance Plot with respect to Speed

Figure 14.4 shows some of the more important performance characteristics for a typical SI engine. In this figure, torque, *ip*, *bp* and *fp* are plotted against engine speed throughout the operating range of the engine, at full throttle and variable load.

The difference between the *ip* produced in the cylinder, and the *bp* realized at the drive-shaft, is the *fp*. At low engine speeds, the *fp* is relatively low, and *bp* is nearly as large as *ip*. As engine speed increases, however, *fp* increases at a greater rate. At engine speeds above the usual operating range, *fp* increases very rapidly. Also, at

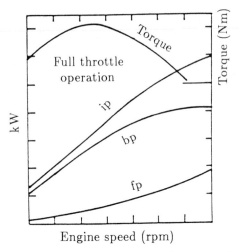

Fig.14.4 Typical SI Engine Performance Curves

these higher speeds, *ip* will reach a maximum and then fall off. At some point, *ip* and *fp* will be equal, and *bp* will then drop to zero. Note that the torque reaches a maximum at approximately 60% of the rated rpm of the engine, while the *ip* has not reached maximum even at the rated speed (3600 rpm).

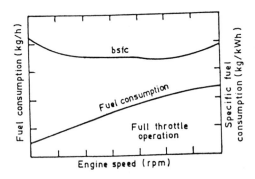

Fig.14.5 Typical Fuel Consumption Curves for an SI Engine

Figure 14.5 shows fuel consumption and *bsfc* plotted against the engine speed, for the same engine operating under the same conditions. The quantity of fuel consumed increases with engine speed. The *bsfc*, on the other hand, drops as the speed is increased in the low speed range, nearly levels off at medium speeds, and increases in the high speed range, At low speeds, the heat loss to the combustion chamber walls is proportionately greater and combustion efficiency is poorer, resulting in higher fuel consumption for the power produced. At the high speeds, the *fp* is increasing at a rapid rate, resulting in a slower increase in *bp* than in fuel consumption, with a consequent increase in *bsfc*.

The *bsfc* curve of Fig.14.5 is for full throttle, variable speed operation. At any one speed, it represents the *bsfc* which will result when the engine is carrying its maximum load at that speed. By reducing throttle opening and load, that same speed may be obtained, but at loads less than the maximum. A family of curves for various speeds can be obtained, each showing the effect on *bsfc* of varying the load at constant speed. Under these conditions of constant speed and variable load, and at a constant air-fuel ratio, the *bsfc* will rise consistently and rapidly as the load (and throttle opening) is decreased. Figure 14.6 illustrates the general shape of the curve for any given rpm. The reason for the rapid increase in *bsfc* with the reduction in throttle opening is that the *fp* remains essentially constant, while the *ip* is being reduced. The *bp* drops more rapidly than fuel consumption, and thereby the *bsfc* rises.

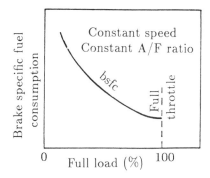

Fig.14.6 *bsfc* Curve at Constant Speed and Variable Load

Performance curves can be constructed for other operating factors such as *imep*, *bmep*, air consumption etc. However, the curves presented are typical, and are among the more important. Probably the most important of these are the curves of torque, *bp* and *bsfc* plotted against engine speed at full throttle operation. These curves are the ones most generally published by engine manufacturers with the descriptive literature on their engine models. Such a plot would look similar to Fig.14.7. If one is looking for an engine of a particular capacity, he can readily determine which engine model approximate meets his needs by referring to these curves as published by the various manufacturers. Once the model which suit the requirements has

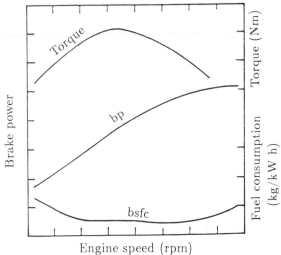

Fig.14.7 Variation of *bsfc*, Torque and *bp*
with respect to Speed for an SI Engine

been located, more comprehensive performance curves may then be studied to determine which engine is the most desirable for selection.

14.5 VARIABLES AFFECTING PERFORMANCE CHARACTERISTICS

In the preceding section, engine performance curves were discussed. The shape of these curves, or the engine performance is determined by the regulation of many design and operating variables. Some of the important variables will be briefly discussed and summarized in this section.

14.5.1 Combustion Rate and Spark Timing

The spark should be timed and the combustion rate controlled such that the maximum pressure occurs as close to the beginning of the power stroke as possible, consistent with a smooth running engine. As a general rule, the spark timing and combustion rate are regulated in such a way that approximately one half of the total pressure rise due to combustion has occurred as the piston reaches TDC on the compression stroke.

14.5.2 Air-Fuel Ratio

This ratio must be set to fulfill engine requirements. Consistent with these requirements, however, it is usually set as close as possible to the best economy proportions during normal cruising speeds, and as close as possible to the best power proportions when maximum performance is required.

14.5.3 Compression Ratio

An increase in compression ratio increases the thermal efficiency, and is, therefore, generally advantageous. The compression ratio in most SI engines is limited by knock, and the use of economically feasible antiknock quality fuels. Increasing compression ratio also increases the friction of the engine, particularly between piston rings and the cylinder walls, and there is a point at which further increase in compression ratio would not be profitable, though this point appears to be rather high.

14.5.4 Engine Speed

At low speeds, a greater length of time is available for heat transfer to the cylinder walls and therefore a greater proportion of heat loss occurs. Up to a certain point, higher speeds produce greater air consumption and therefore greater *ip*. Higher speeds, however, are accompanied by rapidly increasing *fp* and by greater inertia in the moving parts. Consequently, the engine speed range must be a compromise, although most present day designs appear to favour the higher speeds.

14.5.5 Mass of Inducted Charge

The greater the mass of the charge inducted, the higher the power produced. For a given engine, the geometry is fixed, and it is desirable to induct a charge to a maximum possible density giving the highest volumetric efficiency.

14.5.6 Heat Losses

It should be noted that the large proportion of the available energy is lost in a non-usable form, i.e., heat losses. Any method which can be employed to prevent the excessive heat loss and cause this energy to leave the engine is a usable form will tend to increase engine performance. Higher coolant temperatures, for instance, provide a smaller temperature gradient around combustion chamber walls and a reduction in heat loss, but are limited by the possibility of damage to engine parts.

14.6 METHODS OF IMPROVING ENGINE PERFORMANCE

The engine designer is always interested in methods through which engine performance may be improved. By referring to Fig.1.12, it can be seen that there are two general areas in which methods can be utilized to improve performance:

(i) the energy put into the engine at the start may be increased, and/or

(ii) the efficiency with which the fuel energy is converted to mechanical energy may be increased (Areas A and B of Fig.1.12)

Energy supply may be increased by increasing the mass of charge entering the combustion chamber. Supercharging is one method of accomplishing this. Larger piston displacement is another solution, but is limited by engine weight and cooling problems. Improvement in volumetric efficiency would also increase the mass of charge. Higher engine speeds may be utilized, but these result in increased friction losses, and above a certain point, in lowered volumetric efficiency. Improvements in fuels resulting in greater usable energy content without detonation would also help.

The use of higher compression ratios would increase the efficiency of conversion of the energy in the fuel into useful mechanical energy. This requires the development of economically feasible higher anti-knock quality fuels. Even with such fuels, as pointed out earlier, there appears to be a limit to the advantage in increasing the compression ratio. Another solution would be to reduce the losses between the air cycle and the actual cycle, and thereby increase the proportion of energy which can be mechanically utilized.

Also, it is possible to take advantage of the kinetic energy in the exhaust gas to increase the engine output through use of exhaust driven turbines. In this case, the exhaust gas from engine cylinders drives a turbine which is connected to the engine crankshaft, thus increasing engine output. Engines having this type of power booster are known as *turbocompound engines*.

Many of the parameters entering into the performance of four-stroke CI engine are similar to those already analysed for SI engines. Hence, the performance characteristics of CI engines are not discussed separately.

14.7 HEAT BALANCE

Energy supplied to an engine is the heat value of the fuel consumed. As has been repeatedly pointed out, only a part of this energy is transformed into useful work. The rest of it is either wasted or utilized in special application like turbocompounding. The two main parts of the heat not available for work are the heat carried away by the exhaust gases and the cooling medium. Figure 14.8 illustrates the same for spark-ignition engines. A typical heat balance for compression-ignition engines is illustrated in Fig.14.9.

To give sufficient data for the preparation of a heat balance sheet, a test should include a method of determining the friction power and the measurement of speed, load, fuel consumption, air consumption, exhaust temperature, rate of flow of cooling water and its temperature

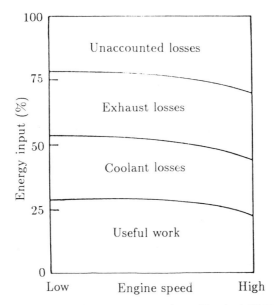

Fig.14.8 Heat Balance Diagram for a Typical SI Engine

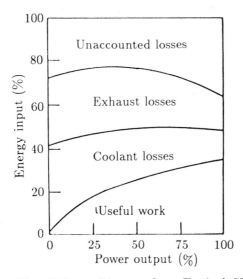

Fig.14.9 Heat Balance Diagram for a Typical CI Engine

rise while flowing through the water jackets. Besides, the small losses, such as radiation and incomplete combustion, the above enumerated data makes it possible to account for the heat supplied by the fuel and indicate its distribution.

It may be argued that same amount of frictional power, is accounted in the rise of cooling water temperature and lubricating oil temperature etc. However, it is taken into account here to show that the frictional losses also include blowdown and pumping losses and therefore it is not appropriate to put it in the heat balance. Since, there are always certain losses which cannot be accounted for, by including fp in the heat balance, the unaccounted losses will reduce.

The heat balance may be external or internal. Typical external heat balance is shown in Fig.14.10. Usually the amount of heat carried by lubricating oil is comparatively small and are normally not included.

A further method of representing heat balance is by means of the Sankey diagram. This is a stream type diagram in which the width of the stream represents the heat quantity being considered, usually as a percentage of the heat supplied, as shown in Fig.14.11.

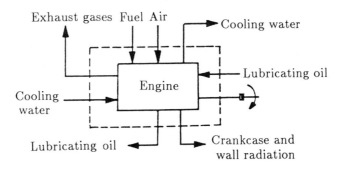

Fig.14.10 External Heat Balance

It may be observed that the diagram starts at the bottom with a stream width which represents the heat input from the fuel which is 100% of the heat input and is marked as such. Moving up the diagram, first the coolant loss stream is let off to the left. The width of this stream represents the percentage loss to the coolant. Still higher the exhaust loss stream is let off to the left and finally the loss to the surroundings appears. The loss streams finally meet a single loss stream as shown of the original vertical stream, only the brake power output stream is left at the top of the diagram. The figures on the diagram are percentages of the heat supplied in the fuel. A more detailed diagram for a spark-ignition engine is shown in Fig.14.12.

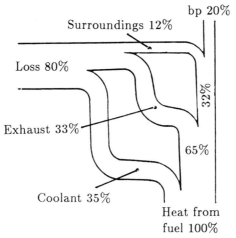

Fig.14.11 Sankey Diagram

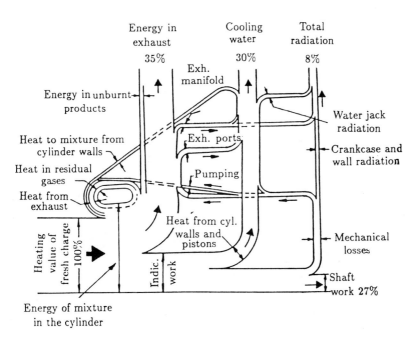

Fig.14.12 Sankey Diagram for an SI Engine

In this case, actually the heat distribution is much more involved. During the suction or scavenging period the entering fresh charge receives heat from the residual gases and from the cylinder walls. Part of the indicated work on the piston is transformed through friction into heat, which goes mostly into the cooling medium but partly into the exhaust gases and lubricating oil and also is dissipated through the crankcase to the surrounding air.

14.8 PERFORMANCE MAPS

For critical analysis the performance of an IC engine under all conditions of load and speed is shown by a performance map. Figure 14.13 shows the performance map of an automotive SI engine and Fig.14.14 the performance map of a four-stroke prechamber CI engine. Figure 14.13 also includes a typical curve of *bmep* vs piston speed for level road operation in high gear. Note that these maps can be used for comparing different sized engines, as performance parameters have been generalized by converting rpm into piston speed and power per unit piston area.

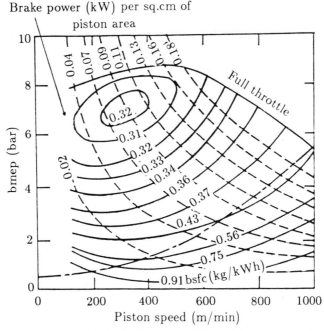

Fig.14.13 Generalized Performance Map of Automotive SI Engine

Generally speaking, all the engines show a region of lowest specific fuel consumption (highest efficiency) at a relatively low piston speed with a relatively high *bmep*.

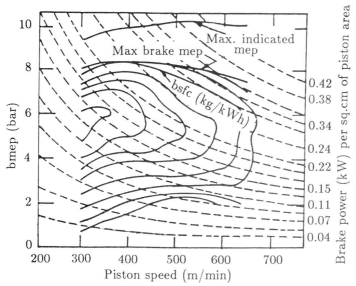

Fig.14.14 Performance Map of a Four-Stroke
Prechamber Diesel Engine

14.8.1 SI Engines

Constant Speed Line : Reduced $bsfc$ results by moving upward along constant speed line, because of mixture enrichment at high load which more than offsets increase in mechanical efficiency. Moving to lower $bmep$s, the $bsfc$ increases because of the reduced mechanical efficiency ($imep$ decreases while $fmep$ remains constant).

Constant $bmep$ Line : Moving from the region of highest efficiency along a line of constant $bmep$, the $bsfc$ increases due to increased friction at higher piston speeds. Moving to the left towards lower piston speed, although friction mep decreases, indicated efficiency falls off owing to poor fuel distribution and increased heat losses.

14.8.2 CI Engines

In the CI engine the $bsfc$ increases at high loads owing to the increased fuel waste (smoke) associated with high fuel-air ratios. At lower load $bsfc$ increases due to decrease in mechanical efficiency (same as in the SI engine).

As the speed is reduced from the point of best economy along a line of constant $bmep$, the product of mechanical and indicated thermal efficiency appears to remain a constant down to the lowest operating speed. The reduction in $fmep$ with speed is apparently balanced by a reduction is indicated thermal efficiency due to poor spray characteristics at very low speeds.

An interesting feature of the performance curves is that they show the power at maximum economy is about half of the maximum power.

14.9 ANALYTICAL METHOD OF PERFORMANCE ESTIMATION

Performance data of engine obtained from theoretical analysis are very useful for the design of a new engine. Due to complex nature of the processes taking place in an engine (such as combustion with variable specific heat and dissociation, mixing of different gases, heat transfer, etc.) the theoretical calculations are rather very difficult and some simplifying assumptions are to be made. Therefore, results obtained from theoretical calculations must be compared with experimental results obtained from engine of similar design for validation. The theoretical results are accepted only when they are reasonably close to the experimental results.

The brake output of an engine depends on brake mean effective pressure, *bmep*, and the piston speed, $\bar{s}_p$. Brake mean effective pressure depends on indicated mean effective pressure, *imep*, and the frictional mean effective pressure, *fmep*. When the speed of the engine changes from N_1 to N_2, the piston speed changes from $\bar{s}_{p1}$ to $\bar{s}_{p2}$ and the brake output changes from bp_1 to bp_2. Now the ratio of bp_2 to bp_1 can be written as

$$\frac{bp_2}{bp_1} = \frac{imep_2 - fmep_2}{imep_1 - fmep_1} \times \frac{\bar{s}_{p2}}{\bar{s}_{p1}} \qquad (14.19)$$

where bp_1 and bp_2 are the power output in kW at speeds N_1 and N_2 respectively.

The ratio of indicated mean effective pressure at N_2 and N_1 is given by

$$\epsilon = \frac{imep_2}{imep_1} = \frac{\rho_{in_2}\,\eta_{v_2}\,(\dot{m}_f\eta_{ith})_2}{\rho_{in_1}\,\eta_{v_1}\,(\dot{m}_f\eta_{ith})_1} \qquad (14.20)$$

Dividing numerator and denominator on the right hand side of Eqn.14.19 by $imep_1$, we get

$$\frac{bp_2}{bp_1} = \frac{\dfrac{imep_2}{imep_1} - \dfrac{fmep_2}{fmep_1}}{1 - \dfrac{fmep_1}{imep_1}} \times \frac{\bar{s}_{p2}}{\bar{s}_{p1}}$$

$$= \frac{\epsilon - \dfrac{fmep_2}{imep_1}}{1 - \dfrac{fmep_1}{imep_1}} \times \frac{\bar{s}_{p2}}{\bar{s}_{p1}} \qquad (14.21)$$

From the above relation the power output at condition 2 (bp_2 at mean effective pressure, $imep_2$ and speed, N_2) can be obtained if the bp_1 at $imep_1$ and N_1 are known. In order to determine the complete performance at condition 2, the brake specific fuel consumption at condition 2 should be determined. The brake specific fuel consumption, $bsfc$, which indicates the economy of the engine operation, is related to the indicated specific fuel consumption, $isfc$ by the following relation:

$$bsfc \;=\; \frac{isfc}{\frac{bmep}{imep}} \;=\; \frac{isfc}{1 - \frac{fmep}{imep}} \tag{14.22}$$

Therefore, the ratio of specific fuel consumption at condition 2 to that at condition 1 is given by the following relation,

$$\frac{bsfc_2}{bsfc_1} \;=\; \frac{isfc_2}{isfc_1}\left(\frac{1 - \frac{fmep_1}{imep_1}}{1 - \frac{fmep_2}{imep_2}}\right)$$

$$=\; \frac{isfc_2}{isfc_1}\left(1 - \frac{\frac{fmep_1}{imep_1}}{\frac{fmep_2}{\epsilon \times imep_1}}\right) \tag{14.23}$$

When engine speed is constant, power developed is proportional to $bmep_2$ and therefore, bp_2 can be obtained from the relation

$$\frac{bp_2}{bp_1} \;=\; \frac{bmep_2}{bmep_1} \;=\; \frac{imep_2 - fmep_2}{imep_1 - fmep_1} \tag{14.24}$$

For four-stroke cycle unsupercharged engine, the change in $fmep$ with change in power output at constant speed is negligibly small.

Therefore, we can assume

$$fmep_1 \;=\; fmep_2 \;=\; fmep$$

Therefore, the ratio of brake outputs is given by

$$\frac{bp_2}{bp_1} \;=\; \frac{\frac{imep_2}{imep_1} - \frac{fmep}{imep_1}}{1 - \frac{fmep}{imep_1}} \;=\; \frac{\epsilon - \frac{fmep}{imep_1}}{1 - \frac{fmep}{imep_1}} \tag{14.25}$$

$$\frac{bsfc_2}{bsfc_1} \;=\; \frac{isfc_2}{isfc_1}\left(\frac{1 - \frac{fmep}{imep_1}}{1 - \frac{fmep}{imep_2}}\right)$$

$$=\; \frac{isfc_2}{isfc_1}\left(\frac{1 - \frac{fmep}{imep_1}}{1 - \frac{fmep}{\epsilon \times imep_1}}\right) \tag{14.26}$$

Value of air capacity is given by the product of inlet density and volumetric efficiency $(\rho_{in} \times \eta_v)$ for a spark-ignition engine and it increases with increase in average piston speed and reaches a maximum value at a particular piston speed. The indicated thermal efficiency is nearly constant with increasing speed if the fuel-air ratio and the spark timing are adjusted to best power condition at each speed. Therefore, the ratio of indicated mean effective pressures, *imep*, is approximately equal to the ratio of air capacities at different speeds. The mean effective pressure, due to mechanical losses, *fmep* changes with change in average piston speed. The variation of *fmep* with variation of piston speed for different designs of engine are obtained from motoring tests. The values of $fmep_1$ and $fmep_2$ are taken from the test data of similar engine at piston speeds $\bar{s}_{p1}$ and $\bar{s}_{p2}$. The air capacities and *fmep* at two condition being known, the power output and brake specific fuel consumption at condition 2 can be calculated (by Eqns.14.21 and 14.23). When the power output of the engine is decreased at a constant speed, the value of $\rho_{in}\eta_v$ decreases appreciably (resistance at inlet increases due to throttling). The indicated thermal efficiency, η_{ith}, also decreases and the change in indicated thermal efficiency being known, the ratio of indicated mean effective pressures, ϵ can be calculated. The power output and brake specific fuel consumption at condition 2 can be calculated using Eqns.14.25 and 14.26.

When the speed of a compression-ignition engine increases the value of air capacity $\rho_{in}\eta_v$ first increases and then decreases after reaching its maximum value. The indicated thermal efficiency η_{ith} increases with increase in speed as long as the fuel feed system operates satisfactorily and the air capacity remains sufficiently high. Therefore, the ratio of indicated mean effective pressures, ϵ increases with speed upto a particular value. The *fmep* also increases with speed of the engine (average piston speed). The variation of *fmep* with piston speed is obtained from motoring tests. Knowing the value of ϵ and $fmep_2$ the power output and brake specific fuel consumption at condition 2 can be calculated.

When the power developed at a constant speed is reduced by decreasing the amount of fuel injected the volumetric efficiency and indicated thermal efficiency increases with decrease in power developed. Therefore, though the fuel-air ratio decreases, the net effect is an increase in the indicated mean effective pressure. Therefore, the value of ϵ increases with decrease in load at constant speed. The *fmep* at constant speed and varying power output (varying load) practically remains constant and knowing the value of ϵ output and brake specific fuel consumption at condition 2 can be determined.

Other factors (except the speed and load of the engine) commonly affecting engine performance are the atmospheric conditions (pressure, temperature and humidity), the fuel-air ratio and the compression ratio. The effects of variation of these factors on the volumetric efficiency, indicated thermal efficiency and mechanical efficiency within the range of speed and output of the engine have been discussed in previous chapters.

If the fuel-air ratio is maintained nearly constant during change of inlet conditions (other factors remain constant), the indicated thermal efficiency may be assumed constant. Therefore, the indicated mean effective pressure will depend on $\rho_a \times \eta_v$. Under this condition, the volumetric efficiency is inversely proportional to $\sqrt{T_0}$ ($p_{in} = p_{ex}$ at all conditions). Therefore, $\rho_a \times \eta_v$ is proportional to $p_0/T_0 \times \sqrt{T_0}$ or $p_0/\sqrt{T_0}$. Therefore,

$$\frac{imep_2}{imep_1} = \frac{p_{02}}{p_{01}} \left(\frac{T_{01}}{T_{02}}\right)^{\frac{1}{2}} = \epsilon \qquad (14.27)$$

Using this value of ϵ the performance of the engine at condition 2 can be obtained if their performance at rated condition is known.

The effect of altitude on engine performance is important when the engine operates at different heights in mountainous region or for the aircraft engines. With change in altitude both the inlet pressure and temperature (also the exhaust pressure) change. Therefore, their performance using constant fuel-air ratio can be obtained as in the previous case.

Fuel-air ratio has a marked effect on the indicated thermal efficiency of an engine. Its effect on volumetric efficiency, is however, small (may be neglected). Therefore,

$$\frac{imep_2}{imep_1} = \frac{(\dot{m}_f \, \eta_{ith})_2}{(\dot{m}_f \, \eta_{ith})_1} = \epsilon \qquad (14.28)$$

when all other conditions remain the same but the fuel-air ratio of the charge is changed. The performance at condition 2 can be obtained if the performance at conditions 1 is known.

Compression ratio has significant effect on indicated thermal efficiency. Therefore, when the compression ratio is changed all other conditions of operation remaining the same, the ratio of indicated mean effective pressures is given by,

$$\frac{imep_2}{imep_1} = \frac{(\eta_{ith})_2}{(\eta_{ith})_1} = \epsilon \qquad (14.29)$$

Worked out Examples

14.1. A four-stroke gas engine has a bore of 20 cm and stroke of 30 cm and runs at 300 rpm firing every cycle. If air-fuel ratio is 4:1 by volume and volumetric efficiency on NTP basis is 80%, determine the volume of gas used per minute. If the calorific value of the gas is 8 MJ/m^3 at NTP and the brake thermal efficiency is 25% determine the brake power of the engine.

SOLUTION:

Swept volume, $V_s \quad = \quad \dfrac{\pi}{4} D^2 L \quad = \quad \dfrac{\pi}{4} \times 20^2 \times 30$

$$= \quad 9424.8 \text{ cc}$$

Total charge taken in per cycle

$$\dot{V}_C \quad = \quad 0.8 \times 9424.8 \quad = \quad 7.54 \times 10^{-3} \text{ m}^3$$

Volume of gas used per minute

$$\dot{V}_g \quad = \quad \frac{7.54 \times 10^{-3}}{4 + 1} \times \frac{300}{2}$$

$$= \quad 0.2262 \text{ m}^3 \text{ at NTP/min}$$

Heat input $\quad = \quad 8000 \times 0.2262 \quad = \quad 1809.6 \text{ kJ/min}$

$bp \quad = \quad \eta_{bth} \times \text{Heat input} \quad = \quad \dfrac{0.25 \times 1809.6}{60}$

$$= \quad \textbf{7.54 kW} \qquad \text{Ans}$$

14.2. The following observations have been made from the test of a four-cylinder, two-stroke gasoline engine. Diameter = 10 cm; stroke = 15 cm; speed = 1600 rpm; Area of the positive loop of the indicator diagram = 5.75 sq cm; Area of the negative loop of the indicator diagram = 0.25 sq cm; Length of the indicator diagram = 55 mm; Spring constant = 3.5 bar/cm; Find the indicated power of the engine.

SOLUTION:

Net area of diagram $\quad = \quad 5.75 - 0.25$

$$= \quad 5.5 \text{ cm}^2$$

Average height of the diagram

$$= \frac{5.5}{5.5} = 1 \text{ cm}$$

$$p_{im} = \text{Average height of the diagram} \times$$

$$\text{Spring constant}$$

$$= 1 \times 3.5 = 3.5 \text{ bar}$$

$$\dot{i}p = \frac{p_{im} L A n K}{60000}$$

$$= \frac{3.5 \times 10^5 \times 0.15 \times \frac{\pi}{4} \times 0.1^2 \times 1600 \times 4}{60000}$$

$$= \textbf{43.98 kW} \qquad \underset{\Longleftarrow}{\textbf{Ans}}$$

14.3. A single-cylinder engine running at 1800 rpm develops a torque of 8 Nm. The indicated power of the engine is 1.8 kW. Find the loss due to friction power as the percentage of brake power.

SOLUTION:

$$\textit{Brake power} = \frac{2\pi NT}{60000} = \frac{2 \times \pi \times 1800 \times 8}{60000}$$

$$= 1.508 \text{ kW}$$

$$\textit{Friction power} = 1.8 - 1.508 = 0.292$$

$$\textit{Percentage loss} = \frac{0.292}{1.8} \times 100$$

$$= \textbf{16.22\%} \qquad \underset{\Longleftarrow}{\textbf{Ans}}$$

14.4. A gasoline engine working on Otto cycle consumes 8 litres of gasoline per hour and develops 25 kW. The specific gravity of gasoline is 0.75 and its calorific value is 44000 kJ/kg. Determine the indicated thermal efficiency of the engine.

SOLUTION:

$$\eta_{ith} = \frac{\text{Heat equivalent of ip}}{\text{Heat input}}$$

$$= \frac{25 \times 60 \times 60}{(8 \times 10^{-3} \times 750) \times 44000} \times 100$$

$$= \quad 34.1\% \qquad\qquad \underset{\Longleftarrow}{\text{Ans}}$$

14.5. A four cylinder engine running at 1200 rpm delivers 20 kW. The average torque when one cylinder was cut is 110 Nm. Find the indicated thermal efficiency if the calorific value of the fuel is 43 MJ/kg and the engine uses 360 grams of gasoline per kW h.

SOLUTION:

$$\text{Average bp for 3 cylinders} = \frac{2\pi NT}{60000}$$

$$= \frac{2\pi \times 1200 \times 110}{60000}$$

$$= \quad 13.82 \text{ kW}$$

$$\text{Average ip with 1 cylinder} = 20 - 13.82$$

$$= \quad 6.18 \text{ kW}$$

$$\text{Total ip} = 4 \times 6.18 = 24.72 \text{ kW}$$

$$isfc = bsfc \times \frac{bp}{ip} = 360 \times \frac{20}{24.72}$$

$$= \quad 291.26 \text{ g/kW h}$$

$$\text{Fuel consumption} = \frac{isfc \times ip}{3600 \times 1000}$$

$$= \frac{291.26 \times 24.72}{3600 \times 1000}$$

$$= \quad 2 \times 10^{-3} \text{ kg/s}$$

$$\eta_{ith} = \frac{ip}{\dot{m}_f \times CV}$$

$$= \frac{24.72}{2 \times 10^{-3} \times 43000} \times 100$$

$$= 28.74\% \qquad \text{Ans}$$

14.6. The bore and stroke of a water-cooled, vertical, single-cylinder, four-stroke diesel engine are 80 mm and 110 mm respectively and the torque is 23.5 Nm. Calculate the brake mean effective pressure of the engine.

SOLUTION:

$$P = \frac{2\pi NT}{60000} = \frac{p_{bm} LAn}{60000}$$

$$p_{bm} = \frac{2\pi NT}{LAn} = \frac{2\pi NT}{L \times \frac{\pi}{4} \times D^2 \frac{N}{2}} = \frac{16T}{D^2 L}$$

$$= \frac{16 \times 23.5}{0.08^2 \times 0.11} = 5.34 \times 10^5 \text{ Pa}$$

$$= 5.34 \text{ bar} \qquad \text{Ans}$$

14.7. Find the mean effective pressure and torque developed by the engine in the previous problem if its rating are 4 kW at 1500 rpm.

SOLUTION:

$$p_{bm} = \frac{P \times 60000}{\frac{\pi}{4} D^2 L \frac{N}{2}} \times 10^{-5} \text{ bar}$$

$$= \frac{4 \times 60000}{\frac{\pi}{4} \times 0.08^2 \times 0.11 \times \frac{1500}{2}} \times 10^{-5}$$

$$= 5.78 \text{ bar} \qquad \text{Ans}$$

$$T = \frac{P \times 60000}{2\pi N} = \frac{4 \times 60000}{2 \times \pi \times 1500}$$

$$= 25.46 \text{ Nm} \qquad \text{Ans}$$

14.8. Find the brake specific fuel consumption in kg/kW h of a diesel engine whose fuel consumption is 5 grams per second when the power output is 80 kW. If the mechanical efficiency is 75%, calculate the indicated specific fuel consumption.

SOLUTION:

$$bsfc = \frac{\dot{m}_f}{bp} = \frac{5}{80} = 0.0625 \text{ g/kW s}$$

$$= \frac{0.0625}{1000} \times 3600$$

$$= \textbf{0.225 kg/kW h} \qquad \underset{\rightleftharpoons}{\textbf{Ans}}$$

$$isfc = bsfc \times \eta_m = 0.225 \times 0.75$$

$$= \textbf{0.169 kg/kW h} \qquad \underset{\rightleftharpoons}{\textbf{Ans}}$$

14.9. For engine in the previous problem find the brake specific energy consumption, *bsec*, given the fuel consumption 5.55 g/s and the lower heating value of the fuel as 43 MJ/kg. Find also the indicated specific energy consumption.

SOLUTION:

$$bsec = \frac{kW \text{ heat input}}{kW \text{ work output}}$$

$$= \frac{CV \times \dot{m}_f}{P} = CV \times bsfc$$

$$bsfc = \frac{5.55}{80} = 0.069 \text{ g/kW s}$$

$$= 0.069 \times 10^{-3} \text{ kg/kW s}$$

$$CV = 43 \text{ MJ/kg} = 43 \times 10^3 \text{ kJ/kg}$$

$$bsec = bsfc \times CV = 43 \times 10^3 \times 0.069 \times 10^{-3}$$

$$= \textbf{2.97} \qquad \underset{\rightleftharpoons}{\textbf{Ans}}$$

$$isec = bsec \times \eta_m = 2.97 \times 0.75$$

$$= \textbf{2.23} \qquad \underset{\rightleftharpoons}{\textbf{Ans}}$$

14.10. Find the air-fuel ratio of a four-stroke, single-cylinder, air-cooled engine with fuel consumption time for 10 cc as 20.4 s and air consumption time for 0.1 m^3 as 16.3 s. The load is 17 kg at the speed of 3000 rpm. Find also brake specific fuel consumption in g/kW h and brake thermal efficiency. Assume the density of air as 1.175 kg/m^3 and specific gravity of fuel to be 0.7. The lower heating value of fuel is 43 MJ/kg and the dynamometer constant is 5000.

SOLUTION:

$$Air\ consumption\ =\ \frac{0.1}{16.3} \times 1.175\ =\ 7.21 \times 10^{-3}\ kg/s$$

$$Fuel\ consumption\ =\ \frac{10}{20.4} \times 0.7 \times \frac{1}{1000}$$

$$=\ 0.343 \times 10^{-3}\ kg/s$$

$$Air\text{-}fuel\ ratio\ =\ \frac{7.21 \times 10^{-3}}{0.343 \times 10^{-3}}$$

$$=\ \mathbf{21} \qquad\qquad \underset{\Leftarrow}{\text{Ans}}$$

$$Power\ output,\ P\ =\ \frac{WN}{5000}\ =\ \frac{7 \times 3000}{5000}\ =\ 4.2\ kW$$

$$bsfc\ =\ \frac{Fuel\ consumption\ (g/h)}{Power\ output}$$

$$=\ \frac{0.343 \times 10^{-3} \times 3600 \times 1000}{4.2}$$

$$=\ \mathbf{294\ g/kW\ h} \qquad\qquad \underset{\Leftarrow}{\text{Ans}}$$

$$\eta_{bth}\ =\ \frac{4.2}{0.343 \times 10^{-3} \times 43000} \times 100$$

$$=\ \mathbf{28.48\%} \qquad\qquad \underset{\Leftarrow}{\text{Ans}}$$

14.11. A six-cylinder, gasoline engine operates on the four-stroke cycle. The bore of each cylinder is 80 mm and the stroke 100 mm. The clearance volume per cylinder is 70 cc. At a speed of 4000 rpm the fuel consumption is 20 kg/h and the torque developed is 150 Nm.

Calculate (i) the brake power (ii) the brake mean effective pressure (iii) brake thermal efficiency if the calorific value of the fuel is 43000 kJ/kg and (iv) the relative efficiency on a brake power basis assuming the engine works on the constant volume cycle. $\gamma = 1.4$ for air.

SOLUTION:

$$bp = \frac{2\pi NT}{60000} = \frac{2 \times \pi \times 4000 \times 150}{60000}$$

$$= \textbf{62.8 kW} \qquad \underset{\Longleftarrow}{\textbf{Ans}}$$

$$p_{bm} = \frac{bp \times 60000}{LAnK}$$

$$= \frac{62.8 \times 60000}{0.1 \times \frac{\pi}{4} \times 0.08^2 \times \frac{4000}{2} \times 6}$$

$$= 6.25 \times 10^5 \text{ Pa}$$

$$= \textbf{6.25 bar} \qquad \underset{\Longleftarrow}{\textbf{Ans}}$$

$$\eta_{bth} = \frac{bp}{\dot{m}_f \times CV} = \frac{62.8 \times 3600}{20 \times 43000} \times 100$$

$$= \textbf{26.3\%} \qquad \underset{\Longleftarrow}{\textbf{Ans}}$$

$$\text{Compression ratio, } r = \frac{V_s + V_{cl}}{V_{cl}}$$

$$V_s = \frac{\pi}{4}D^2 L = \frac{\pi}{4} \times 8^2 \times 10$$

$$= 502.65 \text{ cc}$$

$$r = \frac{502.65 + 70}{70} = 8.18$$

$$\text{Air-std. efficiency, } \eta_{Otto} = 1 - \frac{1}{8.18^{0.4}} = 0.568$$

$$\text{Relative efficiency, } \eta_{rel} = \frac{0.263}{0.568} \times 100$$

$$= \textbf{46.3\%} \qquad \underset{\Longleftarrow}{\textbf{Ans}}$$

14.12. An eight-cylinder, four-stroke engine of 9 cm bore and 8 cm stroke with a compression ratio of 7 is tested at 4500 rpm on a dynamometer which has 54 cm arm. During a 10 minutes test the dynamometer scale beam reading was 42 kg and the engine consumed 4.4 kg of gasoline having a calorific value of 44000 kJ/kg. Air 27 °C and 1 bar was supplied to the carburettor at the rate of 6 kg/min. Find (i) the brake power delivered (ii) the brake mean effective pressure (iii) the brake specific fuel consumption (iv) the brake specific air consumption (v) the brake thermal efficiency (vi) the volumetric efficiency and (vii) the air-fuel ratio.

SOLUTION:

$$bp \quad = \quad \frac{2\pi NT}{60000}$$

$$= \quad \frac{2 \times \pi \times 4500 \times 42 \times 0.54 \times 9.81}{60000}$$

$$= \quad \textbf{104.8 kW} \qquad \underset{=}{\textbf{Ans}}$$

$$bmep \quad = \quad \frac{bp \times 60000}{LAnK}$$

$$= \quad \frac{104.8 \times 60000}{0.08 \times \frac{\pi}{4} \times 0.09^2 \times \frac{4500}{2} \times 8}$$

$$= \quad 6.87 \times 10^5 \text{ Pa}$$

$$= \quad \textbf{6.87 bar} \qquad \underset{=}{\textbf{Ans}}$$

$$bsfc \quad = \quad \frac{\frac{4.4}{10} \times 60}{104.8}$$

$$= \quad \textbf{0.252 kg/kW h} \qquad \underset{=}{\textbf{Ans}}$$

$$bsac \quad = \quad \frac{6 \times 60}{104.8}$$

$$= \quad \textbf{3.435 kg/kW h} \qquad \underset{=}{\textbf{Ans}}$$

$$\eta_{bth} \quad = \quad \frac{bp}{\dot{m}_f \times CV}$$

$$= \frac{104.9 \times 60}{\frac{4.4}{10} \times 44000} \times 100$$

$$= \mathbf{32.5\%} \qquad \underset{\rightleftharpoons}{\mathbf{Ans}}$$

Volume flow rate of air at intake condition

$$\dot{V}_a = \frac{\dot{m}_a RT}{p}$$

$$= \frac{6 \times 287 \times 300}{1 \times 10^5}$$

$$= 5.17 \text{ m}^3/\text{min}$$

Swept volume per minute, $V_s = \frac{\pi}{4} D^2 LnK$

$$= \frac{\pi}{4} \times 0.09^2 \times 0.08 \times \frac{4500}{2} \times 8$$

$$= 9.16 \text{ m}^3/\text{min}$$

Volumetric efficiency, $\eta_v = \frac{5.17}{9.16} \times 100$

$$= \mathbf{56.44\%} \qquad \underset{\rightleftharpoons}{\mathbf{Ans}}$$

Air-fuel ratio, $A/F = \frac{6.0}{0.44}$

$$= \mathbf{13.64} \qquad \underset{\rightleftharpoons}{\mathbf{Ans}}$$

14.13. The following details were noted in a test on a four-cylinder, four-stroke engine, diameter = 100 mm; stroke = 120 mm; speed of the engine = 1600 rpm; fuel consumption = 0.2 kg/min; calorific value of fuel is 44000 kJ/kg; difference in tension on either side of the brake pulley = 40 kg; brake circumference is 300 cm. If the mechanical efficiency is 80%, calculate (i) brake thermal efficiency (ii) indicated thermal efficiency (iii) indicated mean effective pressure and (iv) brake specific fuel consumption

SOLUTION:

$$bp = \frac{2\pi NT}{60000} = \frac{2\pi NWR}{60000} = \frac{WN2\pi R}{60000}$$

$$= \frac{40 \times 9.81 \times 1600 \times 3}{60000} \times 3 \quad = \quad 31.39 \text{ kW}$$

$$\eta_{bth} \quad = \quad \frac{bp}{\dot{m}_f \times CV} \times 100 \quad = \quad \frac{31.39 \times 60}{0.2 \times 44000} \times 100$$

$$= \quad 21.40\% \qquad \qquad \underset{\Longleftarrow}{\text{Ans}}$$

$$\eta_{ith} \quad = \quad \frac{\eta_{bth}}{\eta_m} \times 100 \quad = \quad \frac{21.40}{0.80} \times 100$$

$$= \quad 26.75\% \qquad \qquad \underset{\Longleftarrow}{\text{Ans}}$$

$$imep \quad = \quad \frac{\frac{bp}{\eta_m} \times 60000}{LAnK}$$

$$= \quad \frac{\frac{31.39}{0.8} \times 60000}{0.12 \times \frac{\pi}{4}0.1^2 \times \frac{1600}{2} \times 4} \quad = \quad 7.8 \times 10^5 \text{ Pa}$$

$$= \quad \textbf{7.8 bar} \qquad \qquad \underset{\Longleftarrow}{\text{Ans}}$$

$$bsfc \quad = \quad \frac{\dot{m}_f}{bp} \quad = \quad \frac{0.2 \times 60}{31.39}$$

$$= \quad \textbf{0.382 kg/kW h} \qquad \qquad \underset{\Longleftarrow}{\text{Ans}}$$

14.14. The air flow to a four cylinder, four-stroke oil engine is measured by means of a 5 cm diameter orifice having a coefficient of discharge of 0.6. During a test on the engine the following data were recorded : bore = 10 cm; stroke = 12 cm; speed = 1200 rpm; brake torque = 120 Nm; fuel consumption = 5 kg/h; calorific value of fuel = 42 MJ/kg; pressure drop across orifice is 4.6 cm of water; ambient temperature and pressure are 17 °C and 1 bar respectively. Calculate (i) the thermal efficiency on brake power basis; (ii) the brake mean effective pressure and (iii) the volumetric efficiency based on free air condition.

SOLUTION:

$$bp \quad = \quad \frac{2\pi NT}{60000} \quad = \quad \frac{2 \times \pi \times 1200 \times 120}{60000}$$

$$= \quad 15.08 \text{ kW}$$

$$\eta_{bth} = \frac{15.08 \times 60}{\frac{5}{60} \times 42000} \times 100$$

$$= 25.85\% \qquad \overset{\text{Ans}}{\Longleftarrow}$$

$$p_{im} = \frac{bp \times 60000}{LAnK}$$

$$= \frac{15.08 \times 60000 \times 10^{-5}}{0.12 \times \frac{\pi}{4} \times 0.1^2 \times \frac{1200}{2} \times 4}$$

$$= 4 \times 10^5 \text{ Pa}$$

$$= 4 \text{ bar} \qquad \overset{\text{Ans}}{\Longleftarrow}$$

$$\text{Air flow, } \dot{V}_a = C_d A \sqrt{2g \Delta h_w \frac{\rho_w}{\rho_a}}$$

$$\rho_a = \frac{p}{RT} = \frac{10^5}{287 \times 290}$$

$$= 1.20 \text{ kg/m}^3$$

$$\dot{V}_a = 0.6 \times \frac{\pi}{4} \times 0.05^2$$

$$\times \sqrt{2 \times 9.81 \times 0.046 \times \frac{1000}{1.2}} \times 60$$

$$= 1.938 \text{ m}^3/\text{min}$$

$$\text{Swept volume} = \frac{\pi}{4} d^2 LnK$$

$$= \frac{\pi}{4} \times 0.1^2 \times 0.12 \times \frac{1200}{2} \times 4$$

$$= 2.262 \text{ m}^3/\text{min}$$

$$\eta_v = \frac{1.938}{2.262} \times 100$$

$$= 85.7\% \qquad \overset{\text{Ans}}{\Longleftarrow}$$

14.15. The following observations were recorded during a trial of a four-stroke, single-cylinder oil engine. Duration of trial is 30 min; oil consumption is 4 litres; calorific value of the oil is 43 MJ/kg; specific gravity of the fuel = 0.8; average area of the indicator diagram = 8.5 cm^2; Length of the indicator diagram = 8.5 cm; spring constant = 5.5 bar/cm; brake load = 150 kg; spring balance reading = 20 kg; effective brake wheel diameter = 1.5 m; speed = 200 rpm; cylinder diameter = 30 cm; stroke = 45 cm; jacket cooling water = 10 kg/min; temperature rise is 36 °C. Calculate (i) indicated power (ii) brake power (iii) mechanical efficiency (iv) brake specific fuel consumption in kg/kW h and (v) indicated thermal efficiency.

SOLUTION:

$$p_{im} = \frac{Area \ of \ the \ diagram}{Length \ of \ the \ diagram} \times Spring \ const.$$

$$= \frac{8.5}{8.5} \times 5.5 = 5.5 \ bar$$

$$ip = \frac{p_{im}LAn}{60000}$$

$$= \frac{5.5 \times 10^5 \times 0.45 \times \frac{\pi}{4} \times 0.3^2 \times \frac{200}{2}}{60000}$$

$$= 29.16 \ kW \qquad \underset{=}{\underline{Ans}}$$

$$bp = \frac{2\pi NWR}{60000} = \frac{\pi NWd}{60000}$$

$$= \frac{\pi \times 200 \times (150 - 20) \times 9.81 \times 1.5}{60000}$$

$$= 20.03 \ kW \qquad \underset{=}{\underline{Ans}}$$

$$\eta_m = \frac{20.03}{29.16} \times 100$$

$$= 68.7\% \qquad \underset{=}{\underline{Ans}}$$

$$Fuel \ consumption = \frac{4}{30} \times 60 \times 10^{-3} \times 800$$

$$= 6.4 \ kg/h$$

$$bsfc = \frac{6.4}{20.03}$$

$$= \textbf{0.3195} \text{ kg/kW h} \qquad \underset{\longleftarrow}{\textbf{Ans}}$$

$$\eta_{ith} = \frac{ip}{\dot{m}_f \times CV} = \frac{29.16 \times 3600}{6.4 \times 43000} \times 100$$

$$= \textbf{38.14\%} \qquad \underset{\longleftarrow}{\textbf{Ans}}$$

14.16. A four-stroke cycle gas engine has a bore of 20 cm and a stroke of 40 cm. The compression ratio is 6. In a test on the engine the indicated mean effective pressure is 5 bar, the air to gas ratio is 6:1 and the calorific value of the gas is 12 MJ/m^3 at NTP. At the beginning of the compression stroke the temperature is 77 °C and pressure 0.98 bar. Neglecting residual gases, determine the indicated power, the thermal efficiency and the relative efficiency of the engine at 250 rpm.

SOLUTION:

$$Swept\ volume \ =\ \frac{\pi}{4} D^2 L$$

$$= \frac{\pi}{4} \times 20^2 \times 40$$

$$= 12566.4 \text{ cc}$$

$$Volume\ of\ gas\ in\ the\ cylinder \ =\ \frac{1}{1 + A/F} \times V_1$$

$$V_1 \ =\ V_s + \frac{V_s}{r-1} \ =\ V_s \times \frac{6}{5}$$

$$= \frac{1}{6+1} \times 12566.4 \times \frac{6}{5}$$

$$= 2154.24 \text{ cc/cycle}$$

Since the residual gases are to be neglected, one can assume a volumetric efficiency of 100%

$$Normal\ pressure \ =\ 1 \text{ bar}$$

$$\left(\frac{pV}{T}\right)_{NTP} \ =\ \left(\frac{p_2 V_2}{T_2}\right)_{working}$$

Volume of gas at NTP conditions

$$= \quad 2154.24 \times \frac{0.98}{1} \times \frac{273}{350}$$

$$= \quad 1646.7 \text{ cc}$$

Heat added $\quad = \quad 1646.7 \times 10^{-6} \times 12 \times 10^{3}$

$$= \quad 19.76 \text{ kJ/cycle}$$

$$ip \quad = \quad \frac{p_{im} V_s n}{60000}$$

$$= \quad \frac{5 \times 10^5 \times 12566.4 \times 10^{-6} \times \frac{250}{2}}{60000}$$

$$= \quad \textbf{13.09 kW} \qquad \underset{\Longleftarrow}{\text{Ans}}$$

$$\eta_{ith} \quad = \quad \frac{ip}{\text{Heat added } (in\ kW)} \times 100$$

$$= \quad \frac{13.09}{19.76 \times \frac{250}{2 \times 60}} \times 100$$

$$= \quad \textbf{31.8\%} \qquad \underset{\Longleftarrow}{\text{Ans}}$$

Air-standard efficiency $\quad = \quad 1 - \dfrac{1}{6^{0.4}} \quad = \quad 0.512$

Relative efficiency $\quad = \quad \dfrac{0.318}{0.512} \times 100$

$$= \quad \textbf{62.11\%} \qquad \underset{\Longleftarrow}{\text{Ans}}$$

14.17. A four-stroke, four-cylinder gasoline engine has a bore of 60 mm and a stroke of 100 mm. On test it develops a torque of 66.5 Nm when running at 3000 rpm. If the clearance volume in each cylinder is 60 cc the relative efficiency with respect to brake thermal efficiency is 0.5 and the calorific value of the fuel is 42 MJ/kg, determine the fuel consumption in kg/h and the brake mean effective pressure.

SOLUTION:

$$\text{Swept volume, } V_s \quad = \quad \frac{\pi}{4} \times 0.06^2 \times 0.1$$

$$= \quad 2.83 \times 10^{-4} \text{ m}^3/\text{cylinder}$$

$$= \quad 283 \text{ cc/cylinder}$$

$$\textit{Compression ratio} \quad = \quad \frac{283 + 60}{60} \quad = \quad 5.71$$

$$\textit{Air-standard efficiency} \quad = \quad 1 - \frac{1}{5.71^{0.4}} \quad = \quad 0.50$$

$$\eta_{bth} \quad = \quad \textit{Relative } \eta \times \textit{Air-standard } \eta$$

$$= \quad 0.5 \times 0.5 \quad = \quad 0.25$$

$$bp \quad = \quad \frac{2 \times \pi \times 3000 \times 66.5}{60000} \quad = \quad 20.89 \text{ kW}$$

$$\textit{Heat supplied} \quad = \quad \frac{20.89}{0.25} \quad = \quad 83.56 \text{ kJ/s}$$

$$\textit{Fuel consumption} \quad = \quad \frac{83.56 \times 3600}{42000}$$

$$= \quad \textbf{7.16 kg/h} \qquad \underset{\Longleftarrow}{\textbf{Ans}}$$

$$p_{bm} \quad = \quad \frac{P \times 60000}{V_s n K}$$

$$= \quad \frac{20.89 \times 60000}{2.83 \times 10^{-4} \times \frac{3000}{2} \times 4}$$

$$= \quad 7.38 \times 10^5 \text{ N/m}^2$$

$$= \quad \textbf{7.38 bar} \qquad \underset{\Longleftarrow}{\textbf{Ans}}$$

14.18. The power output of a six cylinder four-stroke engine is absorbed by a water brake for which the law is $WN/20000$ where the brake load, W is in Newton and the speed, N is in rpm. The air consumption is measured by an air box with sharp edged orifice system. The following readings are obtained.

$$\begin{aligned}
\text{Orifice diameter} \quad &= \quad 30 \text{ mm} \\
\text{Bore} \quad &= \quad 100 \text{ mm} \\
\text{Stroke} \quad &= \quad 120 \text{ mm} \\
\text{Brake load} \quad &= \quad 560 \text{ N}
\end{aligned}$$

$$
\begin{aligned}
\text{C/H ratio by mass} &= 83/17 \\
\text{Coefficient of discharge} &= 0.6 \\
\text{Ambient pressure} &= 1 \text{ bar} \\
\text{Pressure drop across orifice} &= 14.5 \text{ cm of Hg} \\
\text{Time taken for 100 cc of fuel consumption} &= 20 \text{ s} \\
\text{Ambient temperature} &= 27\,°\text{C} \\
\text{Fuel density} &= 831 \text{ kg/m}^3
\end{aligned}
$$

Calculate (i) the brake power (ii) the torque (iii) the brake specific fuel consumption (iv) the percentage of excess air and (v) the volumetric efficiency

SOLUTION:

$$
\text{Density of air, } \rho_a = \frac{p}{RT} = \frac{1 \times 10^5}{287 \times 300}
$$

$$
= 1.16 \text{ kg/m}^3
$$

$$
\text{Volume flow rate of air, } \dot{V}_a = C_d A \sqrt{2g\Delta H_g \frac{\rho_{Hg}}{\rho_a}}
$$

$$
= 0.6 \times \frac{\pi}{4} \times 0.03^2
$$

$$
\times \sqrt{2 \times 9.81 \times 0.145 \times \frac{13600}{1.16}}
$$

$$
= 0.077 \text{ m}^3/\text{s}
$$

$$
\text{Swept volume} = \frac{\pi}{4} \times 0.1^2 \times 0.12 \times \frac{2400}{2} \times \frac{6}{60}
$$

$$
= 0.113 \text{ m}^3/\text{s}
$$

$$
\text{Volumetric efficiency} = \frac{0.077}{0.113} \times 100
$$

$$
= \mathbf{68.1\%} \qquad \underleftarrow{\textbf{Ans}}
$$

$$
bp = \frac{WN}{20000} = \frac{560 \times 2400}{20000}
$$

$$
= \mathbf{67.2 \text{ kW}} \qquad \underleftarrow{\textbf{Ans}}
$$

$$
bmep = \frac{bp \times 60000}{LAnK}
$$

$$= \frac{67.2 \times 60000 \times 10^{-5}}{0.12 \times \frac{\pi}{4} \times 0.1^2 \times \frac{2400}{2} \times 6}$$

$$= 5.94 \text{ bar}$$

$$\text{Torque, } T = \frac{bp \times 60000}{2\pi N} = \frac{67.2 \times 60000}{2 \times \pi \times 2400}$$

$$= \textbf{267.3 Nm} \qquad \overset{\text{Ans}}{\Longleftarrow}$$

$$\text{Mass flow rate of fuel} = \frac{100}{20} \times 10^{-6} \times 831 \times 3600$$

$$= 14.96 \text{ kg/h}$$

$$bsfc = \frac{14.96}{67.2}$$

$$= \textbf{0.223 kg/kW h} \qquad \overset{\text{Ans}}{\Longleftarrow}$$

$$O_2 \text{ required/kg of fuel} = 0.83 \times \frac{32}{12} + 0.17 \times \frac{8}{1}$$

$$= 3.57 \text{ kg/kg of fuel}$$

Since, air contains 23.3% of O_2 by weight,

$$\text{Air required/kg of fuel} = \frac{3.57}{0.233} = 15.32 \text{ kg}$$

$$\text{Actual mass flow rate of air} = 0.077 \times 1.16$$

$$= 0.089 \text{ kg/s}$$

$$\text{Actual mass A/F ratio} = \frac{0.089 \times 3600}{14.96}$$

$$= 21.42$$

$$\text{\% of excess air} = \frac{21.42 - 15.32}{15.32} \times 100$$

$$= \textbf{39.8\%} \qquad \overset{\text{Ans}}{\Longleftarrow}$$

14.19. A six-cylinder, four-stroke cycle gasoline engine with a bore of 120 mm and a stroke of 200 mm under test was supplied with gasoline of composition C = 82% and H_2 = 18% by mass. The dry exhaust composition by volume was CO_2 = 11.2%, O_2 = 3.6% and N_2 = 85.2%. Determine the mass of air supplied per kg of gasoline at 17 °C and 1 bar which were the conditions for the mixture entering the cylinder during the test. Also determine the volumetric efficiency of the engine based on intake conditions when the mass of gasoline used per hour was 30 kg and the engine speed was 1400 rpm. The gasoline is completely evaporated before entering the cylinder and the effect of its volume on the volumetric efficiency should be included. Take the density of gasoline vapour as 3.4 times that of air at the same temperature and pressure. 1 kg of air at 0 °C and 1 bar occupies 0.783 m^3. Air contains 23% oxygen by mass.

SOLUTION:

$$\text{Stoichiometric air-fuel ratio} \quad = \quad \frac{0.82 \times \frac{32}{12} + 0.18 \times \frac{8}{1}}{0.23}$$

$$= \quad 15.77$$

Let y mol of air be supplied per kg of fuel and the combustion equation can be written as

$$\frac{0.82}{12}C + \frac{0.18}{2}H_2 + 0.21yO_2 + 0.79yN_2 \quad = \quad aCO_2 + bO_2 + cH_2O + dN_2$$

From carbon balance

$$\frac{0.82}{12} \quad = \quad a \quad = \quad 0.068$$

From hydrogen balance

$$\frac{0.18}{2} \quad = \quad c \quad = \quad 0.09$$

From oxygen balance

$$0.21y \quad = \quad a + b + \frac{c}{2}$$

$$= \quad 0.068 + b + 0.045 \quad = \quad 0.113 + b$$

From nitrogen balance

$$0.79y \quad = \quad d$$

From dry exhaust gas analysis

$$\frac{0.068}{d} = \frac{11.2}{85.2}$$

$$d = \frac{85.2}{11.2} \times 0.068 = 0.517$$

$$0.79\, y = 0.517$$

$$y = 0.655 \text{ mol of air/kg of fuel}$$

$$\text{Mol. wt of air} = 0.23 \times 32 + 0.77 \times 28$$

$$= 28.92 \text{ kg/mol}$$

$$\text{Actual air-fuel ratio} = 0.655 \times 28.92$$

$$= \mathbf{18.94} \qquad \overset{\text{Ans}}{\Longleftarrow}$$

$$\% \text{ excess air} = \frac{18.94 - 15.77}{15.77} \times 100 = 20.1\%$$

$$\text{Volume of air} = v_a \times m$$

$$v_a = 0.783 \times \frac{290}{273} = \mathbf{0.832 \text{ m}^3/\text{kg}}$$

$$= 0.832 \times 18.94 = 15.76 \text{ m}^3$$

$$\text{Volume of fuel} = V_f \times m$$

$$= \frac{0.832}{3.4} \times 1 = 0.245 \text{ m}^3$$

$$\text{Total volume} = 15.76 + 0.245$$

$$= 16.005 \text{ m}^3/\text{kg of fuel}$$

$$\text{Mixture evaporated} = \frac{16.005 \times 30}{60} = 8.002 \text{ m}^3/\text{min}$$

$$\text{Swept volume per minute} = 6 \times \frac{\pi}{4} \times 0.12^2 \times 0.20 \times \frac{1400}{2}$$

$$= 9.5 \text{ m}^3/\text{min}$$

$$\text{Volumetric efficiency, } \eta_v \quad = \quad \frac{8.002}{9.5} \times 100$$

$$= \quad \textbf{84.2\%} \qquad \underset{\Longleftarrow}{\text{Ans}}$$

14.20. An indicator diagram taken from a single-cylinder, four-stroke CI engine has a length of 100 mm and an area of 2000 mm². The indicator pointer deflects a distance of 10 mm for pressure increment of 2 bar in the cylinder. If the bore and stroke of the engine cylinder are both 100 mm and the engine speed is 1000 rpm. Calculate the mean effective pressure and the indicated power. If the mechanical efficiency is 75% what is the brake power developed.

SOLUTION:

Mean height of the indicator diagram

$$= \quad \frac{2000}{100} \quad = \quad 20 \text{ mm}$$

$$\text{Mean effective pressure} \quad = \quad \frac{20}{10} \times 2$$

$$= \quad \textbf{4 bar} \qquad \underset{\Longleftarrow}{\text{Ans}}$$

$$\text{Indicated power, } ip \quad = \quad \frac{p_{im} L A n}{60000}$$

$$= \quad \frac{4 \times 10^5 \times 0.1 \times \frac{\pi}{4} \times 0.1^2 \times \frac{1000}{2}}{60000}$$

$$= \quad \textbf{2.62 kW} \qquad \underset{\Longleftarrow}{\text{Ans}}$$

$$bp \quad = \quad ip \times \eta_m \quad = \quad 2.62 \times 0.75$$

$$= \quad \textbf{1.96 kW} \qquad \underset{\Longleftarrow}{\text{Ans}}$$

14.21. A single-cylinder, four-stroke gas engine has a bore of 180 mm and a stroke of 330 mm and is governed on the hit and miss principle. When running at 400 rpm at full load indicator card are taken which give a working loop mean effective pressure of 6 bar and a pumping loop mean effective pressure of 0.4 bar. Diagrams from the dead cycle give a mean effective pressure of 0.6 bar. When running on no load a mechanical counter recorded 50 firings strokes per minute. Calculate at the full load with regular firing, brake power and the mechanical efficiency of the engine.

SOLUTION:

Assume fp to be constant at a given speed and is independent of load.

$$\text{Net imep} = 6.0 - 0.4 = 5.6 \text{ bar}$$

$$\text{Working cycle/minute} = 50$$

$$\text{Dead cycles/minute} = \frac{400}{2} - 50 = 150$$

In hit and miss governing the working cycle has the same indicated diagram at any load. Since at no load, bp is zero

$$fp = ip - \text{pumping power of dead cycles}$$

$$V_s = \frac{\pi}{4} \times 0.18^2 \times 0.33$$

$$= 8.4 \times 10^{-3} \text{ m}^3$$

$$fp = \frac{p_{im} \times V_s \times n}{60000} - \frac{p_{fm} \times V_s \times n}{60000}$$

$$= \frac{5.6 \times 10^5 \times 8.4 \times 10^{-3} \times 50}{60000}$$

$$- \frac{0.6 \times 10^5 \times 8.4 \times 10^{-3} \times 150}{60000}$$

$$= 3.92 - 1.26 = 2.66$$

At full load, with regular firing $(n = 400/2)$ per minute

$$ip = \frac{p_{im} L A n}{60000}$$

$$= \frac{5.6 \times 10^5 \times 8.4 \times 10^{-3} \times \frac{400}{2}}{60000}$$

$$= 15.68 \text{ kW}$$

$$bp = 15.68 - 2.66$$

$$= 13.02 \text{ kW} \qquad \text{Ans}$$

$$\text{Mechanical efficiency} = \frac{13.02}{15.68} \times 100$$

$$= \textbf{83.03\%} \qquad \underset{\Longleftarrow}{\textbf{Ans}}$$

14.22. A six-cylinder, four-stroke, direct-injection oil engine is to deliver 120 kW at 1600 rpm. The fuel to be used has a calorific value of 43 MJ/kg and its percentage composition by mass is carbon 86.0%, hydrogen 13.0%, non combustibles 1.0%. The absolute volumetric efficiency is assumed to 80%, the indicated thermal efficiency 40% and the mechanical efficiency 80%. The air consumption to be 110% in excess of that required for theoretically correct combustion.

(i) Estimate the volumetric composition of dry exhaust gas

(ii) Determine the bore and stroke of the engine, taking a stroke to bore ratio as 1.5.

Assume the volume of 1 kg of air at the given conditions as 0.77 m^3. Oxygen in air is 23% by mass and 21% by volume.

SOLUTION:

$$\text{Stoichiometric air-fuel ratio} = \frac{0.86 \times \frac{32}{12} + 0.13 \times \frac{8}{1}}{0.23}$$

$$= 14.49$$

$$\text{Actual air-fuel ratio} = \left(1 + \frac{110}{100}\right) \times 14.49$$

$$= 30.43$$

$$\text{Mol wt of air} = 0.23 \times 32 + 0.769 \times 28$$

$$= 28.9 \text{ kg/K mol}$$

Let y mol of air be supplied per kg of fuel and the equation of combustion per kg of fuel is

$$\frac{0.86}{12}C + \frac{0.14}{2}H_2 + 0.21yO_2 + 0.79yN_2 = aCO_2 + bH_2O + cO_2 + dN_2$$

From carbon balance

$$\frac{0.86}{12} = a = 0.0717$$

From hydrogen balance

$$\frac{0.13}{2} = b = 0.065$$

From oxygen balance

$$0.21\,y = a + \frac{b}{2} + c = 0.0717 + 0.025 + c$$

$$= 0.1042 + c$$

Number of kilo moles of air for per kg of fuel

$$\frac{30.43}{28.9} = 1.05 = y$$

Therefore

$$0.21 \times 1.05 = 0.1042 + c$$

$$c = 0.116$$

From nitrogen balance

$$0.79y = d$$

$$d = 0.79 \times 1.05 = 0.8295$$

Volumetric composition of dry exhaust gas

<u>Ans</u>

Constituent	mols	% Vol
CO_2	0.0717	7.05
O_2	0.1160	11.40
N_2	0.8295	81.55
Total	**1.0172**	**100.00**

$$ip = \frac{bp}{\eta_m} = \frac{120}{0.8}$$

$$= 150 \text{ kW}$$

$$\text{Heat input} = \frac{\text{Heat equivalent of } ip}{\eta_{ith}}$$

$$= \frac{150 \times 60}{0.40} = 22500 \text{ kJ/min}$$

$$\dot{m}_f = \frac{22500}{43000} = 0.523 \text{ kg/min}$$

$$\dot{m}_a \;=\; \dot{m}_f \times Actual \; A/F$$

$$=\; 0.523 \times 30.43 \;=\; 15.92 \; \text{kg/min}$$

$$\dot{V}_a \;=\; \dot{m}_a V_a$$

$$=\; 15.92 \times 0.770 \;=\; 12.26 \; \text{m}^3/\text{min}$$

$$Swept \; volume \;=\; \frac{12.26}{0.8}$$

$$=\; 15.32 \; \text{m}^3/\text{min}$$

$$V_s \;=\; \frac{\pi}{4} D^2 L n K$$

$$15.32 \;=\; \frac{\pi}{4} D^2 \times 1.5 D \times \frac{1600}{2} \times 6$$

$$D^3 \;=\; \frac{15.32 \times 4 \times 2}{\pi \times 1.5 \times 1600 \times 6}$$

$$=\; 2.71 \times 10^{-3}$$

$$D \;=\; 0.14 \; \text{m}$$

$$=\; \textbf{14 cm} \qquad\qquad \underset{\Longleftarrow}{\textbf{Ans}}$$

$$L \;=\; 1.5 \times 0.14 \;=\; 0.21 \; \text{m}$$

$$=\; \textbf{21 cm} \qquad\qquad \underset{\Longleftarrow}{\textbf{Ans}}$$

14.23. A six cylinder, four-stroke gasoline engine having a bore of 90 mm and stroke of 100 mm has a compression ratio 7. The relative efficiency is 55% when the indicated specific fuel consumption is 300 gm/kW h. Estimate (i) the calorific value of the fuel and (ii) corresponding fuel consumption, given that imep is 8.5 bar and speed is 2500 rpm.

SOLUTION:

$$Air\text{-}standard \; efficiency \;=\; 1 - \frac{1}{r^{\gamma-1}}$$

$$=\; 1 - \frac{1}{7^{0.4}} \;=\; 0.541$$

$$\text{,Relative efficiency} \quad = \quad \frac{\text{Thermal efficiency}}{\text{Air-standard efficiency}}$$

$$\text{Ind. th. efficiency} \quad = \quad 0.55 \times 0.541$$

$$= \quad 0.297$$

$$\eta_{ith} \quad = \quad \frac{1}{isfc \times CV}$$

$$CV \quad = \quad \frac{1}{\eta_{ith} \times isfc} \quad = \quad \frac{3600}{0.3 \times 0.297}$$

$$= \quad \textbf{40404 kJ/kg} \qquad \underset{\Longleftarrow}{\textbf{Ans}}$$

$$ip \quad = \quad \frac{p_{im}LAnK}{60000}$$

$$= \quad \frac{8.5 \times 10^5 \times 0.1 \times \frac{\pi}{4} \times 0.09^2 \times \frac{2500}{2} \times 6}{60000}$$

$$= \quad 67.6 \text{ kW}$$

$$\text{Fuel consumption} \quad = \quad isfc \times ip \quad = \quad 0.3 \times 67.6$$

$$= \quad \textbf{20.28 kg/h} \qquad \underset{\Longleftarrow}{\textbf{Ans}}$$

14.24. A gasoline engine working on four-stroke develops a brake power of 20.9 kW. A Morse Test was conducted on this engine and the brake power (kW) obtained when each cylinder was made inoperative by short circuiting the spark plug are 14.9, 14.3, 14.8 and 14.5 respectively. The test was conducted at constant speed. Find the indicated power, mechanical efficiency and *bmep* when all the cylinders are firing. The bore of the engine is 75 mm and the stroke is 90 mm. The engine is running at 3000 rpm.

SOLUTION:

$$ip_1 \quad = \quad bp_{1234} - bp_{234}$$

$$= \quad 20.9 - 14.9 \quad = \quad 6.0 \text{ kW}$$

$$ip_2 \quad = \quad bp_{1234} - bp_{134}$$

$$= \quad 20.9 - 14.3 \quad = \quad 6.6 \text{ kW}$$

$$ip_3 = bp_{1234} - bp_{124}$$

$$= 20.9 - 14.8 = 6.1 \text{ kW}$$

$$ip_4 = bp_{1234} - bp_{123}$$

$$= 20.9 - 14.5 = 6.4 \text{ kW}$$

$$ip_1 + ip_2 + ip_3 + ip_4 = ip_{1234} = 6.0 + 6.6 + 6.1 + 6.4$$

$$= \mathbf{25.1} \text{ kW} \qquad \text{Ans}$$

$$\eta_m = \frac{20.9}{25.1} \times 100$$

$$= \mathbf{83.3\%} \qquad \text{Ans}$$

$$p_{bm} = \frac{bp \times 60000}{LAnK}$$

$$= \frac{20.9 \times 60000}{0.09 \times \frac{\pi}{4} \times 0.075^2 \times \frac{3000}{2} \times 4}$$

$$= 5.25 \times 10^5 \text{ Pa}$$

$$= \mathbf{5.25} \text{ bar} \qquad \text{Ans}$$

14.25. A Morse test on a 12 cylinder, two-stroke compression-ignition engine of bore 40 cm and stroke 50 cm running at 200 rpm gave the following readings :

Condition	Brake load (Newton)	Condition	Brake load (Newton)
All firing	2040	7th cylinder	1835
1st cylinder	1830	8th cylinder	1860
2nd cylinder	1850	9th cylinder	1820
3rd cylinder	1850	10th cylinder	1840
4th cylinder	1830	11th cylinder	1850
5th cylinder	1840	12th cylinder	1830
6th cylinder	1855	All firing	2060

The output is found from the dynamometer using the relation

$$bp = \frac{WN}{180}$$

where W, the brake load is in Newton and the speed, N is in rpm.

SOLUTION:

Power output, bp when kth cylinder is cut-off $= \frac{N}{180} \sum_{k=1}^{12} W_k$

Cylinder number cut-off	ip of kth cylinder $ip_k = 2277.8 - \frac{N}{180} \sum W_i$ (kW)
1	244.5
2	222.2
3	222.2
4	244.5
5	233.4
6	216.7
7	238.9
8	211.1
9	255.6
10	233.4
11	222.2
12	244.5

Total indicated power $= ip_1 + ip_2 + \ldots + ip_{12}$

$$= \mathbf{2789.2} \qquad \overset{\textbf{Ans}}{\Longleftarrow}$$

Mechanical efficiency $= \dfrac{2277.8}{2789.2}$

$$= \mathbf{81.66\%} \qquad \overset{\textbf{Ans}}{\Longleftarrow}$$

$$mep = \frac{bp \times 60000}{LAnK}$$

$$= \frac{2277.8 \times 60000}{0.5 \times \frac{\pi}{4} \times 0.4^2 \times 200 \times 12}$$

$$= 9.06 \times 10^5 \text{ Pa}$$

$$= \mathbf{9.06} \textbf{ bar} \qquad \overset{\textbf{Ans}}{\Longleftarrow}$$

14.26. The observations recorded after the conduct of a retardation test on a single-cylinder diesel engine are as follows :

Rated power : 8 kW   Rated speed : 475 rpm

S.No.	Drop in speed (rpm)	Time for fall of speed at no load, t_2 (s)	Time for fall of speed at 50% load, t_3 (s)
1.	475 → 400	7.0	2.2
2.	475 → 350	10.6	3.7
3.	475 → 325	12.5	4.8
4.	475 → 300	15.0	5.4
5.	475 → 275	16.6	6.5
6.	475 → 250	18.9	7.2

Calculate frictional power and mechanical efficiency.

SOLUTION:

First draw a graph of drop in speed versus time taken for the drop.

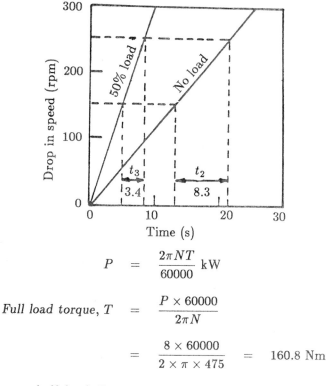

$$P = \frac{2\pi NT}{60000} \text{ kW}$$

$$\text{Full load torque, } T = \frac{P \times 60000}{2\pi N}$$

$$= \frac{8 \times 60000}{2 \times \pi \times 475} = 160.8 \text{ Nm}$$

Torque at half load, $T_{\frac{1}{2}} = 80.4$ Nm

From the graph,

Time for the fall of 100 rpm at no load, $t_2 = 8.3$ s.

Time for the fall of same 100 rpm at half load, $t_3 = 3.4$ s.

$$T_f = \frac{t_3}{t_2 - t_3} T_{\frac{1}{2}} = \left(\frac{3.4}{8.3 - 3.4}\right) \times 80.4$$

$$= 55.8 \text{ Nm}$$

$$\text{Friction power} = \frac{2\pi N T_f}{60000}$$

$$= \frac{2 \times \pi \times 475 \times 55.8}{60000}$$

$$= \textbf{2.77 kW} \qquad \underset{\Longleftarrow}{\textbf{Ans}}$$

$$\eta_m = \frac{bp}{bp + fp} = \frac{8}{8 + 2.77} \times 100$$

$$= \textbf{74.28\%} \qquad \underset{\Longleftarrow}{\textbf{Ans}}$$

14.27. A single-cylinder gas engine, having a bore and stroke of 25 cm and 50 cm respectively and running at 240 rpm fires 100 times per minute. The quantity of coal gas used is 0.3 m^3 per minute at 100 cm of water (gauge) (barometer pressure 1 bar) at 17 °C while the amount of air used is 3 kg/min. Assuming that an extra volume of air is taken in during a missed cycle equal to that of a coal gas normally taken in, if both are measured at NTP, find (i) the charge of air per working cycle as measured at NTP and (ii) the volumetric efficiency.

SOLUTION:

$$\text{Assume 760 mm of Hg} = 1 \text{ bar}$$

$$\text{Gas pressure} = 1 + \frac{100}{13.6} \times \frac{1}{76}$$

$$= 1.097 \text{ bar}$$

$$\text{Volume of coal gas at NTP} = 0.3 \times \frac{1.097}{1} \times \frac{273}{290}$$

$$= 0.31 \text{ m}^3/\text{min}$$

Volume of coal gas used per explosion

$$= \frac{0.31}{100} = 0.0031 \text{ m}^3 \text{ at NTP}$$

Therefore, extra air missed per cycle

$$= \quad 0.0031 \text{ m}^3 \text{ at NTP}$$

Volume of air taken at NTP $\quad = \quad \dfrac{mRT}{p}$

$$= \quad \dfrac{3 \times 287 \times 273}{1 \times 10^5}$$

$$= \quad 2.35 \text{ m}^3/\text{min}$$

The engine is running at 240 rpm and therefore there must be 120 firing cycles per minute. However, there are only 100 cycles per minute. Hence, there are 20 missed cycles. The 2.35 m³ of air per minute at NTP must be made up of 120 normal air charges, V, together with 20 missed cycles each equivalent to 0.0031 m³ at NTP

$$20 \times 0.0031 + 120 \, V \quad = \quad 2.35$$

$$V \quad = \quad 0.019 \text{ m}^3$$

Therefore, total volume of charge

$$= \quad 0.019 + 0.0031$$

$$= \quad \textbf{0.022 m}^3 \textbf{ at NTP} \qquad \underset{\rightleftharpoons}{\text{Ans}}$$

Displacement volume $\quad = \quad \dfrac{\pi}{4} \times 0.25^2 \times 0.5$

$$= \quad 0.0245 \text{ m}^3$$

Volumetric efficiency $\quad = \quad \dfrac{0.022}{0.0245} \times 100$

$$= \quad \textbf{89.8\%} \qquad \underset{\rightleftharpoons}{\text{Ans}}$$

14.28. A trial was conducted on a single-cylinder oil engine having a cylinder diameter of 30 cm and stroke 45 cm. The engine is working on the four-stroke cycle and the following observations were made :

$$\text{Duration of trial} \quad = \quad 45 \text{ minutes}$$
$$\text{Total fuel used} \quad = \quad 7 \text{ litres}$$

$$
\begin{aligned}
\text{Calorific value} &= 42 \text{ MJ/kg} \\
\text{Total number of revolution} &= 12624 \\
\text{Gross imep} &= 7.25 \text{ bar} \\
\text{Pumping imep} &= 0.35 \text{ bar} \\
\text{Net load on the brake} &= 150 \text{ kg} \\
\text{Diameter of the brake wheel drum} &= 1.78 \text{ m} \\
\text{Diameter of the rope} &= 4 \text{ cm} \\
\text{Cooling water circulated} &= 550 \text{ litres} \\
\text{Cooling water temperature rise} &= 48 \,^{\circ}\text{C} \\
\text{Specific heat of water} &= 4.18 \text{ kJ/kg K} \\
\text{Specific gravity of oil} &= 0.8
\end{aligned}
$$

Calculate the mechanical efficiency and also the unaccounted losses.

SOLUTION:

$$
ip = \frac{(7.25 - 0.35) \times 10^5 \times 0.45 \times \frac{\pi}{4} \times 0.3^2 \times \frac{12624}{45 \times 2}}{60000}
$$

$$
= 51.3 \text{ kW}
$$

$$
\text{Heat supplied} = \frac{7 \times 10^{-3} \times 800}{45} \times 42000
$$

$$
= 5226.67 \text{ kJ/min}
$$

$$
bp \doteq \frac{9.8 \times 150 \times \pi \times (1.78 + 0.4) \times 12624}{60000 \times 45}
$$

$$
= 39.34 \text{ kW}
$$

$$
\eta_m = \frac{bp}{ip} \times 100 = \frac{39.34}{51.3} \times 100
$$

$$
= \mathbf{76.68\%} \qquad \underset{\Leftarrow}{\text{Ans}}
$$

Heat equivalent of bp

$$
= 39.34 \times 60 = 2360.4 \text{ kJ/min}
$$

Heat lost in jacket cooling water

$$
= \frac{550 \times 48 \times 4.18}{45}
$$

$$
= 2452.3 \text{ kJ/min}
$$

Unaccounted losses

$$= \quad 5226.67 - (2360.4 + 2452.3)$$

$$= \quad \textbf{413.97 kJ/min} \qquad \underset{\Longleftarrow}{\textbf{Ans}}$$

14.29. A four-stroke gas engine has a cylinder diameter of 25 cm and stroke 45 cm. The effective diameter of the brake is 1.6 m. The observations made in a test of the engine were as follows :

Duration of test	=	40 min
Total number of revolutions	=	8080
Total number of explosions	=	3230
Net load on the brake	=	90 kg
Mean effective pressure	=	5.8 bar
Volume of gas used	=	7.5 m^3
Pressure of gas indicated in meter	=	136 mm water of gauge
Atmospheric temperature	=	17 °C
Calorific value of gas	=	19 MJ/m^3 at NTP
Rise in temp. of jacket cooling water	=	45 °C
Cooling water supplied	=	180 kg

Draw up a heat balance sheet and estimate the indicated thermal efficiency and brake thermal efficiency. Assume atmospheric pressure as 760 mm of Hg.

SOLUTION:

$$ip \quad = \quad \frac{p_{im}LAn}{60000}$$

$$= \quad \frac{5.8 \times 10^5 \times 0.45 \times \frac{\pi}{4} \times 0.25^2 \times \frac{3230}{40}}{60000}$$

$$= \quad 17.25$$

$$bp \quad = \quad \frac{9.81 \times 90 \times \pi \times 1.6 \times \frac{8080}{40}}{60000}$$

$$= \quad 14.94 \text{ kW}$$

Pressure of the gas supplied

$$= \quad 760 + \frac{136}{13.6} \quad = \quad 770 \text{ mm of Hg}$$

Volume of gas used at NTP

$$= 7.5 \times \frac{273}{290} \times \frac{770}{760} = 7.15 \text{ m}^3$$

Heat supplied $= \dfrac{7.15 \times 19000}{40} = 3396.25 \text{ kJ/min}$

Heat equivalent of bp

$$= 14.94 \times 60$$

$$= \textbf{896.4 kJ/min} \qquad \underset{\Longleftarrow}{\text{Ans}}$$

Heat loss in cooling medium

$$= \frac{180 \times 45}{40} \times 4.18$$

$$= \textbf{846.5 kJ/min} \qquad \underset{\Longleftarrow}{\text{Ans}}$$

Heat lost to exhaust, radiation, etc. (by difference)

$$= 3396.25 - 896.4 - 846.5$$

$$= \textbf{1653.35 kJ/min} \qquad \underset{\Longleftarrow}{\text{Ans}}$$

$$\eta_{ith} = \frac{ip \times 60}{Heat\ supplied/min} \times 100$$

$$= \frac{17.25 \times 60}{3396.25} \times 100$$

$$= \textbf{30.47\%} \qquad \underset{\Longleftarrow}{\text{Ans}}$$

$$\eta_{bth} = \frac{bp \times 60}{Heat\ supplied} = \frac{14.94 \times 60}{3396.25} \times 100$$

$$= \textbf{26.39\%} \qquad \underset{\Longleftarrow}{\text{Ans}}$$

Heat input (per minute)	(kJ)	Heat expenditure (per minute)	(kJ)
Heat supplied by fuel	3396.25	1. Heat equivalent to *bp*	896.40
		2. Heat lost to cooling medium	846.50
		3. Heat lost in exhaust	1653.35
		Total	**3396.25**

14.30. The following observations were made during a trial of a single-cylinder, four-stroke cycle gas engine having cylinder diameter of 18 cm and stroke 24 cm.

Duration of trial	=	30 min
Total number of revolution	=	9000
Total number of explosion	=	4450
Mean effective pressure	=	5 bar
Net load on the brake wheel	=	40 kg
Effective diameter of brake wheel	=	1 m
Total gas used at NTP	=	2.4 m^3
Calorific value of gas at NTP	=	19 MJ/m^3
Total air used	=	36 m^3
Pressure of air	=	720 mm Hg
Temperature of air	=	17 °C
Density of air at NTP	=	1.29 kg/m^3
Temperature of exhaust gas	=	350 °C
Room temperature	=	17 °C
Specific heat of exhaust gas	=	1 kJ/kg K
Cooling water circulated	=	80 kg
Rise in temperature of cooling water	=	30 °C

Draw up a heat balance sheet and estimate the mechanical and indicated thermal efficiencies of the engine. Take $R = 287$ J/kg K.

SOLUTION:

$$ip = \frac{p_{im} L A n}{60000}$$

$$= \frac{5 \times 10^5 \times 0.24 \times \frac{\pi}{4} \times 0.18^2 \times 4450}{60000 \times 30}$$

$$= 7.55 \text{ kW}$$

$$bp = \frac{\pi N d W}{60000}$$

$$= \frac{\pi \times 9000 \times 1 \times 40 \times 9.81}{60000 \times 30}$$

$$= 6.16 \text{ kW}$$

Heat supplied at NTP

$$= \frac{2.4}{30} \times 19000 = 1520 \text{ kJ/min}$$

$$\text{Heat equiv. of bp} \quad = \quad 6.16 \times 60$$

$$= \quad \textbf{369.6 kJ/min} \qquad \underset{\Leftarrow}{\textbf{Ans}}$$

Heat lost to cooling medium

$$= \quad \frac{80}{30} \times 30 \times 4.18$$

$$= \quad \textbf{334.4 kJ/min} \qquad \underset{\Leftarrow}{\textbf{Ans}}$$

$$\text{Total air used} \quad = \quad 36 \text{ m}^3 \text{ at } 720 \text{ mm of Hg}$$

Volume of air used at NTP

$$= \quad 36 \times \frac{273}{290} \times \frac{720}{760} \quad = \quad 32.1 \text{ m}^3$$

Mass of air used

$$= \quad \frac{32.1 \times 1.29}{30} \quad = \quad 1.38 \text{ kg/min}$$

Mass of gas at NTP,

$$m_g \quad = \quad \frac{pV}{RT} \quad = \quad \frac{1 \times 10^5 \times 2.4}{287 \times 273}$$

$$= \quad 3.06 \text{ kg}$$

$$\text{Mass of gas per/min} \quad = \quad \frac{3.06}{30} \quad = \quad 0.102 \text{ kg}$$

Total mass of exhaust gas

$$= \quad 1.38 + 0.102 \quad = \quad 1.482 \text{ kg}$$

Heat lost to exhaust gas

$$= \quad 1.482 \times (350 - 17) \times 1$$

$$= \quad \textbf{493.5 kJ/min} \qquad \underset{\Leftarrow}{\textbf{Ans}}$$

Heat lost by radiation

$$= \quad 1520 - (369.6 + 334.4 + 493.5)$$

$$= \quad \textbf{322.5 kJ/min} \qquad \underset{\Leftarrow}{\textbf{Ans}}$$

$$\eta_m \quad = \quad \frac{bp}{ip} \times 100 \quad = \quad \frac{6.16}{7.55} \times 100$$

$$= \quad 81.6\% \qquad \text{Ans}$$

$$\eta_{ith} \quad = \quad \frac{ip \times 60}{Heat\ supplied} \quad = \quad \frac{7.55 \times 60}{1520} \times 100$$

$$= \quad 29.8\% \qquad \text{Ans}$$

Heat input (per minute)	(kJ)	Heat expenditure (per minute)	(kJ)
Heat supplied by fuel	1520	1. Heat equivalent to *bp*	369.6
		2. Heat lost to cooling medium	334.4
		3. Heat lost in exhaust	493.5
		4. Unaccounted losses	322.5
		Total	**1520.0**

14.31. The following results were obtained in a test on a gas engine :

$$\begin{aligned}
\text{Gas used} &= 0.16 \text{ m}^3/\text{min at NTP} \\
\text{Calorific value of gas at NTP} &= 14 \text{ MJ/m}^3 \\
\text{Density of gas at NTP} &= 0.65 \text{ kg/m}^3 \\
\text{Air used} &= 1.50 \text{ kg/min} \\
\text{Specific heat of exhaust gas} &= 1.0 \text{ kJ/kg K} \\
\text{Temperature of exhaust gas} &= 400\ ^\circ\text{C} \\
\text{Room temperature} &= 20\ ^\circ\text{C} \\
\text{Cooling water per minute} &= 6 \text{ kg} \\
\text{Specific heat of water} &= 4.18 \text{ kJ/kg K} \\
\text{Rise in temperature of cooling water} &= 30\ ^\circ\text{C} \\
ip &= 12.5 \text{ kW} \\
bp &= 10.5 \text{ kW}
\end{aligned}$$

Draw a heat balance sheet for the test on per hour basis in kJ.

SOLUTION:

$$Heat\ supplied\ at NTP \quad = \quad 0.16 \times 14000 \times 60$$

$$= \quad 134400 \text{ kJ/h}$$

$$Heat\ equivalent\ of\ bp \quad = \quad 10.5 \times 60 \times 60$$

$$= \quad \textbf{37800 kJ/h} \qquad \text{Ans}$$

Heat lost in cooling medium

$$= 6 \times 30 \times 4.18 \times 60$$

$$= \mathbf{45144} \text{ kJ/h} \qquad \underline{\underline{\mathbf{Ans}}}$$

$$\text{Mass of gas used} = 0.16 \times 0.65 = 0.104 \text{ kg/min}$$

$$\text{Mass of air used} = 1.50 \text{ kg/min}$$

$$\text{Mass of exhaust gas} = 0.104 + 1.50 = 1.604 \text{ kg/min}$$

Heat carried away in exhaust gases

$$= 1.604 \times 1 \times (400 - 20) \times 60$$

$$= \mathbf{36571.2} \text{ kJ/h} \qquad \underline{\underline{\mathbf{Ans}}}$$

$$\text{Unaccounted losses} = 134400 - (37800 + 45144 + 36571.2)$$

$$= \mathbf{14884.8} \text{ kJ/h} \qquad \underline{\underline{\mathbf{Ans}}}$$

Heat input (per hour)	(kJ)	Heat expenditure (per hour)	(kJ)
Heat supplied by fuel	134400	1. Heat equivalent to *bp*	37800.0
		2. Heat lost to cooling medium	45144.0
		3. Heat lost in exhaust	36571.2
		4. Unaccounted losses	14884.8
		Total	**134400.0**

14.32. A test on a two-stroke engine gave the following results at full load

$$\begin{aligned}
\text{Speed} &= 350 \text{ rpm} \\
\text{Net brake load} &= 65 \text{ kg} \\
\text{mep} &= 3 \text{ bar} \\
\text{Fuel consumption} &= 4 \text{ kg/h} \\
\text{Jacket cooling water flow rate} &= 500 \text{ kg/h} \\
\text{Jacket water temperature at inlet} &= 20 \text{ °C} \\
\text{Jacket water temperature at outlet} &= 40 \text{ °C} \\
\text{Test room temperature} &= 20 \text{ °C} \\
\text{Temperature of exhaust gases} &= 400 \text{ °C} \\
\text{Air used per kg of fuel} &= 32 \text{ kg} \\
\text{Cylinder diameter} &= 22 \text{ cm}
\end{aligned}$$

$$
\begin{aligned}
\text{Stroke} &= 28 \text{ cm} \\
\text{Effective brake diameter} &= 1 \text{ m} \\
\text{Calorific value of fuel} &= 43 \text{ MJ/kg} \\
\text{Proportion of hydrogen in fuel} &= 15\% \\
\text{Mean specific heat of dry exhaust gas} &= 1 \text{ kJ/kg K} \\
\text{Mean specific heat of steam} &= 2.1 \text{ kJ/kg K} \\
\text{Sensible heat of water at room temp.} &= 62 \text{ kJ/kg} \\
\text{Latent heat of steam} &= 2250 \text{ kJ/kg}
\end{aligned}
$$

Find *ip*, *bp* and draw up a heat balance sheet for the test in kJ/min and in percentage.

SOLUTION:

$$
\begin{aligned}
ip &= \frac{p_{im} L A n}{60000} \\[2mm]
&= \frac{3 \times 10^5 \times 0.28 \times \frac{\pi}{4} \times 0.22^2 \times 350}{60000} \\[2mm]
&= \textbf{18.63 kW} \qquad \underset{\Longleftarrow}{\textbf{Ans}}
\end{aligned}
$$

$$
\begin{aligned}
bp &= \frac{W \pi N d}{60000} = \frac{65 \times 9.81 \times \pi \times 350 \times 1}{60000} \\[2mm]
&= \textbf{11.68 kW} \qquad \underset{\Longleftarrow}{\textbf{Ans}}
\end{aligned}
$$

$$
\begin{aligned}
\text{Heat supplied/min} &= \frac{4 \times 43000}{60} \\[2mm]
&= \textbf{2866.7 kJ} \quad = \quad \textbf{100\%} \; (let) \qquad \underset{\Longleftarrow}{\textbf{Ans}}
\end{aligned}
$$

Heat equivalent of *bp*

$$
\begin{aligned}
&= 11.68 \times 60 \\[2mm]
&= \textbf{700.8 kJ/min} \; = \; \textbf{24.4\%} \qquad \underset{\Longleftarrow}{\textbf{Ans}}
\end{aligned}
$$

Heat lost to cooling water

$$
\begin{aligned}
&= \frac{500}{60} \times (40 - 20) \times 4.18 \\[2mm]
&= \textbf{696.7 kJ/min} \; = \; \textbf{24.3\%} \qquad \underset{\Longleftarrow}{\textbf{Ans}}
\end{aligned}
$$

1 kg of H_2 produces 9 kg of H_2O. Therefore,

H_2O *produced per kg of fuel burnt*

$$= \quad 9 \times \% \ H_2 \times mass \ of \ fuel/min$$

$$= \quad 9 \times 0.15 \times \frac{4}{60} \quad = \quad 0.09 \ kg/min$$

$$\begin{matrix} Mass \ of \ wet \ exhaust \\ gases/min \end{matrix} \quad = \quad \begin{matrix} Mass \ of \ air/min \\ +Mass \ of \ fuel/min \end{matrix}$$

$$= \quad \frac{(32+1) \times 4}{60} \quad = \quad 2.2 \ kg/min$$

$$\begin{matrix} Mass \ of \ dry \ exhaust \\ gases/min \end{matrix} \quad = \quad \begin{matrix} Mass \ of \ wet \ exhaust gases/min \\ - \quad Mass \ of \ H_2O \ produced/min \end{matrix}$$

$$= \quad 2.2 - 0.09 \quad = \quad 2.11 \ kg$$

Heat lost to dry exhaust gases/minute

$$= \quad 2.11 \times 1 \times (400 - 20)$$

$$= \quad 801.8 \ kJ/min \quad = \quad 28\% \qquad \underset{=}{\text{Ans}}$$

Assuming that steam in exhaust exists as super heated steam at atmospheric pressure and at exhaust gas temperature, total heat of 1 kg of steam at atmospheric pressure (1 bar) and 400° reckoned above room temperature,

$$Heat \ in \ steam \quad = \quad H_{sup} - h$$

where h is the sensible heat of water at room temperature.

$$C_{pw}(100 - T_R) + C_{ps}(T_{sup} - 100) + L - h$$

$$= \quad [h + 100C_{pw} + L + C_{ps}(T_{sup} - 100)] - h$$

$$= \quad 100C_{pw} + L + C_{ps}(T_{sup} - 100)$$

$$= \quad [100 \times 4.18 + 2250 + 2.1 \times (400 - 100)]$$

$$= \quad 3298 \ kJ/kg$$

Heat carried away by steam in exhaust gases per minute

$$= \quad 0.09 \times 3298$$

$$= \quad 296.8 \ kJ/min \quad = \quad 10.4\% \qquad \underset{=}{\text{Ans}}$$

$$\text{Unaccounted losses} \quad = \quad 2866.7 - (700.8 + 696.7 + 801.8 + 296.8)$$

$$= \quad \textbf{370.6 kJ} \quad = \quad \textbf{12.9\%} \qquad \underset{\Longleftarrow}{\textbf{Ans}}$$

Heat input (per minute)	kJ	%	Heat expenditure (per minute)	kJ	%
Heat supplied by fuel	2866.7	100	1. Equivalent of *bp*	700.8	24.4
			2. Lost to cooling water	696.7	24.3
			3. Lost to dry exhaust gas	801.8	28.0
			4. Carried away by steam	296.8	10.4
			5. Unaccounted losses	370.6	12.9
			Total	2866.7	100.0

14.33. During the trial of a single-cylinder, four-stroke oil engine, the following results were obtained.

$$
\begin{aligned}
\text{Cylinder diameter} &= 20 \text{ cm} \\
\text{Stroke} &= 40 \text{ cm} \\
\text{Mean effective pressure} &= 6 \text{ bar} \\
\text{Torque} &= 407 \text{ Nm} \\
\text{Speed} &= 250 \text{ rpm} \\
\text{Oil consumption} &= 4 \text{ kg/h} \\
\text{Calorific value of fuel} &= 43 \text{ MJ/kg} \\
\text{Cooling water flow rate} &= 4.5 \text{ kg/min} \\
\text{Air used per kg of fuel} &= 30 \text{ kg} \\
\text{Rise in cooling water temperature} &= 45 \text{ °C} \\
\text{Temperature of exhaust gases} &= 420 \text{ °C} \\
\text{Room temperature} &= 20 \text{ °C} \\
\text{Mean specific heat of exhaust gas} &= 1 \text{ kJ/kg K} \\
\text{Specific heat of water} &= 4.18 \text{ kJ/kg K}
\end{aligned}
$$

Find the *ip*, *bp* and draw up a heat balance sheet for the test in kJ/h.

SOLUTION:

$$\dot{ip} \quad = \quad \frac{p_{im}LAn}{60000}$$

$$= \quad \frac{6 \times 10^5 \times 0.4 \times \frac{\pi}{4} \times 0.2^2 \times \frac{250}{2}}{60000}$$

$$= \quad \textbf{15.7 kW} \qquad \underset{\Longleftarrow}{\textbf{Ans}}$$

$$bp \quad = \quad \frac{2\pi NT}{60000}$$

$$= \frac{2 \times \pi \times 250 \times 407}{60000}$$

$$= \text{ } \textbf{10.6 kW} \qquad \qquad \underset{\Longleftarrow}{\text{Ans}}$$

Heat supplied $= 4 \times 43000 = 172000 \text{ kJ/h}$

Heat equivalent of bp $= 10.6 \times 60 \times 60$

$$= \textbf{38160 kJ/h} \qquad \qquad \underset{\Longleftarrow}{\text{Ans}}$$

Heat carried away by cooling water

$$= 4.5 \times 60 \times 45 \times 4.18$$

$$= \textbf{50787 kJ/h} \qquad \qquad \underset{\Longleftarrow}{\text{Ans}}$$

Heat carried away by exhaust

$$= 4 \times (30 + 1) \times 1 \times (420 - 20)$$

$$= \textbf{49600 kJ/h} \qquad \qquad \underset{\Longleftarrow}{\text{Ans}}$$

Unaccounted loss by difference

$$= \textbf{33453 kJ/h} \qquad \qquad \underset{\Longleftarrow}{\text{Ans}}$$

Heat input (per hour)	(kJ)	Heat expenditure (per hour)	(kJ)
Heat supplied by fuel	172000	1. Heat equivalent to *bp*	38160
		2. Heat lost to cooling medium	50787
		3. Heat lost in exhaust	49600
		4. Unaccounted losses	33453
		Total	**172000**

14.34. In a test of an oil engine under full load condition the following results were obtained.

$$\begin{aligned}
ip &= 33 \text{ kW} \\
\text{brake power} &= 27 \text{ kW} \\
\text{Fuel used} &= 8 \text{ kg/h} \\
\text{Rate of flow of water through gas calorimeter} &= 12 \text{ kg/min} \\
\text{Cooling water flow rate} &= 7 \text{ kg/min} \\
\text{Calorific value of fuel} &= 43 \text{ MJ/kg} \\
\text{Inlet temp. of cooling water} &= 15 \text{ °C} \\
\text{Outlet temp. of cooling water} &= 75 \text{ °C}
\end{aligned}$$

Inlet temp. of water to exhaust gas calorimeter $=$ 15 °C
Outlet temp. of water to exhaust gas calorimeter $=$ 55 °C
Final temp. of the exhaust gases $=$ 80 °C
Room temperature $=$ 17 °C
Air-fuel ratio on mass basis $=$ 20
Mean specific heat of exhaust gas $=$ 1 kJ/kg K
Specific heat of water $=$ 4.18 J/kg K

Draw up a heat balance sheet and estimate the thermal and mechanical efficiencies.

SOLUTION:

$$Heat\ supplied \quad = \quad \frac{8 \times 43000}{60} \quad = \quad 5733.3 \text{ kJ/min}$$

$$Heat\ equivalent\ of\ bp \quad = \quad 27 \times 60$$

$$= \quad \mathbf{1620} \text{ kJ/min} \qquad \underset{\rightleftharpoons}{\mathbf{Ans}}$$

Heat carried away by cooling water

$$= \quad 7 \times 4.18 \times (75 - 15)$$

$$= \quad \mathbf{1755.6} \text{ kJ/min} \qquad \underset{\rightleftharpoons}{\mathbf{Ans}}$$

Heat lost by exhaust gases in exhaust calorimeter in kJ/min

$$= \quad 12 \times 4.18 \times (55 - 15) \quad = \quad 2006.4$$

$$Heat\ lost\ in\ exhaust\ gases \quad = \quad \dot{m}_{ex} C_{p_{ex}} (t_{ex} - t_R)$$

$$= \quad 6.96 \times (320 - 17)$$

$$= \quad 2110.9 \text{ kJ/min}$$

Total heat carried away by the exhaust gases

$$Unaccounted\ loss \quad = \quad \mathbf{246.8} \text{ kJ} \qquad \underset{\rightleftharpoons}{\mathbf{Ans}}$$

$$Indicated\ thermal\ efficiency \quad = \quad \frac{ip \times 60}{Heat\ supplied/min} \times 100$$

$$= \quad \frac{33 \times 60}{5733.3} \times 100$$

$$= \quad \mathbf{34.5\%} \qquad \underset{\rightleftharpoons}{\mathbf{Ans}}$$

$$\text{Brake thermal efficiency} \quad = \quad \frac{27 \times 60}{5733.3} \times 100$$

$$= \quad \textbf{28.3\%} \qquad \underset{\Longleftarrow}{\textbf{Ans}}$$

$$\text{Mechanical efficiency} \quad = \quad \frac{bp}{ip} \times 100 \quad = \quad \frac{27}{33} \times 100$$

$$= \quad \textbf{81.8\%} \qquad \underset{\Longleftarrow}{\textbf{Ans}}$$

Heat input (per minute)	(kJ)	Heat expenditure (per minute)	(kJ)
Heat supplied by fuel	5733.3	1. Heat equivalent of *bp*	1620
		2. Heat lost to cooling medium	1755.6
		3. Heat lost in exhaust	2110.9
		4. Unaccounted losses	246.8
		Total	5733.3

14.35. A gasoline engine has a stroke volume of 0.0015 m³ and a compression ratio of 6. At the end of the compression stroke, the pressure is 8 bar and temperature 350°C. Ignition is set so that the pressure rises along a straight line during combustion and attains its highest value of 25 bar after the piston has traveled 1/30 of the stroke. The charge consists of a gasoline-air mixture in proportion by mass 1 to 16. Take R = 287 J/kg K, calorific value of the fuel = 42 MJ/kg and C_p = 1 kJ/kg K. Calculate the heat lost per kg of charge during combustion.

SOLUTION:

$$V_1 - V_2 \quad = \quad 0.0015 \times 10^6 \quad = \quad 1500 \text{ cc}$$

$$V_2 \quad = \quad \frac{1500}{6-1} \quad = \quad 300 \text{ cc}$$

$$V_3 \quad = \quad \frac{1500}{30} + 300 \quad = \quad 350 \text{ cc}$$

$$T_2 \quad = \quad 350 + 273 \quad = \quad 623 \ K$$

$$T_3 \quad = \quad T_2 \frac{p_3 V_3}{p_2 V_2}$$

$$= \quad 623 \times \frac{25}{8} \times \frac{350}{300}$$

$$= \quad 2271 \ K$$

In order to estimate the heat added to the mixture during the process $2 \to 3$, it is required to calculate the work done and the increase in internal energy between 2 and 3.

$$W_{2-3} = \left(\frac{25+8}{2}\right) \times 10^5 \times (350 - 300) \times 10^{-6}$$

Area of trapezoid $2'23'3$

$$= 82.5 \text{ J}$$

$$\text{Mixture mass, } m = \frac{pV}{RT}$$

$$= \frac{8 \times 10^5 \times 300 \times 10^{-6}}{287 \times 623}$$

$$= 1.342 \times 10^{-3} \text{ kg}$$

$$\Delta E = E_3 - E_2$$

$$= 1.342 \times 10^{-3} \times 1 \times (2271 - 623)$$

$$= 2.2121 \text{ kJ}$$

$$Q = 0.0825 + 2.2121 = 2.2946 \text{ kJ}$$

This is the quantity of heat actually given to the mixture in one cycle. But the heat liberated in one cycle must have been

$$\frac{1}{17} \times 1.342 \times 10^{-3} \times 42000 = 3.3155 \text{ kJ}$$

$$\text{Heat lost during explosion} = 3.3155 - 2.2946$$

$$= 1.0209 \text{ kJ}$$

$$= \frac{1.0209}{1.342 \times 10^{-3}}$$

$$= \textbf{760.75 kJ/kg of charge} \quad \underline{\underline{\text{Ans}}}$$

14.36. The air flow to a four-cylinder four-stroke gasoline engine was measured by means of a 8 cm diameter sharp edged orifice with $C_d = 0.65$. During a test the following data were recorded:

$$
\begin{aligned}
\text{Bore} &= \text{10 cm} \\
\text{Stroke} &= \text{15 cm} \\
\text{Engine speed} &= \text{2500 rpm} \\
\text{Brake power} &= \text{36 kW} \\
\text{Fuel consumption} &= \text{10 kg/h} \\
\text{Calorific value of fuel} &= \text{42 MJ/kg} \\
\text{Pressure drop across the orifice} &= \text{4 cm of water}
\end{aligned}
$$

Atmospheric temperature and pressure are 17 °C and 1 bar respectively. Calculate

    (i) brake thermal efficiency

   (ii) brake mean effective pressure

  (iii) volumetric efficiency based on free air condition

SOLUTION:

Brake thermal efficiency,

$$
\eta_{bth} = \frac{36 \times 60}{\frac{10}{60} \times 42000} \times 100
$$

$$
= \mathbf{30.86\%} \qquad\qquad \underset{\Longleftarrow}{\mathbf{Ans}}
$$

$$
p_{bm} = \frac{bp \times 60000}{LAnK}
$$

$$
= \frac{36 \times 60000}{0.15 \times \frac{\pi}{4} \times 0.1^2 \times \frac{2500}{2} \times 4}
$$

$$
= \text{3.67} \times 10^5 \text{ Pa}
$$

$$
= \mathbf{3.67 \text{ bar}} \qquad\qquad \underset{\Longleftarrow}{\mathbf{Ans}}
$$

$$
V_s = \frac{\pi}{4} d^2 LK = \frac{\pi}{4} \times 10^2 \times 15 \times 4
$$

$$
= \text{4712.4 cc}
$$

Velocity of air through the orifice,

$$
C_a = \sqrt{2g\Delta h_a}
$$

$$
\Delta h_w \rho_w = \Delta h_a \rho_a
$$

$$
\rho_a = \frac{p}{RT} = \frac{1 \times 10^5}{287 \times 290}
$$

$$
= \text{1.20 kg/m}^3
$$

$$C_a = \sqrt{2g \times \frac{\Delta h_w}{1000} \times \frac{1000}{\rho_a}}$$

where Δh_w is the pressure drop in mm of water column

$$C_a = \sqrt{2 \times 9.81 \times \frac{40}{1.2}} = 25.57 \text{ m/s}$$

$$\dot{m}_a = C_d A \rho_a C_a$$

$$= 0.65 \times \frac{\pi}{4} \times 0.08^2 \times 1.20 \times 25.57 \times 3600$$

$$= 360.9 \text{ kg/h}$$

$$\text{Volume of air} = \frac{mRT}{p} = \frac{360.9 \times 287 \times 290}{1 \times 10^5}$$

$$= 300.4 \text{ m}^3/\text{h}$$

$$\text{Swept volume} = 4712.4 \times 10^{-6} \times \frac{2500}{2} \times 60$$

$$= 353.43 \text{ m}^3/\text{h}$$

$$\text{Volumetric efficiency, } \eta_v = \frac{300.4}{353.43} \times 100$$

$$= 85\% \qquad \underset{\Longleftarrow}{\text{Ans}}$$

14.37. A test on a single-cylinder, four-stroke oil engine having a bore of 15 cm and stroke 30 cm gave the following results: speed 300 rpm; brake torque 200 Nm; indicated mean effective pressure 7 bar; fuel consumption 2.4 kg/h; cooling water flow 5 kg/min; cooling water temperature rise 35 °C; air-fuel ratio 22; exhaust gas temperature 410 °C; barometer pressure 1 bar; room temperature 20 °C. The fuel has a calorific value of 42 MJ/kg and contains 15% by weight of hydrogen. Take latent heat of vapourisation as 2250 kJ/kg. Determine,

(i) the indicated thermal efficiency

(ii) the volumetric efficiency based on atmospheric conditions

Draw up a heat balance in terms of kJ/min. Take C_p for dry exhaust gas = 1 kJ/kg K and super heated steam C_p = 2.1 kJ/kg K; R=0.287 kJ/kg K.

SOLUTION:

$$ip = \frac{p_{im} LAn}{60000}$$

$$= \frac{7 \times 10^5 \times 0.3 \times \frac{\pi}{4} \times 0.15^2 \times \frac{300}{2}}{60000}$$

$$= 9.28 \text{ kW}$$

Indicated thermal efficiency $= \dfrac{9.28 \times 3600}{2.4 \times 42000} \times 100$

$$= \textbf{33.1\%} \qquad \underset{\Longleftarrow}{\textbf{Ans}}$$

Air consumption $= \dfrac{2.4 \times 22}{60} = 0.88 \text{ kg/min}$

Density of air $= \dfrac{p}{RT} = \dfrac{1 \times 10^5}{287 \times 293}$

$$= 1.19 \text{ kg/m}^3$$

Volume of air consumed $= \dfrac{0.88}{1.19} = 0.739 \text{ m}^3/\text{min}$

Swept volume per minute $= \dfrac{\pi}{4} \times 0.15^2 \times 0.30 \times \dfrac{300}{2}$

$$= 0.795 \text{ m}^3/\text{min}$$

Volumetric efficiency $= \dfrac{0.739}{0.795} \times 100$

$$= \textbf{93\%} \qquad \underset{\Longleftarrow}{\textbf{Ans}}$$

$$bp = \frac{2\pi NT}{60000}$$

$$= \frac{2\pi \times 300 \times 200}{60000} = 6.28 \text{ kW}$$

Heat input $= \dfrac{2.4 \times 42000}{60}$

$$= \textbf{1680 kJ/min} \qquad \underset{\Longleftarrow}{\textbf{Ans}}$$

Heat equivalent of bp $=$ 6.28×60

$\qquad\qquad\qquad\qquad = \quad$ **376.8 kJ/min** $\overset{\text{Ans}}{\Longleftarrow}$

Heat in cooling water $=$ $5 \times 4.18 \times 35$

$\qquad\qquad\qquad\qquad = \quad$ **731.5 kJ/min** $\overset{\text{Ans}}{\Longleftarrow}$

1 kg of H_2 in the fuel will be converted to 9 kg of H_2O during combustion. Assuming the steam in the exhaust is in the superheated state, heat carried away by steam can be calculated as

$$C_{pw}(100 - T_a) + h_{fg} + C_{pa}(T_{sup} - 100)$$

where h_{fg} is the latent heat of vapourisation = 2250 kJ/kg.

Enthalpy of steam

$\qquad = \quad 4.18 \times (100 - 20) + 2250 + 2.1 \times (410 - 100)$

$\qquad = \quad 3235.4 \text{ kJ/kg}$

Heat carried away by steam

$\qquad = \quad \dot{m}_s(h - C_{pw}t_R)$

$\qquad = \quad 9 \times 0.15 \times \dfrac{2.4}{60} \times (3225.4 - 4.18 \times 20)$

$\qquad = \quad$ **170.2 kJ/min** $\overset{\text{Ans}}{\Longleftarrow}$

Heat carried away by exhaust gas

$\qquad = \quad \left(0.88 + \dfrac{2.4}{60} - 9 \times 0.15 \times \dfrac{2.4}{60}\right) \times 1 \times (410 - 20)$

$\qquad = \quad$ **337.7** $\overset{\text{Ans}}{\Longleftarrow}$

Unaccounted loss (by difference)

$\qquad = \quad 1680 - (376.8 + 731.5 + 170.2 + 337.7)$

$\qquad = \quad$ **63.8 kJ/min** $\overset{\text{Ans}}{\Longleftarrow}$

Heat input (per minute)	(kJ)	Heat expenditure (per minute)	(kJ)
Heat supplied by fuel	1680	1. Heat equivalent to *bp*	376.8
		2. Heat lost to cooling medium	731.5
		3. Heat carried away by steam	170.2
		4. Heat lost in exhaust	337.7
		5. Unaccounted losses	63.8
		Total	**1680.0**

14.38. A four-stroke cycle gasoline engine has six single-acting cylinders of 8 cm bore and 10 cm stroke. The engine is coupled to a brake having a torque radius of 40 cm. At 3200 rpm, with all cylinders operating the net brake load is 350 N. When each cylinder in turn is rendered inoperative, the average net brake load produced at the same speed by the remaining 5 cylinders is 250 N. Estimate the indicated mean effective pressure of the engine. With all cylinders operating the fuel consumption is 0.33 kg/min; calorific value of fuel is 43 MJ/kg; the cooling water flow rate and temperature rise is 70 kg/min and 10 °C respectively. On test, the engine is enclosed in a thermally and acoustically insulated box through which the output drive, water, fuel, air and exhaust connections pass. ventilating air blown up through the box at the rate of 15 kg/min enters at 17 °C and leaves at 62 °C. Draw up a heat balance of the engine stating the items as a percentage of the heat input.

SOLUTION:

$$bp = \frac{2\pi NT}{60000} = \frac{2 \times \pi \times 3200 \times 350 \times 0.4}{60000}$$

$$= 46.91 \text{ kW}$$

bp of engine when each cylinder is cut-off in turn

$$= \frac{2 \times \pi \times 3200 \times 250 \times 0.4}{60000}$$

$$= 33.51 \text{ kW}$$

$$ip \text{ of the engine} = 6 \times (46.91 - 33.51) = 80.4 \text{ kW}$$

$$imep = \frac{ip \times 60000}{LAnK}$$

$$= \frac{80.4 \times 60000}{0.1 \times \frac{\pi}{4} \times 0.08^2 \times \frac{3200}{2} \times 6}$$

$$= \quad 10 \times 10^5 \text{ Pa} \quad = \quad 10 \text{ bar}$$

Heat input $\quad = \quad 0.33 \times 43000$

$$= \quad \textbf{14190} \text{ kJ/min} \quad = \quad \textbf{100\%} \text{ (let)} \quad \underset{=}{\textbf{Ans}}$$

Heat equivalent of bp $\quad = \quad 46.91 \times 60$

$$= \quad \textbf{2814.6} \text{ kJ/min} \quad = \quad \textbf{19.8\%} \quad \underset{=}{\textbf{Ans}}$$

Heat in cooling water $\quad = \quad 70 \times 4.18 \times 10$

$$= \quad \textbf{2926} \text{ kJ/min} \quad = \quad \textbf{20.6\%} \quad \underset{=}{\textbf{Ans}}$$

Heat carried away by ventilating air

$$= \quad 15 \times 1.005 \times (62 - 17)$$

$$= \quad \textbf{678.4} \text{ kJ/min} \quad = \quad \textbf{4.8\%} \quad \underset{=}{\textbf{Ans}}$$

Unaccounted loss (by difference)

$$= \quad 14190 - (2814.6 + 2926 + 678.4)$$

$$= \quad \textbf{7771} \text{ kJ/min} \quad = \quad \textbf{54.8\%} \quad \underset{=}{\textbf{Ans}}$$

Heat input (per minute)	kJ	%	Heat expenditure (per minute)	kJ	%
Heat supplied by fuel	14190	100	1. Equivalent of *bp*	2814.6	19.8
			2. Lost to cooling water	2926.0	20.6
			3. Lost to dry exhaust gas	678.4	4.8
			4. Unaccounted losses	7771.0	54.8
			Total	14190.0	100.0

14.39. A full load test on a two-stroke engine yielded the following results: speed 440 rpm; brake load 50 kg; imep 3 bar; fuel consumption 5.4 kg/h; rise in jacket water temperature 36 °C; jacket water flow 440 kg/h; air-fuel ratio by mass 30; temperature of exhaust gas 350 °C; temperature of the test room 17 °C; barometric pressure 76 cm of Hg; cylinder diameter 22 cm; stroke 25 cm; brake diameter 1.2 m; calorific value of fuel is 43 MJ/kg; proportion of hydrogen by mass in the fuel 15%; R = 0.287 kJ/kg of mean specific heat of dry exhaust gases = 1 kJ/kg K; specific heat of dry steam 2 kJ/kg K. Assume enthalpy of super heated steam to be 3180 kJ/kg. Determine,

(i) the indicated thermal efficiency
(ii) the specific fuel consumption in g/kW h
(iii) volumetric efficiency based on atmospheric conditions

Draw up a heat balance for the test on the percentage basis indicating the content of each item in the balance.

SOLUTION:

$$bp = \frac{2\pi NT}{60000}$$

$$= \frac{2\pi \times 440 \times 50 \times 9.81 \times 0.6}{60000}$$

$$= 13.56 \text{ kW}$$

$$ip = \frac{p_{im}LAn}{60000}$$

$$= \frac{3 \times 10^5 \times 0.25 \times \frac{\pi}{4} \times 0.22^2 \times 440}{60000}$$

$$= 20.91 \text{ kW}$$

$$\eta_{ith} = \frac{20.91 \times 60}{\frac{5.4}{60} \times 43000} \times 100$$

$$= \mathbf{32.4\%} \qquad \underset{\Longleftarrow}{\text{Ans}}$$

$$bsfc = \frac{5.4 \times 1000}{13.56}$$

$$= \mathbf{398.2 \text{ g/kW h}} \qquad \underset{\Longleftarrow}{\text{Ans}}$$

$$\text{Swept volume} = \frac{\pi}{4}D^2Ln$$

$$= \frac{\pi}{4} \times 0.22^2 \times 0.25 \times 440$$

$$= 4.18 \text{ m}^3/\text{min}$$

$$\dot{m}_a = A/F \times \dot{m}_f = 30 \times \frac{5.4}{60} = 2.7 \text{ kg/min}$$

$$\rho \;=\; \frac{p}{RT} \;=\; \frac{1 \times 10^5}{287 \times 290}$$

$$=\; 1.20 \text{ m}^3/\text{kg}$$

Air consumption in m^3

$$=\; \frac{2.7}{1.2} \;=\; 2.25 \text{ m}^3/\text{min}$$

$$\eta_v \;=\; \frac{2.25}{4.18} \times 100$$

$$=\; \mathbf{53.8\%} \qquad\qquad\qquad \text{Ans}$$

$$\text{Heat input} \;=\; \frac{5.4}{60} \times 43000$$

$$=\; \mathbf{3870 \text{ kJ/min}} \;=\; \mathbf{100\%} \text{ (let)} \qquad \text{Ans}$$

Heat equivalent of bp

$$=\; 13.56 \times 60$$

$$=\; 813.6 \text{ kJ/min}$$

$$=\; \mathbf{21\%} \qquad\qquad\qquad \text{Ans}$$

Heat lost to cooling water

$$=\; \frac{440}{60} \times 4.18 \times 36 \;=\; 1103.5 \text{ kJ/min}$$

$$=\; \mathbf{28.5\%} \qquad\qquad\qquad \text{Ans}$$

Heat carried away by dry exhaust gas

$$=\; \frac{5.4}{60} \times (30 + 1 - 9 \times 0.15) \times 1 \times (350 - 17)$$

$$=\; 888.6 \text{ kJ/min}$$

$$=\; \mathbf{23\%} \qquad\qquad\qquad \text{Ans}$$

Heat carried away by the steam in the exhaust gas

$$=\; 9 \times 0.15 \times \frac{5.4}{60} \times (4.18 \times 17)$$

$$= \quad 377.8 \text{ kJ/min}$$

$$= \quad 9.8\% \qquad\qquad \underset{\Longleftarrow}{\text{Ans}}$$

Unaccounted loss (by difference)

$$= \quad 3870 - (813.6 + 1103.5 + 888.6 + 377.8)$$

$$= \quad 686.5 \text{ kJ/min}$$

$$= \quad \textbf{17.7\%} \qquad\qquad \underset{\Longleftarrow}{\text{Ans}}$$

Heat input	%	Heat expenditure	%
Heat supplied	100	1. Heat equivalent of hp	21.0
by fuel		2. Heat lost in cooling water	28.5
		3. Heat carried away by dry exhaust	23.0
		4. Heat lost in steam	9.8
		5. Unaccounted losses	17.7
		Total	**100**

14.40. A gas engine working on the constant-volume cycle gave the following results during a one-hour test run. Cylinder diameter 24 cm; stroke 48 cm; torque 770 Nm; average speed 220 rpm; average explosion per minute 77; mep 7.5 bar; volume of gas used 12 m³ at 17 °C and 770 mm of mercury pressure; lower calorific value of gas 21 MJ/m³ at NTP; inlet and outlet temperature of cooling water are 25 °C and 60 °C respectively; cooling water used 600 kg. Determine (i) the mechanical efficiency (ii) the indicated specific gas consumption in m³/kw h and (iii) the indicated thermal efficiency.

Draw up a heat balance for the engine on minute basis, explaining why friction power has been included in or omitted from your heat balance. NTP conditions are 760 mm of Hg and 0 °C.

SOLUTION:

$$bp \quad = \quad \frac{2 \times \pi \times 220 \times 770}{60000}$$

$$= \quad 17.74 \text{ kW}$$

$$ip \quad = \quad \frac{7.5 \times 10^5 \times 0.48 \times \frac{\pi}{4} \times 0.24^2 \times 77}{60000}$$

$$= \quad 20.9 \text{ kW}$$

$$\eta_m = \frac{17.74}{20.9} \times 100 = \mathbf{84.9\%} \qquad \underset{\Leftarrow}{\text{Ans}}$$

Gas consumption at NTP, $\dot{V}_f$

$$= 12 \times \frac{770}{760} \times \frac{273}{290}$$

$$= \mathbf{11.44\ m^3/h} \qquad \underset{\Leftarrow}{\text{Ans}}$$

$$\text{Ind. th. efficiency} = \frac{ip}{\dot{V}_f \times CV}$$

$$= \frac{20.9 \times 60 \times 60}{11.44 \times 21000} \times 100$$

$$= \mathbf{31.3\%} \qquad \underset{\Leftarrow}{\text{Ans}}$$

$$\text{Heat input} = \frac{11.44 \times 21000}{60}$$

$$= \mathbf{4004\ kJ/min}$$

$$\text{Heat equivalent of } bp = 17.74 \times 60$$

$$= \mathbf{1064.4\ kJ/min} \qquad \underset{\Leftarrow}{\text{Ans}}$$

$$\text{Heat in cooling water} = \frac{600}{60} \times 4.18 \times (60 - 25)$$

$$= \mathbf{1463\ kJ/min} \qquad \underset{\Leftarrow}{\text{Ans}}$$

Heat in exhaust, radiation etc. (by difference)

$$= 4004 - (1064.4 + 1463)$$

$$= \mathbf{1476.6\ kJ/min} \qquad \underset{\Leftarrow}{\text{Ans}}$$

Heat input (per minute)	(kJ)	Heat expenditure (per minute)	(kJ)
Heat supplied by fuel	4004	1. Heat equivalent to *bp*	1064.4
		2. Heat lost to cooling medium	1463.0
		3. Heat lost in exhaust	1476.6
		Total	4004.0

The friction heat is not included since it is assumed that friction heat is rejected to cooling water, exhaust, radiation etc.

Questions :-

14.1 Mention the basic aspects covered by the engine performance.

14.2 List the parameters by which performance of an engine is evaluated.

14.3 Define mean effective pressure and distinguish between brake mean effective pressure and indicated mean effective pressure.

14.4 Develop an expression for the calculation of indicated mean effective pressure.

14.5 Derive the expression for calculating indicated power of an engine.

14.6 How is the torque of an engine is evaluated from the nomenclature of the engine?

14.7 Distinguish between the power and specific output.

14.8 Briefly discuss the various efficiency terms associated with an engine.

14.9 Define air-fuel ratio and briefly state its effect on power output, fuel consumption and combustion pressure.

14.10 What is the importance of specific fuel consumption?

14.11 Schematically explain the use of the study of the heat balance of an engine.

14.12 Explain the effect of the following factors on the performance of an engine: (i) compression ratio (ii) air-fuel ratio (iii) spark timing (iv) engine speed (v) mass of inducted charge and (vi) heat losses.

14.13 Briefly explain the reason of poor part load thermal efficiency of the SI engine compared with CI engine.

14.14 What limits the brake power and brake mean effective pressure of a CI engine?

14.15 What are the methods available for improving the performance of an engine?

14.16 Draw a typical performance map for a four-stroke fast burn SI engine showing contours of constant *bsfc*.

14.17 Draw a typical performance map showing the effect of fuel injection parameters, air swirl and bowl in piston design and fraction of exhaust gas recirculation.

14.18 By means of a Sankey diagram explain the energy flow through an engine.

14.19 Draw a detailed Sankey diagram for SI engine and explain.

14.20 Explain the details of the analytical method of performance estimation.

Exercise:-

14.1. Find the bore of the single-cylinder diesel engine working on the four-stroke cycle and delivers 40 kW at 200 rpm from the following data :

$$
\begin{aligned}
\text{Compression ratio} &= 14{:}1 \\
\text{Fuel cut-off} &= 5\% \text{ of the stroke} \\
\text{Index of compression curve} &= 1.4 \\
\text{Index for expansion curve} &= 1.3 \\
\text{Pressure at the beginning of the compression} &= 1 \text{ atm} \\
\text{Ratio of stroke to bore} &= 1.5 \text{ to } 1
\end{aligned}
$$

Ans: 34.5 cm

14.2. Determine the diameter of a gas engine cylinder to develop 18 kW when making 100 explosions per minute, gives clearance volume 1/3 swept volume, law of compression and explosion $pV^{1.3}$, absolute maximum pressure is 3 times the absolute pressure at the end of the stroke. Take length of stroke to be twice the bore.

Ans: 24.7 cm

14.3. A four-stroke gas engine having a cylinder of 250 mm diameter and stroke 450 mm has a volumetric efficiency of 80%, ratio of air to gas is 8 to 1, calorific value of gas is 20 MJ/m^3 at NTP. Find the heat supplied to the engine per working cycle. If the compression ratio is 5, what is the heating value of the mixture per working stroke per m^3 of total cylinder volume?

Ans: (i) 40 kJ (ii) 1428.6 kJ/m^3

14.4. The swept volume of a gas engine is 9 l and the clearance volume is 2.25 l. The engine consumes 3500 l of gas per hour, when running at 160 rpm firing every cycle and developing 4 kW. It has a mechanical efficiency of 75%. What is the relative efficiency of the engine compared with air-standard cycle if the calorific value of gas is 20 MJ/m^3. Assuming a volumetric efficiency of 0.87, find the ratio of air to gas used by 1 m^3 of the mixture in the cylinder. *Ans*: (i) 58% (ii) 9.7

14.5. In order to study the effect of mixture strength on the thermal efficiency of the engine, tests were made on a four-stroke gas engine and the following heat balance results were obtained.

Test	Engine A Lean mixture	Engine B Rich mixture
Indicated work	37	33
Heat in exhaust gas	42	39
Heat lost	21	28
	100	1000

In the test A the supply of gas was 0.23 m^3 per minute and in test B was 0.30 m^3 per minute. The calorific value of the gas used was 20 MJ/m^3. Calculate the indicated power and the heat loss to the cylinder walls and piston per minute in the two cases. Calculate the indicated thermal efficiencies of the two engines and comment on the result.

Ans: (i) 966 kJ; 1600 kJ (ii) 37%; 33%

Reasons for higher thermal efficiency on lean mixtures

(i) *With a weak mixture the temperature rise for a given heat input is greater than with rich mixture since there is a little or no dissociation and the specific heat is less at the lower temperature.*

(ii) *The smaller maximum temperature reduces the heat flow to the cylinder walls.*

The more nearly the maximum temperature approaches the temperature at the end of compression will the Otto efficiency approaches the Carnot.

14.6. An engine of 175 mm bore and 375 mm stroke is governed by hit and miss type governor to 220 rpm. With a fixed setting of the gas supply and ignition advance, indicator diagrams gave the following values of mean effective pressure. Firing, positive loop 5.7 bar, negative loop 0.25 bar; missing negative loop 0.42 bar. When developing 6 kW the explosions were 100 per minute and the gas used was 0.1 m^3 per minute. Calculate the friction power and assuming uniform gas supply per explosion, find the gas consumption at no load.   *Ans*: (i) 2.13 kW (ii) 0.0023 m^3

14.7. Calculate the maximum brake power which can be developed in the cylinder of a four-stroke gas engine which runs at 200 rpm. The diameter of the piston is 300 mm and stroke 400 mm. Clearance volume is 25% of the swept volume. The gas supplied consists of CO = 19.7%; H_2 = 28.8%; CO_2 = 14.4%; N_2 = 37.1%. Assume that the total mixture at NTP admitted per suction stroke is 0.875 of total volume behind the piston at the end of the stroke and that the thermal efficiency is 35%. Calorific value of H_2 = 121.4 MJ/kg; Calorific value of carbon burning from CO to CO_2 = 23.4 MJ/kg; Density of air = 1.2 kg/m³.

Ans: 47 kW

14.8. The following observations were made in a test of a gas engine in which a waste heat boiler served as an exhaust gas calorimeter. Gross calorific value of gas 20.0 MJ/m³ at NTP; gas consumption 9.35 m³/h at NTP; density of gas = 0.706 kg/m³; mass of water vapour of combustion produced per m³ of gas (at NTP) = 0.72 kg; air consumption = 77.4 kg/h; temperature of air and gas = 17 °C; rate of flow of water through boiler = 168 kg/h; inlet and outlet temperature of water is 20 °C and 80 °C. Temperature of exhaust gases leaving the boiler 127 °C. Calculate the heat per hour leaving the engine and express as a percentage of heat supplied.

Assume the dew point of the exhaust gases as 50 °C; the total heat of dry saturated steam at 50 °C = 2580 kJ/kg; the mean specific heat of steam as 2 kJ/kg K and the specific heat of flue gas as 1 kJ/kg K. Take atmospheric temperature = 17 °C as datum. *Ans:* (i) 187000 kJ/h (ii) 36.9%

14.9. In olden days in order to compare the various forms and sizes of the engine a factor known as Tookey factor is defined. It is given by

$$TF = \frac{mep \ of \ the \ cycle \ in \ bar}{Heat \ equivalent \ of \ 1 \ m^3 \ of \ cyl. \ mixture \ in \ MJ}$$

Calculate the value of the Tookey factor for an engine developing 10 kW using the following data. Piston diameter = 200 mm; stroke = 300 mm; explosion per minute = 100; calorific value of gas = 19 MJ/m³ gas per hour 6 m³; clearance volume = 25% of swept volume; volumetric efficiency = 87.5%. (Note : Modern engines will have a TF > 4). *Ans:* 3.44

14.10. In a test of a gas engine, the gas used had the following composition by volume : CH_4 = 65%; H_2 = 2%; N_2 = 2%; CO_2 = 31%. The dry exhaust gas analysis gave O_2 = 5.3%; N_2 = 83%; CO = 0.3% and CO_2 = 11.4%. Find (i) the air-fuel ratio by volume, to give complete combustion (ii) the percentage of excess air actually used in the test. Air contains 79% by volume of nitrogen. *Ans:* (i) 6.235 m³ (ii) 36.7%

14.11. Assume that an oil engine cylinder is not cooled. One kg of air and 0.00518 of fuel are introduced into the cylinder at 17 °C and

1 bar and compressed to 4 bar. Taking the specific heat of the product to 0.717 kJ/kg K, find the maximum temperature and pressure at the end of explosion. The calorific value of the fuel is 46 MJ/kg. Take $\gamma = 1.4$.   *Ans*: (i) 761.6 K (ii) 7.07 bar

14.12. Assuming a volumetric efficiency of 75%, estimate the probable indicated power of a four-cylinder petrol engine, given the following data, diameter of cylinder = 180 mm; stroke = 210 mm; speed = 1000 rpm; air-fuel ratio = 16:1. Engine works on a four-stroke cycle, net calorific value of the fuel = 44 MJ/kg; thermal efficiency is 30%.   *Ans*: 140.8 kW

14.13. A nine-cylinder petrol engine of bore 150 mm and stroke 200 mm has a compression ratio of 6:1 and develops 360 kW at 2000 rpm when running on a mixture of 20% rich. The fuel used has a calorific value of 43 MJ/kg and contains 85.3 carbon and 14.7% hydrogen. Assuming volumetric efficiency of 70% at 17 °C and mechanical efficiency of 90%, find the indicated thermal efficiency of the engine. Air contains 23.3% by mass of oxygen.

Ans: 30.84%

14.14. In a test the percentage analysis of petrol is found to be C = 83.2%; H = 14.3% O_2 = 2.5%. Calculate the mixture strength theoretically required for complete combustion of this fuel. Tests were conducted on a petrol engine at full throttle and constant speed, the quantity of fuel supplied to the engine being varied by means of an adjustable needle valve fitted to the jet of the carburettor. The results obtained were as follows :

Speed (rpm)	Torque (Nm)	Fuel (kg/min)	Air (kg/min)	Air-fuel ratio
1570	122.9	0.097	1.570	16.20
1570	128.5	0.100	1.560	15.60
1573	131.5	0.103	1.570	15.20
1569	138.0	0.110	1.560	14.20
1572	139.3	0.113	1.560	13.80
1563	141.5	0.118	1.560	13.20
1560	142.7	0.123	1.560	12.70
1572	141.2	0.129	1.560	12.10

Calorific value of fuel is 42 MJ/kg. Find the mixture strength for maximum brake power and maximum thermal efficiency.

Ans: (i) 12.1 (ii) 15.75

14.15. A four-cylinder automobile engine of 90 mm bore and 115 mm stroke was tested at constant speed over the complete practical range of mixture strength. The speed, the brake loads and the fuel consumption are as follows :

Test No.	Brake load (kg)	Speed (rpm)	Fuel consumption (kg/h)
1	17.35	1510	10.99
2	17.39	1500	10.80
3	17.44	1510	10.53
4	17.48	1512	10.35
5	17.48	1510	9.99
6	17.48	1510	9.72
7	17.21	1509	9.17
8	16.35	1493	8.54
9	15.08	1513	8.08

The brake arm length was 1 m. Plot *sfc* vs *bmep* curve and find the mixture strength for maximum power and maximum economy.   *Ans:* (i) 0.37 kg/kW h (ii) 0.341 kg/kW h

14.16. Calorific value of kerosene = 42 MJ/kg. Plot the curves of indicated mean effective pressure and thermal efficiency with respect to a compression ratio and obtain the values at compression ratio of 5.5. Also find the relative efficiency at the above compression ratio.   *Ans:* 62.3%

14.17. In a test of a four-cylinder, four-stroke petrol engine of 75 mm bore and 100 mm stroke, the following results were obtained at full throttle at a constant speed and with a fixed setting of the fuel supply of 0.082 kg/min.

bp with all cylinders working	=	15.24 kW
bp with cylinder number 1 cut-off	=	10.45 kW
bp with cylinder number 2 cut-off	=	10.38 kW
bp with cylinder number 3 cut-off	=	10.23 kW
bp with cylinder number 4 cut-off	=	10.45 kW

Estimate the indicated power of the engine under these conditions. If the calorific value of the fuel is 44 MJ/kg, find the indicated thermal efficiency of the engine. Compare this with the air-standard efficiency, the clearance volume of one cylinder being 115 cc.

Ans: (i) $ip_1 = 4.79; ip_2 = 4.86; ip_3 = 5.01; ip_4 = 4.79;$
$ip_{1234} = 19.45;$ (ii) 32.35% (iii) 69.1%

14.18. A six-cylinder petrol engine of 100 mm bore and 125 mm stroke was run at full throttle at a constant speed of 1500 rpm over the practical range of air-fuel ratio, and the following results were deduced from the series :

bmep (bar)	sfc (kg/kW h)	Air-fuel ratio (mass basis)
6.19	0.555	11.0
6.53	0.494	11.5
6.67	0.438	12.9
6.63	0.383	14.7
6.60	0.352	16.1
6.26	0.339	17.6
5.71	0.352	19.2
5.07	0.407	20.8

The engine has a compression ratio of 5. The fuel used has a calorific value of 44300 kJ/kg and the stoichiometric air-fuel ratio is 14.5. Plot on a base of air-fuel ratio, curves of brake mean effective pressure and the specific fuel consumption. Point out these characteristics of petrol engine in general and this engine in particular, revealed by these curves. Calculate the highest brake thermal efficiency given by these tests.   *Ans:* **24%**

14.19. A four-cylinder, four-stroke automobile engine of 60 mm bore and 115 mm stroke was tested at a constant speed over the complete range of mixture strength. The arm of the brake was 1 m. The following data were recorded.

Speed (rpm)	Brake load (kg)	Fuel consumption (kg/h)
1510	8.67	5.50
1500	8.70	5.40
1510	8.72	5.27
1510	8.74	5.18
1510	8.74	5.00
1510	8.74	4.86
1509	8.60	4.59
1493	8.17	4.27
1513	7.10	4.04

Plot a diagram showing the relation between fuel consumption in kg/kW h and *bmep*. Discuss the information which this diagram provides with respect to the performance of the engine. Calculate the power output at the most economized point and also the corresponding brake thermal efficiency if the calorific value of the fuel is 44 MJ/kg h.   *Ans:* (i)12.56 kW (ii) 24.1%

14.20. The fuel supplied to a Diesel engine has a gross calorific value of 44800 kJ/kg and contains 85.4% C and 12.3% H_2. The average temperature of the exhaust gases is 260 °C and their volumetric analysis gives CO_2 : 5.77%, CO : 0.12%, O_2 : 13.09%, N_2 (by difference) : 81.02%. Find (i) the heat carried away by the exhaust expressed as a percentage of the heat supplied and (ii) the mass of air per kg of fuel in excess of that theoretically required for complete combustion. Take mean specific heat of the dry exhaust gases as 1 kJ/kg K and atmospheric temperature as 17 °C. Air contains 23% oxygen on mass basis.

Ans: (i) 26.21% (ii) 21.62 kg

14.21. The following set of observations refer to a trial on a single-cylinder, four-stroke solid injection diesel engine of 200 mm bore and 400 mm stroke; gross mep = 6.20 bar, pumping mep = 0.44 bar, speed 262 rpm, brake torque = 468 Nm; fuel used 3.85 kg of oil per hour of gross calorific value 46600 kJ/kg, cooling water 6 kg/min raised 47 °C.

Draw up a heat balance sheet for the trial expressing various quantitative in kJ/min and calculate the mechanical efficiency of the engine. If the fuel contains 13.5% H_2 (by mass) and the air supply to the engine 1.71 kg/min at 17 °C. Estimate the heat carried away per minute by the exhaust gases when their temperature is 280 °C. Assume a mean specific heat of 1 kJ/kg and specific heat of fuel to be 4 kJ/kg K. Take sensible heat of air and fuel also in the heat balance.   *Ans:* (i) 81.2% (ii) 678.7 kJ/min

14.22. A four-cylinder, four-stroke diesel engine develops 28 kW at 2000 rpm. Its *bsfc* is 0.26 kg/kW h. Calculate the power output of the engine and its *bsfc* when the fuel rate is reduced by 40% at the same speed. Mechanical efficiency is 0.80. Assume that the indicated thermal efficiency changes linearly with equivalence ratio and there is 1% increase in indicated thermal efficiency for 6% increase in the equivalence ratio. Equivalence ratio (ϕ) at higher fuel flow rate is 0.6666.   *Ans:* (i) 16.1 kW (ii) 0.271 kg/kW h

14.23. An eight-cylinder SI engine with 90 mm bore and 110 mm stroke produces 100 kW at a mean piston speed of 660 m/min at full throttle and the *bsfc* is 0.39 kg/kW h. What will be the power produced by the engine and the *bsfc* when the engine runs with a mean piston speed of 440 m/min with the throttle set at the same position. The fuel-air ratio and spark timing are adjusted for best power at each speed. From experimental curves of similar engine it is found that for the given percentage reduction in piston speed the air capacity of the engine is reduced by 20%. Friction mep at piston speeds of 660 and 440 m/min are 1.9 and 1.2 bar respectively.   *Ans*: (i) 56.3 kW (ii) 0.369 kg/kW h

14.24. An eight-cylinder automobile petrol engine of 100 mm bore and 90 mm stroke has a compression ratio of 7. The engine develops 136 kW at 4000 rpm. The engine operates at 20% rich in fuel and the atmospheric conditions are 27 °C and 760 mm of Hg barometer. The *bsfc* is 0.34 kg/kW h. Estimate the power output and *bsfc* when (i) barometer is 740 mm of Hg and temperature is 47 °C. (ii) barometer is 775 mm of Hg and temperature is 7 °C. *fmep* at 4000 rpm is 2.10 bar.

Ans: (i) 125.8 kW; 0.346 kg/kW h (ii) 145.7 kW; 0.335 kg/kW h

14.25. A four-cylinder, four-stroke cycle diesel engine produces 60 kW at 2200 rpm (under maximum fuel delivery position) with a *bsfc* of 0.28 kg/kW h at sea level (barometer 760 mm of Hg and 27 °C). Cylinder bore of the engine is 110 mm and stroke is 140 mm. The friction mean effective pressure obtained from motoring test at 2200 rpm is 2 bar. Estimate the *bsfc* of the engine at 2200 rpm, when it is taken to a hill top at 3000 m altitude. Effect of humidity may be neglected. The fuel-air ratio is maintained at a constant value. At 3000 m altitude barometer is 540 mm of Hg and temperature is 270 K.   *Ans*: 0.314 kg/kW h

15

TWO–STROKE ENGINES

15.1 INTRODUCTION

A two-stroke engine is one which completes its cycle of operation in one revolution of the crankshaft or in two strokes of the piston. In this engine the functions of the intake and exhaust processes of the four-stroke engine are taken care of by the incoming fresh charge which is compressed either in the crankcase or by means of a separate blower while the engine piston is near the bottom dead center. The engine piston needs only to compress the fresh charge and expand the products of combustion. Since a two-stroke engine will have twice as many cycles per minute as a four-stroke engine operating at the same speed and with the same number of cylinders, theoretically it will develop twice the power when operating at the same mean effective pressure. As with the four-stroke engine, the power output of this engine also depends upon the number of kilograms of air per minute available for combustion.

In many two-stroke engines the mechanical construction is greatly simplified by using the piston as a slide valve in conjunction with intake and exhaust ports cut in the side of the cylinder.

If the engine is running on Otto cycle the charge consists of the correct mixture of fuel and air whereas for engines running on the Diesel or dual combustion cycles the charge will consist of pure air. The application of two-stroke principle to small high speed compression ignition engines involves many difficulties, although these are by no means insurmountable. In case of spark-ignition engines, the two-stroke cycle principle has been applied to a large number of engines ranging from tiny single cylinder model engines developing a fraction of a kW to the largest of the aircraft units developing 2500 kW or more.

15.2 TYPES OF TWO–STROKE ENGINES

Depending upon the scavenging method used there are basically two types of two-stroke engines:

(i) crankcase scavenged engine

(ii) separately scavenged engine

The details of the above two types of two-stroke engines are discussed briefly in the following sections.

15.2.1 Crankcase Scavenged Engine

One of the simplest types of two-stroke engines is shown in Fig.15.1. In this engine, the charge (fuel-air mixture in SI engine and air in CI engine) is compressed in the crankcase by the underside of the piston during the expansion stroke. There are three ports in this engine.

(i) intake port at the crankcase

(ii) transfer port

(iii) exhaust port

The compressed charge passes through the transfer port into the engine cylinder flushing the products of combustion. This process is called scavenging and this type of engines are called the crankcase-scavenged engines.

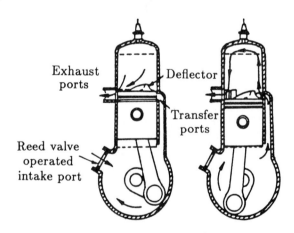

Fig.15.1 Crankcase-Scavenged Two-Stroke Engine

As the piston moves down, it first uncovers the exhaust ports, and the cylinder pressure drops to atmospheric level as the combustion products escape through these ports. Further, downward motion of the piston uncovers the transfer ports, permitting the slightly compressed mixture or air (depending upon the type of the engine) in the crankcase to enter the engine cylinder. The top of the piston and the ports are usually shaped in such a way that the fresh air is directed towards the top of the cylinder before flowing towards the exhaust ports. This is for the purpose of scavenging the upper part of the cylinder of the combustion products and also to minimize the flow of the fresh fuel-air mixture directly through the exhaust ports. The projection on the piston is called the deflector. As the piston returns from bottom center, the transfer ports and then the exhaust ports are closed and compression of the charge begins. Motion of the piston during compression lowers the pressure in the crankcase so that the fresh mixture or air is drawn into the crankcase through the inlet reed valve. Ignition and expansion take place in the usual way, and the cycle is repeated. Due to the flow restriction in the inlet reed valve and the transfer ports the engine gets charged with less than one cylinder displacement volume.

15.2.2 Separately Scavenged Engine

Another type of engine which uses an external device like a blower to scavenge the products of combustion is called the externally or separately scavenged engine. The details of this type of engine are shown in Fig.15.2. In these engines air-fuel mixture is supplied at a slightly higher pressure at the inlet manifold. As the piston moves down on the expansion stroke it uncovers the exhaust ports at approximately 60° before bottom center. About 10° later, when the cylinder pressure has been considerably reduced, the inlet ports are uncovered and the scavenging process takes place. The inlet ports are shaped in such a way that most of the air flows to the top of the cylinder on the inlet side and move down on the exhaust side forming a loop (Fig.15.2) before reaching the exhaust ports. This ensures scavenging of the upper cylinder volume. Piston deflectors are not used as they are heavy and tend to become overheated at high output. The scavenging process is more efficient in these types of engines than in the usual crankcase scavenged engine with deflector piston.

Opposed Piston or End-to-End Scavenged Engine : Another type of two-stroke engine is the opposed piston type or sometime called *end-to-end* scavenged engine. The details of this type of engine are shown in Fig.15.3. In this type the exhaust ports are opened first. The inlet ports are shaped in such a way that the air or mixture enters tangentially, thereby imparting a swirling motion (Fig.15.3). The swirl helps to prevent mixing of fresh charge and combustion products during the scavenging process. Early on the compression stroke, the exhaust ports close first. Because of the flow through the

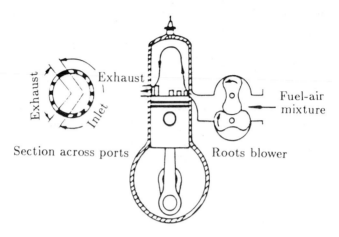

Fig.15.2 Loop-Scavenged Two-Stroke Engine

inlet port the cylinder pressure start rising and when the pressure is close to inlet manifold pressure the inlet ports close. In this opposed piston engine it may be noted that two pistons are a little out of phase (Fig.15.3).

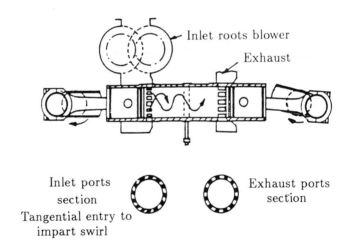

Fig.15.3 Port Arrangement of a Two-Stroke End-to-End
Scavenged Compression-Ignition Engine

In the loop-scavenged engine the port timing is symmetrical, so that the exhaust port must close after the inlet port closes. The timing prevents this type of engine from filling its cylinder at full inlet pressure. In the end-to-end scavenged engines, counterflow within the cylinder is eliminated, and there is less opportunity for mixing of fresh and spent gases. The scavenging should therefore be more efficient.

15.3 TERMINOLOGIES AND DEFINITIONS

In the absense of a terminology that is commonly accepted or at least commonly understood, it is necessary to explain and define the terms used in connection with two-stroke engines. In this section a list of terminologies used are explained. The terminologies can be easily understood from the diagram showing the charging process of the engine. The flow of gases through a two-stroke cycle engine is diagrammatically represented in Fig.15.4. The hatched areas represent fresh air or mixture and the cross hatched areas represent combustion gases. The width of the channels represent the quantity of the gases expressed by volume at NTP condition.

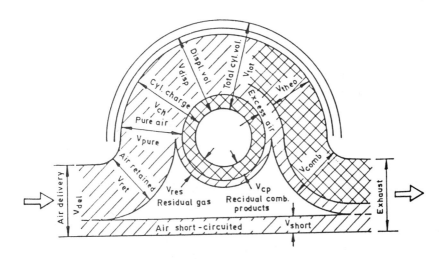

Fig.15.4 Scavenging Diagram for Two-stroke Cycle SI Engine

15.3.1 Delivery Ratio (R_{del})

In two-stroke cycle engines, air or mixture is supplied to the cylinders either by a separate pump or blower, or by the piston of the working cylinder acting as a pump. Depending on the relative capacity of the blower, the air delivered, V_{del}, may be either more or less than the swept volume.. The delivery ratio is usually defined as

$$R_{del} = \frac{V_{del}}{V_{ref}} \qquad (15.1)$$

The reference volume, V_{ref}, has been variously chosen to be displacement volume, effective displacement volume, total cylinder volume, or total effective cylinder volume. Since, it is only the quantity of charge in the remaining total cylinder volume at exhaust port closure that enters into the combustion, it is recommended that the total effective cylinder volume should be preferred.

The delivery ratio, nowadays, is defined on mass basis. According to it, the delivery ratio is mass of fresh air delivered to the cylinder divided by a reference mass, i.e.,

$$R_{del} \quad = \quad \frac{M_{del}}{M_{ref}} \tag{15.2}$$

Thus the delivery ratio is a measure of the air supplied to the cylinder relative to the cylinder content. If $R_{del} = 1$, it means that the volume of the scavenging air supplied to the cylinder is equal to the cylinder volume (or displacement volume whichever is taken as reference).

Delivery ratio usually varies between 1.2 and 1.5 except for crankcase scavenged engines where it is less than unity.

15.3.2 Trapping Efficiency

The air or mixture delivery, V_{del}, is split into two parts (Fig.15.4): the short circuiting air, V_{short} which leaves through the exhaust port or valve without remaining in the cylinder, and the retained air or mixture, V_{ret}, which is trapped in the cylinder and takes part in the subsequent combustion. Therefore, an additional term, trapping efficiency, is used to indicate the ability of the cylinder to retain the fresh charge. It is defined as the ratio of the amount of charge retained in the cylinder to the total charge delivered to the engine, i.e.,

$$\eta_{trap} \quad = \quad \frac{V_{ret}}{V_{del}} \tag{15.3}$$

Trapping efficiency indicates that fraction of the fresh air or mixture supplied to the cylinder which is retained in the cylinder the rest being wasted through the exhaust. Short circuiting is naturally equal to $(1 - \eta_{trap})$. The trapping efficiency, therefore, is a measure of the success of trapping the supplied air or mixture with minimum waste. It is mainly controlled by the geometry of the ports and the overlap time.

15.3.3 Relative Cylinder Charge

The air or mixture retained, V_{ret}, together with the residual gas, V_{res}, remaining in the cylinder after flushing out the products of combustion constitutes the cylinder charge, V_{ch}. Relative cylinder charge is a measure of the success of filling the cylinder irrespective of the composition of charge and is defined as

$$C_{ret} = \frac{V_{ch}}{V_{ref}} \tag{15.4}$$

$$= \frac{V_{ret} + V_{res}}{V_{ref}} \tag{15.5}$$

The relative cylinder charge may be either more or less than unity depending upon the scavenging pressure and the port heights.

It must be noted that all volumes referred to are at NTP condition. However, some authors, recommend the use of inlet temperature and exhaust pressure as the reference. *In view of the use of different reference volumes as well as different reference conditions, it should always be kept in mind that the published values may not be directly comparable.*

The cylinder charge may also be considered as composed of pure air and residual combustion products. By denoting

$$p_{ar} = \frac{V_{pure}}{V_{ret}}$$

as pure air ratio and

$$\frac{V_{cp}}{V_{ret}}$$

as residual combustion products ratio.

$$C_{rel} = p_{ar} + \frac{V_{cp}}{V_{ret}}$$

During combustion, part of the air (or, rather of the oxygen in the air) contained in the cylinder charge burns, while the remainder, the excess air, is not involved in the attendant chemical reactions. Part of this excess air escapes through the exhaust with the combustion products, and another part, $V_{res} - V_{cp}$, remains in the cylinder and participates in the subsequent cycle, where V_{cp} represents combustion products in the residual gas. Therefore, the cylinder charge consists of three parts: the retained portion of the air delivered, part of the combustion products from the preceding cycle, and part of the excess air from the preceding cycle.

As the cylinder charge, V_{ch} is contaminated by combustion products, V_{cp},

$$\eta_p = \frac{V_{pure}}{V_{ch}} = \frac{V_{ch} - V_{cp}}{V_{ch}} = 1 - \frac{V_{cp}}{V_{ch}}$$

represents the purity of the charge and $\frac{V_{cp}}{V_{ch}}$ constitutes the pollution. It should be realized that V_{pure} is more than that part of the air delivered which is retained in the cylinder; it includes some pure air contained in the residual gas remaining in the cylinder from the previous cycle.

15.3.4 Scavenging Efficiency

The purity of charge is a measure of the success of scavenging the cylinder from the combustion products of the preceding cycle. It is largely controlled by the shape of the combustion chamber and the scavenging arrangement (cross, loop or uniflow). Experimental determination of the purity is relatively simple, consisting only of analysis of a gas sample taken during the compression stroke. The difficulty lies in representative sample.

In German technical literature the term *scavenging efficiency (spuelwirkungsgrad)* has been widely used. It is defined as

$$\eta_{sc} = \frac{V_{ret}}{V_{ch}} = \frac{V_{ret}}{V_{ret} + V_{res}}$$

It can be shown that the delivery ratio, scavenging and trapping efficiencies are related by the following equation

$$R_{del} = \frac{C_{rel} \, \eta_{sc}}{\eta_{trap}} \tag{15.6}$$

Scavenging efficiency is a term somewhat similar to purity and expresses the measure of the success in clearing the cylinder of residual gases from the preceding cycle.

Scavenging efficiency indicates to what extent the residual gases in the cylinder are replaced with fresh air. If it is equal to unity, it means that all the gases existing in the cylinder at the beginning of scavenging have been swept out.

The scavenging efficiency in diesel engines can be measured by drawing off a small sample of the products of combustion just before the exhaust valve opens or during the earlier part of blowdown. This sample is analyzed and the results obtained are compared with standard curves of exhaust products versus fuel-air ratio. This determines the fuel-air ratio that must have existed in the cylinder before combustion. Knowing the quantity of fuel injected per cycle, the quantity of fresh air retained in the cylinder per cycle can be calculated. Air in the residual gas is not considered as it represents a constant quantity which does not take part in combustion.

15.3.5 Charging Efficiency

The amount of fresh charge in the cylinder is a measure of the power output of the engine. The useful fresh charge divided by the displacement volume is the charging efficiency defined as

$$\eta_{ch} \;=\; \frac{V_{ret}}{V_{ref}}$$

Charging efficiency is a measure of the success of filling the cylinder with fresh air. Naturally

$$\eta_{ch} \;=\; R_{del}\,\eta_{trap}$$

as can be seen from Eqn.15.1 and 15.3. The charging efficiency is an important index as it most directly affects the power output of the engine.

15.3.6 Pressure Loss Coefficient (P_l)

The pressure loss coefficient is the ratio between the main upstream and downstream pressures during the scavenging period and represents the loss of pressure to which the scavenge air is subjected to when it crosses the cylinder.

15.3.7 Index for Compressing the Scavenge Air (N)

This index is proportional to the power required for compressing the scavenge air and is calculated on the basis of the delivery ratio and pressure loss coefficient measurements. It allows, along with scavenging efficiency evaluation, a comparison of different types of scavenging systems.

15.3.8 Excess Air Factor (λ)

The value $(R_{del} - 1)$ is called the excess air factor, λ. For example, if the delivery ratio is 1.7, the excess air factor is 0.7.

15.4 TWO-STROKE AIR CAPACITY

Except when large valve overlaps are used, a four-stroke engine retains almost all the charge entering the cylinder during the intake process, i.e., nothing is lost through the exhaust valve. An air meter at the engine inlet will therefore measure $\dot{m}_a$, the amount of air available for combustion, and the engine output will be proportional to $\dot{m}_a$.

The exhaust ports of a two-stroke engine would remain open during most of the scavenging process. This makes it impossible to avoid some loss of fresh air or mixture at that time through the open exhaust port. With the increasing quantity of charge coming into the cylinder, increased amount of charge would be lost through the exhaust port. The power output of a two-stroke engine is therefore proportional not to the amount of charge entering the engine but to the amount of charge retained in the cylinder. The amount of charge that can be retained in the cylinder is called the air capacity of the two-stroke engine.

15.5 THEORETICAL SCAVENGING PROCESSES

The amount of charge retained in the cylinder depends upon the scavenging ability of the engine. Three theoretical scavenging processes are illustrated in Fig.15.5. These are perfect scavenging, perfect mixing and complete short circuiting. The details are explained in the following sections.

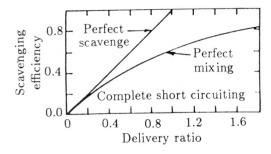

Fig.15.5 Theoretical Scavenging Processes

15.5.1 Perfect Scavenging

Ideally the fresh fuel-air mixture should remain separated from the residual products of combustion with respect to both mass and heat transfer during the scavenging process. Fresh charge is pumped into the cylinder by the blower through the inlet ports at the lower end of the cylinder. This in turn, pushes the products of combustion in the cylinder out through the exhaust port/valve at the other end. Here, it is assumed that there is no mixing of fresh charge and the products and as long as any products remain in the cylinder the flow through the exhaust ports is considered to be products of combustion only. However, when sufficient fresh air has entered to fill the entire cylinder volume the flow abruptly changes from one of products to one of air. This ideal process would represent perfect scavenging without any short circuiting loss.

15.5.2 Perfect Mixing

The second theoretical scavenging process is perfect mixing, in which the incoming fresh charge mixes completely and instantaneously with the cylinder contents, and a portion of this mixture passes out through the exhaust ports at a rate equal to that of the charge entering the cylinder. This homogeneous mixture consists initially of products of combustion only and then gradually changes to pure air. This mixture flowing through the exhaust ports is identical with that momentarily existing in the cylinder and changes with it.

For the case of perfect mixing the scavenging efficiency can be represented by the following equation

$$\eta_{sc} \quad = \quad 1 - e^{-R_{del}} \qquad (15.7)$$

where η_{sc} and R_{del} are scavenging efficiency and delivery ratio respectively. This is plotted in Fig.15.5. The result of this theoretical process closely approximates the results of many actual scavenging processes, and is thus often used as a basis of comparison.

15.5.3 Short Circuiting

The third type of scavenging process is that of short circuiting in which the fresh charge coming from the scavenge manifold directly goes out of the exhaust ports without removing any residual gas. This is a dead loss and its occurrence must be avoided.

15.6 ACTUAL SCAVENGING PROCESS

The actual scavenging process is neither one of perfect scavenging nor of perfect mixing. It probably consists partially of perfect scavenging, mixing and short circuiting. Figure 15.6 shows the variation of delivery ratio and trapping efficiency with crank angle for three different scavenging modes, i.e., perfect scavenging, perfect mixing and intermediate scavenging.

The scavenging parameters for the intermediate scavenging is shown in Fig.15.7. This represents the actual scavenging process. It can be seen from this figure that a certain amount of combustion products is initially pushed out of the cylinder without being diluted by fresh air. Gradually, mixing and short circuiting causes the outflowing products to be diluted by more and more fresh air until ultimately the situation is the same as for perfect mixing, i.e., the first phase of the scavenging process is a perfect scavenging process which then gradually changes into a complete mixing process.

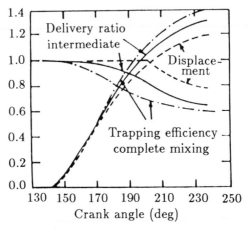

Fig.15.6 Delivery Ratio and Trapping Efficiency Variation for a Crankcase Scavenged Engine (Different Scavenging Modes)

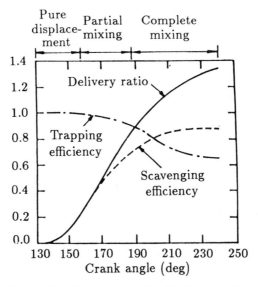

Fig.15.7 Scavenging Parameters for the Intermediate Scavenging

15.7 CLASSIFICATION BASED ON SCAVENGING PROCESS

The simplest method of introducing the charge into the cylinder is to employ crankcase compression as shown in Fig.15.1. This type of engine is classified as the crankcase scavenged engine. In another type,

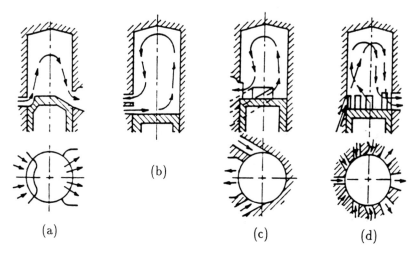

Fig.15.8 Methods of Scavenging

(a) *Cross Scavenging*   (b) *Loop Scavenging, M.A.N. Type*
(c) *Loop Scavenging,*   (d) *Loop Scavenging, Curtis Type*
     Schnuerle Type

a separate blower or a pump (Fig.15.2) may be used to introduce the charge through the inlet port. They are classified as the separately scavenged engines.

Another classification of two-stroke cycle engines is based on the air flow. The most common arrangement is cross scavenging, illustrated in Fig.15.8(a). The incoming air is directed upward, and then down to force out the exhaust gases through the oppositely located exhaust ports. With loop or reverse scavenging, the fresh air first sweeps across the piston top, moves up and then down and finally out through the exhaust. In the M.A.N. type of loop scavenge, Fig.15.8(b), the exhaust and inlet ports are on the same side, the exhaust above the inlet. In the Schnuerle type, Fig.15.8(c), the ports are side by side. The Curtis type of scavenging, Fig.15.8(d), is similar to the Schnuerle type, except that upwardly directed inlet ports are placed also opposite the exhaust ports.

The most perfect method of scavenging is the uniflow method, where the fresh air charge is admitted at one end of the cylinder and the exhaust escapes at the other end. The air flow is from end to end, and little short-circuiting between the intake and exhaust openings is possible. The three available arrangements for uniflow scavenging are shown in Fig.15.9. A poppet valve is used in *a* to admit the inlet air or for the exhaust, as the case may be. In *b* the inlet and exhaust

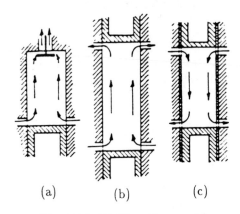

(a)   (b)   (c)

Fig.15.9 Uniflow Scavenging

(a) Poppet Valve   (b) Opposed Piston   (c) Sleeve Valve

ports are both controlled by separate pistons that move in opposite directions. In *c* the inlet and exhaust ports are controlled by the combined motion of piston and sleeve. In an alternative arrangement one set of ports is controlled by the piston and the other set by a sleeve or slide valve. All uniflow systems permit unsymmetrical scavenging and supercharging.

Fig.15.10 Reverse Flow Scavenging

Reverse flow scavenging is shown in Fig.15.10. In this type the inclined ports are used and the scavenging air is forced on to the opposite wall of the cylinder where it is reversed to the outlet ports. One obvious disadvantage of this type is the limitation on the port area. For long stroke engines operating at low piston speeds, this arrangement has proved satisfactory.

15.8 COMPARISON OF SCAVENGING METHODS

An interesting comparison of the merits of two cycle engine air scavenging methods is illustrated in Fig.15.11. In fact, specific output of the engine is largely determined by the efficiency of the scavenging system and is directly related to the brake mean effective pressure.

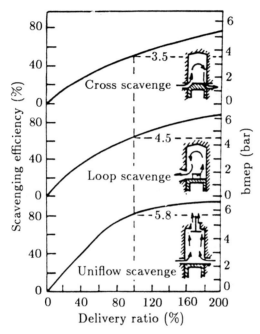

Fig.15.11 Scavenging Efficiency

As shown in Fig.15.11 scavenging efficiency varies with the delivery ratio and the type of scavenging. In this respect cross scavenging is least efficient and gives the lowest brake mean effective pressure. The main reason for this is that the scavenging air flows through the cylinder but does not expel the exhaust residual gases effectively. Loop scavenging method is better than the cross scavenging method. Even with a delivery ratio of 1.0 in all cases the scavenging efficiencies are about 53, 67 and 80 per cent for cross scavenging, loop scavenging and uniflow scavenging systems with corresponding values of *bmep* as 3.5, 4.5 and 5.8 bar.

15.9 SCAVENGING PUMPS

The performance of a two-stroke engine will depend largely on the characteristics of the type of compressor used as a scavenging pump. Figure 15.12 illustrates various types of scavenging pumps used.

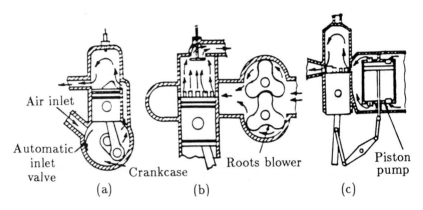

Fig.15.12 Scavenging Pumps

Displacement type of pumps include crankcase scavenging, piston and roots type blowers. An important characteristic of the displacement pumps is that their volumetric efficiency, and therefore, the mass of air delivered per unit time is not seriously reduced even by considerable increase in the outlet pressure. Thus it tends to maintain constant scavenging ratio even though ports may become partially clogged with deposits etc.

In case of centrifugal pumps, however, the mass flow is heavily influenced by the engine and exhaust system resistance. The pressure ratio in this case varies almost linearly with the rpm squared. Thus, the centrifugal type scavenging pump is a very satisfactory type provided the resistance of the flow system is not suddenly altered.

15.10 ADVANTAGES AND DISADVANTAGES OF TWO–STROKE ENGINES

Two-stroke engines have certain advantages as well as disadvantages compared to four-stroke engines. In the following sections the main advantages and disadvantages are discussed briefly.

15.10.1 Advantages of Two-stroke Engines

(i) As there is a working stroke for each revolution, the power developed will be nearly twice that of a four-stroke engine of the same dimensions and operating at the same speed.

(ii) The work required to overcome the friction of the exhaust and suction strokes is saved.

(iii) As there is a working stroke in every revolution, a more uniform turning moment is obtained on the crankshaft and therefore, a lighter flywheel is required.

(iv) Two-stroke engines are lighter than four-stroke engines for the same power output and speed.

(v) For the same output, two-stroke engines occupy lesser space.

(vi) The construction of a two-stroke cycle engine is simple because it has ports instead of valves. This reduces the maintenance problems considerably.

(vii) In case of two-stroke engines because of scavenging, burnt gases do not remain in the clearance space as in case of four-stroke engines.

15.10.2 Disadvantages of Two-Stroke Engines

(i) High speed two-stroke engines are less efficient owing to the reduced volumetric efficiency.

(ii) With engines working on Otto cycle, a part of the fresh mixture is lost as it escapes through the exhaust port during scavenging. This increases the fuel consumption and reduces the thermal efficiency.

(iii) Part of the piston stroke is lost with the provision of the ports thus the effective compression is less in case of two-stroke engines.

(iv) Two-stroke engines are liable to cause a heavier consumption of lubricating oil.

(v) With heavy loads, two-stroke engines get heated due to excessive heat produced. Also at light loads, the running of engine is not very smooth because of the increased dilution of charge.

15.11 COMPARISON OF TWO–STROKE SI AND CI ENGINES

The two-stroke SI engine suffers from two big disadvantages—fuel loss and idling difficulty. The two-stroke CI engine does not suffer from these disadvantages and hence CI engine is more suitable for two-stroke operation.

If the fuel is supplied to the cylinders after the exhaust ports are closed, there will be no loss of fuel and the indicated thermal efficiency of the two-stroke engine will be as good as that of a four-stroke engine. However, in an SI engine using carburettor, the scavenging is done with fuel-air mixture and only the fuel mixed with the retained air is used for combustion. To avoid the fuel loss, fuel-injection just before the exhaust port closure may be used instead of a carburettor.

The two-stroke SI engine picks up only gradually and may even stop at low speeds when mean effective pressure is reduced to about 1.2 bar. This is because a large amount of residual gas (more than in four-stroke engine) mixing with small amount of charge. At low speeds there may be backfiring due to slow burning rate. Fuel-injection improves idling and also eliminates backfiring as there is no fuel present in the inlet system.

In CI engines there is no loss of fuel as the charge is only air and there is no difficulty at idling because the fresh charge (air) is not reduced.

Questions :-

15.1 What is a two-stroke engine and how does it differ from a four-stroke engine?

15.2 Explain with neat sketches the two different types of two-stroke engines.

15.3 What is an opposed piston engine? Explain.

15.4 Define the following :
    (i) delivery ratio
    (ii) trapping efficiency
    (iii) relative cylinder charge
    (iv) scavenging efficiency
    (v) charging efficiency
    (vi) pressure loss coefficients
    (vii) excess air factor
    (viii) index of compression

15.5 Explain with a graph the three possible theoretical scavenging processes.

15.6 How does the actual scavenging process differ from the theoretical one? Explain by means of suitable graphs.

15.7 Briefly explain the classification of two-stroke engines based on scavenging process.

15.8 Compare the various scavenging methods.

15.9 Explain the various scavenging pumps used in a two-stroke engine.

15.10 What are the advantages and disadvantages of a two-stroke engine? Compare two-stroke SI and CI engines.

INDEX